T0258120

Crystallization: Advanced Methods

Crystallization: Advanced Methods

Edited by **Garry Hollis**

New York

Published by NY Research Press,
23 West, 55th Street, Suite 816,
New York, NY 10019, USA
www.nyresearchpress.com

Crystallization: Advanced Methods
Edited by Garry Hollis

International Standard Book Number: 978-1-63238-106-4 (Hardback)

Printed in the United States of America.

Contents

Preface

In my initial years as a student, I used to run to the library at every possible instance to grab a book and learn something new. Books were my primary source of knowledge and I would not have come such a long way without all that I learnt from them. Thus, when I was approached to edit this book; I became understandably nostalgic. It was an absolute honor to be considered worthy of guiding the current generation as well as those to come. I put all my knowledge and hard work into making this book most beneficial for its readers.

Recently, numerous novel applications have started counting on crystallization for their successful implementation, such as the crystallization of nano and amorphous materials. A common processing industry procedure for production, purification or recovery of solid materials, crystallization is applicable to multiple fields. The content of this book has been compiled by some highly respected experts in the field and pertains to cutting-edge areas of study and advancements in crystallization procedures. This book documents current work advancements in various facets of bulk crystallization from aqueous solutions and discusses general issues in crystallization. This book will prove to be invaluable for both basic study and expert practice, while adding substantially to the knowledge reservoir of all chemical engineers, students and academicians.

I wish to thank my publisher for supporting me at every step. I would also like to thank all the authors who have contributed their researches in this book. I hope this book will be a valuable contribution to the progress of the field.

Editor

Section 1

Bulk Crystallization from Aqueous Solutions

Crystallization, Alternation and Recrystallization of Sulphates

Joanna Jaworska
Adam Mickiewicz University
Poland

1. Introduction

Sulphates as well as silicates and carbonates are one of the most common minerals on the Earth's surface. They cover about 25% of continents surface (Blatt et al. 1980; Ford &Williams, 1989). Their recent sedimentary environments are the terrains of: the southern Mediterranean coast – coastal salt lakes of Marocco, Libya, Tunisia and Egypt, Gulf of Kara Bogaz (Caspian Sea), Persian Gulf – coastal sabkhas of UAE (special Abu Dhabi Emirate) and Qatar, Texas and California (Death Valley), salt lakes of South and Central Australia and salt lakes, salinas and salares of South America.

Annual total world production of gypsum in 2010 exceeded 146 million metric tones (http://minerals.usgs.gov/minerals/pubs/commodity/).

First of all, the sulphates are represented by two kinds of calcium sulphate - gypsum ($CaSO_4 \cdot 2H_2O$) and anhydrite ($CaSO_4$); mainly the first one creates deposits that are of economical value; it is used in the construction industry as bond material and to control the bonding speed, in casting and modelling and also in medicine (surgery and stomatology), during the production of paper. Its properties influence the parameters and quality of materials which it consist in. In construction/building industry the semi-hydrated gypsum is used as a result of frying in temperatures about 160°C (150-190°C) with sufficient amount of added water, the material bonds and hardens – the reaction is exothermic and the gypsum's volume increases of about 1%. Bassanite ($CaSO_4 \, \frac{1}{2} \, H_2O$), calcium sulphate semi-hydrate, is also known.

Rarely we can find the sulphates of: strontium (celestine), barium (barite), potassium (e.g. polyhalite), sodium (e.g. mirabilite, glauberite), magnesium (e.g. epsomite, kieserite), copper (e.g. brochantite, chalcanthite) and others. Most of gypsum and anhydrite on Earth are of evaporate origin, they are formed in specific order as a result of precipitation of the calcium sulphate inside the gradually drying sea basin (deep or shallow), lake, by the coastal lagoons, bays or sabkhas (indications of hot and arid climate). They are also the products of volcanic exhalations or low temperature hydrothermal processes, as well as of oxidation of sulphide deposits. The sulphates are also found above the salt mirror of diapirs, where they form the secondary deposit as the harder soluble residuum after the salt leaching – they constitute the main component of so-called gypsum or anhydrite-gypsum cap-rock.

The average precipitation rate of sulphates (gypsum and anhydrite) in the evaporite basin is ca. 0.5-1.2 mm/year and requires the evaporation of few to few tens cm high (2 m) column of water.

Probably, the oldest documented sulphate pseudomorphs are 3. 45 billion years old and come from West Australia (Pilbara), cm-size growth and interpreted to replace gypsum (Barley et al., 1979; Buick & Dunlop, 1990); only slightly younger are pseudomorphs after swallowtail gypsum – 3.4 billion years old – from S. Africa, Kaapvall Craton (Wilson & Versfeld, 1994).

1.1 Gypsum: $CaSO_4 \cdot 2H_2O$

Crystal system: monoclinic, hardness: 2, density: 2.3-2.4 g/cm^2
soluble: in water, in HCl and in concentrated solution of H_2SO_4
contains impurities: Ba, Sr, deposit grains where it crystallizes, bituminous substances
habit: platy, columnar, fibrous, needle-like, lenticular; forms massive aggregates and twins - swallowtail (figs. 1.,2.,3.,7. and 10.), usually colourless, might be coloured by Fe compounds
particular varieties:
- alabaster – fine-grained, sugar-like variety used in sculpture (fig. 4.),
- selenite – large well-crystallized varieties with dimensions reaching few m (fig. 5.); usually colourless
- spar – fibrous variety with semi-gloss, filling fissures and fractures (fig. 7.)
- desert rose – flower-like form of rounded gypsum aggregates (fig. 8.), occurring in the deserts as a result of ascent of the underground water rich in sulphates; it contains embedded sand grains built-in during the fast crystal growth.

Fig. 1. Platy gypsum (Petunia Bukta, Spitsbergen) phot. J. Jaworska

Fig. 2. Fibrous gypsum (Germany) phot. J. Jaworska

Fig. 3. Columnar to needle-like gypsum (Polkowice, Poland) phot. J. Jaworska

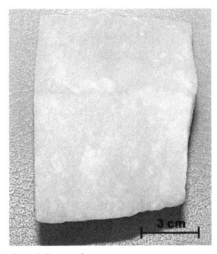

Fig. 4. Alabaser (Ukraine) phot. J. Jaworska

Primarily, gypsum that crystallizes in the evaporite basins forms usually medium or coarse grains; sometimes the lamination occurs, reflecting the changes in the basin (water composition, water level). Among the gypsum laminas, biolaminae appear; they are formed in the neritic zones and can be either deformed by periodical droughts (mudcraks) or ruptured by crystallizing sulphates (teepee-like structures, see fig. 9). In deeper zones of the basin, sabre-like gypsum (fig. 14.) can crystallize; these are elongated gypsum crystals, 20-30 cm long, distorted in one direction due to the demersal current activity (they constitute the perfect indicators of paleocurrents). Selenite gypsum is an exceptional feature; it forms under stable conditions at the depth of few to several m (figs. 16. and 17.) and reaches the dimensions of 3.5-4 m usually, but even up to 10 m. In deeper zones, laminated gypsum forms; sometimes with the ripplemark remains or even turbidites and slump structure with fragments of older, more lithified gypsum.

Fig. 5. Selenite (Busko-Zdrój, Poland) phot. J. Jaworska

Fig. 6. Gypsum twins – swallowtail (Dymaczewo Stare, Poland) phot. J. Jaworska

1.2 Anhydrite: $CaSO_4$

crystal system: orthorhombic; hardness: 3.5, density: 2.98 g/cm^2
hardly soluble in: HCl and concentrated H_2SO_4
contains impurities: Ba, Sr

habit: platy, columnar, fibrous; the crystal size rarely exceeds 0.5-1 mm (fig. 11.); sometimes crystals grown in caverns and fractures appear; massive aggregates (fig. 13.), rare radiant aggregates exceptionally reach the length of few cm

usually colourless crystals
particular varieties:
-   enterolithic anhydrite
-   bluish, fibrous variety, resembling twisted viscera (regional mining name, see fig. 12.).

Recently, the gypsum precipitates from among calcium sulphates; whereas anhydrite crystallizes very rarely – the only locations of its recent crystallization are: the Persian Gulf coast, lakes: Elton and Inger, Death Valley and Clayton Playa (Nevada).

Fig. 7. Fibrous –spar gypsum in clay-slate (Niwnice, Poland) phot. J. Jaworska

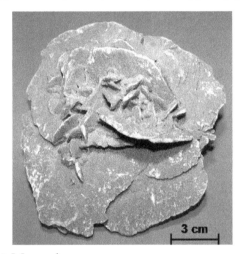

Fig. 8. Desert rose; phot. J. Jaworska

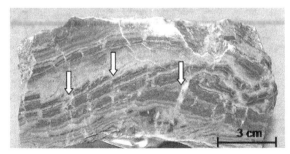

Fig. 9. Biolaminas deformed by crystallizing sulphates (near Ostrówka quarry, Poland) phot. J. Jaworska

Fig. 10. Lenticular gypsum (Wapno cap-rock, Poland) phot. J. Jaworska

Fig. 11. Anhydrite crystals (from Dębina salt dome, Poland), phot. A. Kyc

12. Enterolithic anhydrite (Wieliczka mine, Poland) phot. J. Jaworska

2. Crystallization and alternation: Hydratation and dehydratation (gypsification, anhydritization)

2.1 Crystallization

In most of the cases during evaporation processes, the gypsum crystallizes first, than the anhydrite (higher concentration of solution, 5-6 times higher than the normal sea water

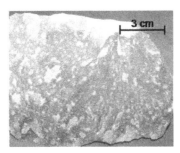

Fig. 13. Anhydrite rock - massie agreggates (Niwnice, Poland) phot. J. Jaworska

Fig. 14. Sabre-like gypsum (Nida region, Poland) phot. J. Jaworska

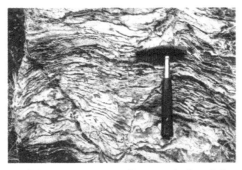

Fig. 15. Laminated gypsum (Niemeyer quarry, Germany) phot. J. Jaworska

salinity and in temperature about 40°C). Not until the concentration of solution reaches values close to NaCl concentration, the only phase of calcium sulphate which crystallizes and accompanies the rock salts is anhydrite; even if the temperature does not exceed 18°C. The thick rock salt deposits seldom form salt pillows together with salt swells and diapirs; their roof surfaces are located close to the Earth's surface (at the boundary of the salt mirror) and easily undergo leaching, leaving less soluble residue of - among the others - anhydrite grains and next - the anhydrite sandstone (fig. 20.), forming so-called cap-rock that forms the natural cover of the salt deposit. The anhydrite sandstone can – depending on the conditions - undergo further transformation typical for this very mineral (fig. 21.).

Fig. 16. and 17. Outcrop of 2.5-3 m senlenite gypsums, regional named szklica (Nida region, Poland) phot. J. Jaworska

Fig. 18. Second native sulphur in gypsum rock (Niemeyer quarry, Germany) phot. J. Jaworska

In the recent evaporation basins mainly the gypsum precipitates; anhydrite crystallizing under more extreme conditions occurs more rarely. Whereas among the sediments – particularly at the depths of few hundreds to few thousand meters – the anhydrite dominates. In many cases the anhydrite occurs as a product of the dehydration of gypsum; usually it is easily recognized pseudomorph of gypsum (e.g. selenite gypsum). The primary anhydrite, as well as the secondary one (dehydrate), as a result of tectonic processes, intense weathering of the overburden, climate changes etc., can be placed within the range of the underground or subsurface water (ground, meteoric) – where the hydration processes occur resulting in substitution of anhydrite by gypsum.

Fig. 19. Gypsum-karst (Nida region, Poland). phot. J. Jaworska

Fig. 20. Anhydrite sandstone; phot. J. Jaworska

2.2 Alternation

The sulphates – mainly the products of the hypergenic processes – very easily undergo the diagenetic processes, in which the dominant role is played by: hydration (gypsification) of anhydrite and dehydration (anhydritization) of gypsum; both processes are reversible and the reaction takes place as follows:

$$CaSO_4 \bullet 2H_2O \leftrightarrows CaSO_4 + 2H_2O$$

gypsum ⇆ anhydrite + water

There are many factors affecting the start and course of this reaction:

1. temperature and environmental pressure – depending on:
-   climate (for sulphates on the surface or close below it)
-   depth of the deposits – thickness of the overburden,
-   geothermal gradient of the area where the deposits occur – geotectonic environment and lithology of the overburden (thermal conductivity of the overburden),
2. chemical composition and concentration of solution, pore fluid pressure and the activity of water,
3. presence of micro-organisms and organisms (changes in Eh),
4. presence of cracks and pores in the sulphates as well as in the surrounding rocks.

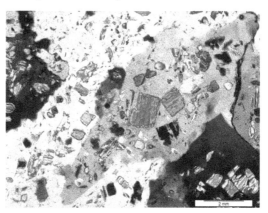

Fig. 21. Lenticular gypsum with anhydrite inclusions; phot. J. Jaworska

Fig. 22. Fine-crystalline gypsum; phot. J. Jaworska

2.2.1 Conditions

Anhydrite under surface conditions or close to the surface can be formed as a result of intense heating (over 50°C) of primary gypsum by the sun under hot and arid conditions.

When the gypsum deposits are buried, their transformation into anhydrite can theoretically start at the depth of about 450-500 m (Murray, 1964; Hardie, 1967; Jowett et al. 1993); those are the depths where temperature reaches 20°C, so the dehydration should not appear, however it is compensated by high overburden pressure (10 MPa; Kubica, 1972); on the other hand, according to Sonnenfeld (1984), gypsum can be found at the depth of 1200 m; and according to Ford and Williams (2007) even at 3000 m. The depth of the gypsum dehydration among others is modified by the geotectonic environment and the lithology of the overburden. The weakly heat conducting overburden, e.g. schists and gneisses, upon the areas seismically active, volcanic, causes the increase of the hydration speed – anhydrite can substitute the gypsum already at the depth of about 400 m; whereas well conducting overburden, e.g. rock salt of the cratonic areas, causes the process of transformation of the gypsum into anhydrite to occur hypothetically at the depth of even 4 km (Jowett et al., 1993). But the anhydrite gypsification process during the exhumation occurs usually at the depth of about 100-150 m (Murrey, 1964; Klimchouk &Andrejchuk, 1996). It starts either when the anhydrite appears in the area of influence of the ground water, or when it is exposed to rain water.

Fig. 23. Lenticular gypsum; phot. J. Jaworska

The crystallization process of calcium sulphates, as well as their gypsification or anhydritization are affected by the solutions (and their pressure). The NaCl solution occurring in the pore fluids plays special role; it modifies the temperature of the gypsum-anhydrite phase transformation. If the composition of pore fluids corresponds to the composition of sea water, the water activity (aH_2O) is 0.93 and the transformation of gypsum into anhydrite occurs at the temperature of 52°C; however if the pore fluids are NaCl saturated, then the water activity reaches 0.75 and the transformation occurs at 18°C (Jowett et al., 1993). The temperature of gypsum-anhydrite transformation is increased by: the presence of alkaline metal ions (Conley and Bundy, 1958) up to 98°C and the solution of $CaSO_4$ up to 95°C, but with lack of the anhydrite nuclei (Posnjak, 1940). Additionally it is necessary to take into account the regime of pore fluids pressure; if it is hydrostatic, then the temperature of the gypsum transformation decreases along with depth from 52°C under surface conditions to about 40°C at the depth of 3 km, and in the case of the lithostatic regime – rises to about 58°C at 2 km (Jowett et al., 1993).

Fig. 24. Grain boundary migration between two gypsum crystals; phot. J. Jaworska

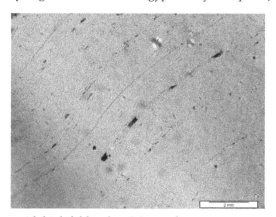

Fig. 25. Large gypsum with kink folds; phot. J. Jaworska

Fig. 26. 'Kink bands' and result of subgrain rotation in gypsum; phot. J. Jaworska

Fig. 27. 'Kink bands' and result of subgrain rotation in gypsum; phot. J. Jaworska

Shahid et al. (2007) comparing the crystallization and transformation conditions of sulphates in salt lakes and sabkhas in north Africa (Libia) and those from the Persian Gulf (Abu Dhabi) noted that while the climate is comparable, in the first case the anhydrite occurs very rarely, unlike in the area of the Arabian Peninsula. The main causes of this difference are the geochemical environment conditions: in the African sabkhas and salt lakes, the environment is more reducing and there is an occurrence of the organic material, the hydrogen sulphide releases and the sediment is dark; while sabkhas from the Persian gulf are more oxidised with lack of hydrogen sulphide - the sediment is light. The presence of fractures and joints in sediments/rocks surrounding the sulphates, as well as the microfractures and pores in the sulphates themselves strongly affect the start of the gypsification and anhydritization. Those free spaces allows the water to migrate and solutions to start and catalyse the course of processes.

2.2.2 Time

The anhydritization and gypsification (dehydration and hydration) under natural conditions can occur very quickly: within few years (Farnsworth, 1925) or even within one year (Moiola & Glover, 1965); and experiments showed that even within several/several dozen of days (i.e. Sievert et al., 2005), what depends on physical and chemical conditions under which the process occurs. We can see for ourselves the speed of these processes, when inside a brick (ceramic material) we note the anhydrite grains, which with infiltrating water are being gypsificated and expand destroying the material   the damage of walls occurs even within several years.

2.2.3 Volume

The volumetric change comes along with hydration and dehydration processes of the sulphates – the increase of volume of anhydrite by its gypsification is about 30-50% according to Petijohn (1957), and according to Azam (2007) - close to 63%. Whereas the gypsum anhydritization decreases its volume of about 39% (Azam, 2007); sometimes it occurs together with many alterations, especially of the primary rock structure. The different situation takes place in case of sulphate deposits which already contain water; according to Farnsworth (1924), 1000g of gypsum fills 431 cm^3, while the sum of anhydrite and water needed to form the same amount of gypsum fills 473 cm³, 9% more – then under natural conditions, when the anhydrite deposit is porous/fractured and water supersaturated, the gypsification process can result not in increase but decrease of volume of the newly formed rock.

2.2.4 Models of gypsification and anhydritisation

According to Hardie (1967) there are three models describing transformation of gypsum into anhydrite (or backwards – anhydrite into gypsum):

1. dissolution of gypsum, and furthermore precipitation of anhydrite (during anhydritisation) or dissolution of anhydrite and later precipitation of gypsum (during gypsification);
2. direct dehydration of gypsum, that is loosing of the crystallization water (during anhydritisation) or adding the water – hydration of anhydrite (during gypsification). This mechanism results in change of the rock volume;
3. dehydration or hydration with mid stage, with participation of bassanite (mineral rarely occurring in nature). During the hydration, the reaction (occurring very slowly) is as follows:

$$4CaSO_4 \cdot \frac{1}{2}H_2O \rightarrow CaSO_4 \cdot 2H_2O + 3CaSO_4$$

bassanite   gypsum anhydrite

2.2.4.1 Anhydritization

Petrichenko (1989) stated that the process of anhydritisation of gypsum began with its dissolution. This process is accompanied by the appearance of the nuclei of the new mineral phase – bassanite. During the second stage, bassanite transforms into anhydrite. The structural rearrangement of this mineral occurs, resulting in increase of thickness at the cost of length. Sheets (plates) of anhydrite crystals form with corroded edges. However in case of the presence of anhydrite "nuclei", the bassanite does not form, but anhydrite continues its crystallization at the cost of the calcium sulphate from dissolved gypsum. On the basis of examination of the inclusions in minerals, Petrichenko (1989) determined the conditions of the origin of anhydrite: this process takes place in the presence of concentrated brine solutions and under the conditions of high pressure and temperature, but not above 40-50°C.

Depending on time and speed of the sulphates transformation there are three kinds of the process: syndepositional, early- and late-diagenetic. The syndepositional anhydritisation occurring during the deposit formation, in shallow basin, sabkhas, in the subsurface environment, causes the substitution of gypsum to take place so fast that the anhydrite remain in its primary form. Anhydritisation during the later stages, according to the solutions of lower salinity, causes the primary sedimentary structures to disappear and the nodular structures to form – gypsum is substituted by incohesive mass of fine anhydritic strips and water, whereas the anhydritisation under the influence of highly concentrated brines can lead to the preservation of the primary gypsum pseudomorphs (Peryt, 1996; Warren, 1999), especially apparent in the coarse-crystalline gypsum forming "the grass-like selenite".

2.2.4.2 Gypsification

The process of hydration was described in detail by Sievert et al. (2005):

1. during the first quick phase, there is an initial partial dissolution of $CaSO_4$ and adsorption of hydratem Ca^{2+} and SO_4^{2-} ions at the surface of anhydrite;
2. during the second – the slower one, there is an increase of thickness of adsorbed layer
3. during the third phase, there is a crack formation in the absorber layer and counter migration of H_2O and Ca^{2+}, SO_4^{2-} ions;
4. during the fourth phase - the formation of gypsum nuclei at the surface of anhydrite occurs and in the end gypsum crystals are formed.

This process takes place in the presence of water (in the active phreatic zone), in temperatures below 40°C (process takes place faster in lower temperatures), and its speed depends on the presence of chemical activators, for example K_2SO_4, $MgSO_4 \bullet 7H_2O$ or H_2SO_4 (Sievert et al., 2005) and CO_2, which speeds and eases the hydration. At first, it covers the most fractured parts of the rock, taking place along the cracks and grain boundaries. As a result of hydration, the anhydrite rock transforms into gypsum rock with fine-grained (alabaster), fibrous, porphyroblastic texture (Warren, 1999), coarse/lenticular-crystalline gypsum (sometimes with preserved relic of the anhydritic precursor) – they result from the dissolution of primary sulphates (fine-crystalline anhydrites); see fig. 22. and 23. The secondary gypsum can also be formed as a pseudomorph of the primary anhydrite (e.g. the floor of the cap-rock) or the coarse-crystalline gypsum (selenitic gypsum), which underwent anhydritisation and furthermore gypsification – in this case, despite the multi-stage characteristics of the diagenetic processes, the primary rock structure is preserved. There is an example of the Zechstein (Permian) sulphates, which were uplifted close to the surface as a result of diapirism, and further incorporated into a cap-rock, while being anhydritised and later gypsificated (Jaworska & Ratajczak, 2008).

2.3 Inclusions

Inclusions or remains of the primary precursor minerals (e.g. the remains of anhydrite in gypsum) can appear in the primary as well as in the secondary sulphates. Particularly valuable are the inclusions in the primary minerals which can be liquid, solid, gaseous, or even organic. They reach diameters between few and several hundred of µm. Sometimes they are arranged zonally, rhythmically – as the crystal grew. Among the inclusions:

a. solid – most often occur: clay minerals, quartz, chalcedony, barite, halite, carbonates - calcite, dolomite, magnesite
b. liquid – mainly the chlorine-sulphate solutions of various mineralization,

Part of the solutions can be saturated with gases (CO_2, N_2, CH_4, H_2 and H_2S), e.g. originating from the organic decomposition (Petrichenko et al., 1995). For example, in the badenian gypsum of Carpathian Foredeep, the presence of: fragments of characean algas, filamentous algas, and colony of unicellular cyanobacterium, insects, coccoids, and multicellular organisms – most probably fungi, has been confirmed. The good state of preservation of these microorganism tissues indicates anaerobic conditions during gypsum precipitation (Petrichenko et al., 1995). The detailed inclusion analyses led to a series of conclusions on the environment, chemical (basin type: open sea or inland ?; brine type: e.g. Na- (Ca)-SO_4-Cl or Mg-Na-(Ca)-SO_4-Cl or Na-CO_3-SO_4-Cl ?) and biochemical conditions during the sulphate sedimentation; the variations of the solution chemical composition (e.g. indication of the fresh sea water inflow direction). In addition, the analyses of one-phase liquid inclusions provide information on the water temperature in the crystallization basin.

2.4 Calcitization

Sulphates, as well as gypsum and anhydrite can undergo calcification by:

a.   bacterial reduction in deposits rich in organic substances – the most effective process,
-    sulphates are altered by S-reducing bacteria to form H_2S, pyrite and other sulphides, native S and calcite (Holster, 1992)
b.   infiltration of meteoric water rich in carbonate ions – occurs during the sulphates exposition onto water activity (Warren, 1999),
c.   thermal reduction of sulphates – late diagenetic process, occurs in temperatures over 100°C, under atoxic conditions and with presence of the hydrocarbons (Machel, 1987).

Dissolution of sulphates in presence of hydrocarbons leads to biogenic SO_4 reductions and calcite precipitation according to reactions:

$$CH_4 + CaSO_4 \rightarrow CaCO_3 + H_2S + H_2O$$

Calcitization of the sulphates can be a multi-stage process (Scholle et al., 1992), which begins with (1) dissolution (or at least corrosion) of anhydrite, (2) hydration of anhydrite and gypsum formation, (3) dissolution of gypsum (this process can be accompanied by the formation of collapse breccia), and afterwards (4) precipitation of calcite inside free spaces and pores arisen after leached sulphates. Sometimes sulphur is the secondary product of calcitization of sulphates (see fig. 18.).

Generally, the gypsum – more easily than the anhydrite – can be substituted by calcite. In case where this process occurs in bigger scale, the post-gypsum limestones form. They can occur in the highest parts of the cap-rock, covering the upper parts of some diapirs - upon the area of Costal Gulf the shallowest subsurface cap-rock levels are usually formed as the calcitic deposits and therefore named as calcitic cap-rocks. However, the microscopic analyses of the cap-rock deposits demonstrated that among the secondary coarse-grained gypsum with the anhydrite remains, the calcification process starts exactly with these anhydrite inclusions, not with the gypsum.

2.5 Polyhalitization

The sulphate rocks can also undergo the polyhalitization process. It proceeds during the early stages of the diagenesis of evaporites as a result of infiltration of hot brines into the sulphate deposits (in the peripheral zones of the evaporite basins): halite saturated, with high contents of Mg^{2+} and K^+ (originating from the dissolution of the potassium salts in the local salt pans), sulphate-rich (Peryt, 1995 and 1996). This process starts from the edges of the grain/crystal and proceeds with deep embayments into the core – the anhydrite/gypsum grain disintegrates into smaller parts that undergo polyhalitization more easily (Stańczyk, 1970).

2.6 Dissolution and Karst

Sulphates – gypsum in particular – are common ingredients of the lithosphere and often occur close to the Earth's surface. Additionally, the gypsum easily undergo physical weathering (is soft and has ductile rheology), as well as chemical (dissolves in water). Gypsum dissolution rates reach 29 mm/year and have been measured in Ukraine (Klimchouk & Aksem, 2005). Therefore upon the areas of gypsum deposits karst processes and forms occur (fig. 19.). Gypsum-karst features commonly develop along bedding planes, joint or fractures; sometimes up to 30 m below the Earth's surface. The evidence is the presence of: caves, sinkholes, karren, disappearing streams and springs, collapse structures (Johnson, 2008). One of the longest reported gypsum caves is D.C. Jaster Cave (SW Oklahoma, USA) where main passage is 2,413 m long but total length of all the passages reaches 10,065 m (Johnson, 2008). Speleothems in gypsum caves may provide information about paleoclimate and climate changes in the past, because in arid or semi-arid climates, the speleothems in gypsum cave are mainly composed of gypsum, whereas in contrast, in humid or tropical climate – of carbonate (calcite). The dating of speleothems could provide the paleoclimatic data relating to:

a. dry periods, when gypsum speleothems were deposited,
b. wet periods in arid zone, when calcite speleothems were deposited (Calaforra et al., 2008).

Gypsum-karst area could be dangerous and should be monitored due to the risk of danger. Some sinkholes and collapse structures, commonly being few hundreds m wide and tens of m deep, may cause the loss of human lives and damages, e.g. in Spain in Oviedo and Calatayud situated on cavernous gypsum area, direct economic losses by collapse events were estimated to be 18 mln euro in 1998 and 4.8 mln euro in 2003 (Gutiérrez et al., 2004 and 2008).

The process of the sulphates dissolution is visible not only in developement of karst features; it reveals itself in the smaller scale for example in development of stylolites as a result of pressure solution. The development of the stylolitization process has been usually described among the carbonate rocks - mainly limestones; in the evaporites the stylolites are exceptional. Bäuerle et al. (2000) took under consideration the problem of stylolites genesis in the main anhydrite deposits located in the salts of the Gorleben diapir (Germany). Detailed studies of these forms led to estimation of the amount of dissolved material thanks to the measurements of the maximum amplitudes of the stylolitic sutures visible inside the core. The calculations showed that over 26% rock mass were dissolved. Moreover the microscopic observations indicated the gaps in the sutures – the sutures were 'cut' by the anhydrite crystals formed as pseudomorphs after gypsum. This fact proves that the stylolitization had developed before the gypsum underwent anhydritization. In the article summary, the authors plotted the conditions of the stylolites formation in sulphates, especially in gypsum as the primary deposit where such forms appear. The process requires:

a. the presence of interbeds different than the sulphate rocks; the lithological heterogeneity,
b. the presence of overburden in which the increase of thickness and its chemical characteristics favour the conditions where the lower gypsum is under conditions balancing between pressure solution and the gypsum-anhydrite transformation.

3. Geochemystry of sulphates

The analysis of chemical (including isotopes) contents of the sulphate rocks leads to the conclusions regarding their genesis and diagenetic transformations; e.g. strontium and boron.

3.1 Strontium (Sr)

Sr can substitute Ca ions in minerals (mainly in carbonates and sulphates) or create their own minerals (celestine or strontianite), which very often occur dispersed in the marine sediments. High level of Sr characterizes the rocks formed during the final stage of carbonates' sedimentation and during the first stage of calcium sulphates crystallization; generally in sulphates the Sr content increases in direct proportion to the brine concentration (Rosell et al., 1988). The primary gypsum precipitated from evaporated seawaters is expected to have a Sr content of 0.1-0.2% (Ichikuni & Setsuko Musha, 1978) and the one from K-Mg brines - the content of 0.97% Sr (Usdowsky, 1973). Butler (1973) thinks that gypsum precipitated from the celestine saturated solution should consist about 0.09% Sr, and anhydrite – about 0.24% Sr, but primary selenitic gypsum from the Eastern Betics basin contents strontium only between 493-625 ppm Sr (Warren, 2006) and primary Zechstein (Permian) anhydrites content 0.61% Sr (Polański & Smulikowski 1969).

Multimodal distribution of Sr compound in the primary sulphates (gypsum in particular) profiles indicates various sources of this element and multi-stage process of its concentration. The Miocene selenite gypsum from the southern border of the Holy Cross Mts. shows high Sr content (averagely 1300-1500 up to max. 2575 ppm); and scarce variations of the content indicate only episodic salinity fluctuations of the basin, probably connected with the inflow of fresh sea or meteoric water; the gypsum was formed in the sub-aqueous environment characterized by high salinity, whereas the laminated stromatolitic gypsum is characterized by high variations of Sr content (from max. 3695 to 179 ppm), simultaneously indicating high salinity fluctuations (Kasprzyk, 1993).

Strontium can also originate from diagenetic processes: bacterial sulphates reduction, dissolution and recrystallization – they may favour the liberation of strontium ions from the sulphate and could locally form higher concentrations within the other sulphate rocks (Kasprzyk, 1994).

Apparent decrease of Sr concentration occurs during rock transformation in the open system with unbounded circulation of the solution in free pore spaces, whereas the residual products of these transformations are often enriched in strontium. During the hydration of anhydrite, gypsum shows limited ability of Sr ions incorporation into its crystal lattice and is not able to incorporate them completely. Dissolution and recrystallization purify gypsum and anhydrite from impurities, and activate strontium lowering its content in newly created mineral comparing to the primary mineral, i.e. some secondary gypsums from Wapno Salt Dome consist only 159 ppm Sr (Jaworska & Ratajczak, 2008), primary anhydrite from which it has been created consist 1700 ppm Sr.

3.2 Boron (B)

B likewise Sr is a sensitive indicator of changing conditions in the evporite sedimentary environment, as its concentration in the sediment depends on the salinity.

Systematic increase of B content in the profile of sulphate sediments indicates progressive increase of basin salinity during the crystallization of successive generations of sulphates – evaporites containing the highest amounts of B originate from the most concentrated solutions. Any decrease/variation/fluctuation of this element concentration indicates fresh (sea or meteoric) water supply to the evaporite basin.

Sea water contains 4,45 ppm of boron, mainly in the form of undissociated ortho-boric acid. Solutions of this element deriving from the terrigenic sediments, submarine exhalations and decomposing clay minerals (especially illite) constitute the source of borate ions in the sedimentary basins. Ions of BO_3^{2-} can isomorphically replace SO_4^{2-} and form their own minerals (borates, e.g. boracite).

The highest B concentrations are noted during the latest stages of evaporation – when the K-Mg salts precipitate accompanied (under favour conditions) by borates' crystallization. The B content in sulphate rocks (gypsum, as well as anhydrite) can fluctuate between 2 and 5500 ppm; in the Zechstein anhydrite the content ranges from 16 to 500 ppm, and in polyhalite reaches 800 ppm (Pasieczna, 1987) – generally, there are high B contents noted in polyhalite.

Sulphates can be analysed from the point of view of Mn and Fe contents; increased concentrations of both elements usually indicate the terrigenic deposit (siliciclastic sediments, clay minerals) supply into the sedimentary basin.

3.3 Isotopes

Another indication of the genesis and diagenesis of sulphates are the isotopic analyses of $^{87}Sr/^{86}Sr$ ratio, S ($\delta^{34}S$) and O ($\delta^{18}O$) in SO_4, and in the case of gypsum, also O ($\delta^{18}O$) of the crystallization water. $\delta^{34}S$ and $\delta^{18}O$ in SO_4 does not change despite of many transformations, the sulphate molecule maintain its primary isotopic composition, what allows to determine the primary sedimentary conditions, but dynamic and multiple transformation can affect the $\delta^{18}O$ of crystallization water, so in gypsum we have to indicate two $\delta^{18}O$ – in SO_4 and H_2O.

3.3.1 Sulphur (S)

The present-day $^{34}S/^{32}S$ ($\delta^{34}S$) ratio of sulphates in oceanic water is constant and reaches +20±0.5‰ with respect to V-CDT (Pierre, 1988) and the fractionation between dissolved sulphates in oceanic water and crystallized sulphates is negligible (Thode & Monster, 1965; Raab & Spiro, 1991). $\delta^{34}S$ was changing in the geological past and its general trends are known as the sulphur-isotope age curve (Claypool et al., 1980). This curve allows to define the time of evaporate crystallization.

3.3.2 Oxygen (O)

The present-day $^{18}O/^{16}O$ ($\delta^{18}O$) ratio of sulphates in oceanic water reaches 9.5±0.5‰ with respect to V-SMOW (Longinelli & Craig, 1967) but during crystallization of the oceanic sulphates, the $\delta^{18}O$ is raised up to 3.5‰ (Lloyd, 1968; Pierre, 1988) and $\delta^{18}O$ value of this sulphates reaches 13.0±0.5‰.

Primary gypsum and its crystallization water are formed in isotopic equilibrium with the mother brine (Sofer, 1978), but gypsum can easy loose its original crystallization water during further dehydration and hydration. During hydration sulphates interact with meteoric-, ground-, residual or sea water and gypsum absorbs this new, fresh or sometimes mixed primary water. In the areas of several-, several dozen of m long profiles consisting gypsum rocks, basing on the determination of $\delta^{18}O$ of their crystallization water, it is possible to indicate the type and range of individual water types which affected the sulphates. E.g. in profiles of the cap-rock of the Wapno and Mogilno salt diapirs (Jaworska, 2010) there is gypsum, which shows $\delta^{18}O$ of crystallization water indicating the influence of: cold period - post-glacial water – $\delta^{18}O$ reaches values from -11 up to -13‰ in the lowest part of the profile (Wapno and Mogilno), recent (or similar to) meteoritic water - $\delta^{18}O$ reaches values of -9 to -10‰ (Wapno), cap-rock water - $\delta^{18}O$ reaches -4.3 to -6.6‰ (Mogilno), „mixing" water or warmer period water - $\delta^{18}O$ is -5.6‰ (Wapno) and from -6.9 to -8.7‰ (Mogilno).

The presence of water described as recent or originated from the colder periods inside the lowest and the middle parts of the cap-rock is very important for further management plan of such salt structure. The influence of present day water or the water from colder periods in the lowest part of the cap-rock indicates free flow of surface water into the area of so called salt mirror; the presence of this water in the middle part of the cap-rock indicates the occurrence of cracks, fractures and karst forms in cap-rock body. In consequence it means, that such cap-rock is not a hermetic cover and does not fulfil the requirements for a seal which protects the rock salt and salt mirror against inflow of freshwater. This information is of great importance for salt structures which are prepared for underground disposal of radioactive waste or for the storage of hydrocarbons, as well as salt mine.

3.3.3 Strontium (Sr)

The $^{87}Sr/^{86}Sr$ ratio of modern oceanic water is uniform and reaches 0.70901 (Burke et al., 1982) but has been changing in time. Main reasons of these irregular changes were contribution of Sr with high $^{87}Sr/^{86}Sr$ ratios from continents and input of Sr with low $^{87}Sr/^{86}Sr$ ratios from active mid –oceanic ridges (Veizer, 1989; Chaudhuri & Clauer, 1992). The general trends and variations of the marine Sr isotopes during the Phanerozoic carbonates are known (Burke et al., 1982) and this curve (the same as S-curve) allows us to study the age of evaporates precipitation. In evaporites the $^{87}Sr/^{86}Sr$ ratios reflect the isotopic composition of the brines or diagenetic fluids. Strontium does not fractionate (Holster, 1992).

Present-day strontium isotope ratio equilibrated between [87]Sr-depleted young oceanic basalts and hydrothermal activity along mid-oceanic ridges (ca. 0.7035) and [87]Sr-enriched continental sediments (from old continental granites) transported into the basin by wind and rivers (ca. 0.7119 and more; Chaudhuri & Clauer, 1992; Dickin, 2005). It is the same reason why primary Sr isotopic ratio of evaporites could not be the same as that of contemporaneous sea water – e.g. sediments may have deposited in closed basin with inflow of continental water and continental Sr - the Sr ratio of such sulphates is higher than the one of contemporaneous ocean water, so any variation of Sr isotopic composition may relate to the paleohydrology of the basin. Additionally, variations of Sr isotopic ratio may be explain by contamination with more radiogenic Sr or by diagenesis (Hess et al., 1986; Saunders et al., 1988; Chaudhuri & Clauer, 1992).

4. Recrystallization

In the classic approach recrystallization means the transformation of fine-crystalline minerals/rocks into coarse-crystalline ones and makes sometimes the continuation of the recovery process, when the mineral/rock or the whole material tries to loose the excess of the internal energy generated during the deformation/strain, when the crystal lattice defects occur. During those processes the shape and size of grains change and the crystallographic axes rotate; they are also accompanied by progressive loss and disappearance of the primary rock texture/structure.

In the case of recrystallization of cap-rock gypsum, a reverse process can be generally observed (looking upwards) - the size reduction of the mineral grains (dominant or subordinate components).

The boundaries between adjacent fine gypsum grains are usually blurred and irregular, what results from transformation of the larger grains into smaller ones, which successively become individual.

The recrystallization of gypsum can occur via: grain boundary migration or subgrain rotation. The grain boundary migration is characteristic for the mineral grains with large variety of lattice defects density, whereas the subgrain rotation occurs in grains with uniformly dispersed defects (Passchier & Trouw, 1998).

4.1 Grain boundary migration

If the adjacent grains differs in defects density, the defect-poor one bulges into the defect-rich one; see fig. 24. It results in the removal of grains with many dislocations. It also enables the spontaneous crystallization and the growth of new grains - "nuclei" (either defectless or with few dislocations) inside the defect-rich grain; these fine new grains are called 'subgrains' as well.

4.2 Subgrain rotation

The deformation bands formed during the recovery tighten progressively, creating a grid determined by subgrain walls that developed successively within the grain. The subgrains are fragments of larger grain with fine boundaries. As a result of rotation, the crystalline axis of the subgrain becomes slightly misoriented relating to the axes of the adjacent subgrains or the main grain/crystal; the misorientation angle usually reaches max. 5° (FitzGerald et al., 1983; White & Mawer, 1988 fide Passchier & Trouw, 1998). During the rotation recrystallization the mylonitic and porphyroblastic/porphyroclastic rocks are formed.

Another (however not so common) mechanism of subgrain development can be observed in the rocks of the gypsum cap-rock - the process is called kinking and leads to formation of 'kink bands' (Means & Ree, 1988 fide Passchier & Trouw, 1998), which are represented by narrow accumulation of kink folds; see fig. 25. They are formed in brittle-ductile system and correspond to the initial shearing along the planes oblique to the dense anisotropic planes (sedimentary, metamorphic, lattice anisotropy) under the influence of parallel (to those planes) or close to parallel compression at rather high surrounding pressure (Dadlez & Jaroszewski, 1994). This process has been observed in few mm to few cm lenticular, cigar-shaped gypsum crystals; see fig. 26. and 27.

5. Summary

Sulphates are common minerals; they are easy crystallized, alternated and recrystallized.

Distinct variation of isotope ratios of sulphur, oxygen and strontium in the sea water sulphates in time enables their use to determine:

- the age of evaporite deposits;
- the sulphates' origin (marine or non-marine?, and primary or secondary minerals?) and
- in the case of gypsum (oxygen analysis of crystallization water), the determination of paleoclimatic conditions (also the time) when the gypsification occurred due to water particles accretion or isotopic composition exchange of water in gypsum.

The liquid inclusions analysis in the primary evaporites enables determination of chemical composition of primary solutions/brines from which the sulphates crystallized, as well as the temperature of water.

The analysis of the primary minerals remains constituting the impurities in the secondary crystals enables determination of the diagenetic processes taking place in the evaporite deposits (including the mineral precursor for the secondary crystal), and the direction and cause of diagenetic transformations (e.g. anhydrite gypsification: primary mineral – anhydrite, cause – presence of fresh or low-mineralized water in the deposit, e.g. as a result of tectonic uplift and exposition to the activity of shallow underground water).

The crystal shape, form and texture of gypsum and anhydrite sediments indicate the environmental conditions of their formation such as: basin bathymetry (shallow or deep zones of the basin), water oxygenation, either stability or dynamics of the environment (e.g. turbidity currents, sea-level fluctuations – in case of high variability and low thickness of separate sulphate lithotypes in the profile).

Trace elements analysis in sulphates:

1. Sr, B contents: constant increase of their contents in the profile indicate stable evaporation conditions; their variations episodes connected with the fresh water inflows to the evaporite basin and its dilution;

2. Mn and Fe contents: elevated concentrations of both elements indicate the supply of terrigenic sediments to the basin.

6. References

Azam S. 2007 - Study on the geological and engineering aspects of anhydrite/gypsum transition in the Arabian Gulf coastal deposits. Bull. Eng. Geol. Env., 66: 177–185.

Barley M.E., Dunlop J.S.R., Glover J.E. & Groves D.I. 1979 – Sedimentary evidence for an Archaean shallow-water volcanic-sedimentary facies, eastern Pilbara Block, Western Australia. Earth and Planetary Science Letters, 43: 74-84.

Blatt H., Middleton G. & Murray R. 1980 - *Origin of Sedimentary Rocks* (2nd ed.). Prentice-Hall, Inc. Englewood Cliffs, 782 pp.

Buick R. & Dunlop J.S.R. 1990 – Evaporitic sediments of early Archaean age from the Warrawoona Group, North Pole, Western Australia. Sedimentology, 37: 247-277.

Burke W.H., Denison R.E., Hetherington E.A., Koepnick R.E., Nelson H.F. & Otto J.B. 1982 – Variations of sea water $^{87}Sr/^{86}Sr$ throughout Phanerozoic time. Geology, 10: 516-519.

Butler G.P. 1973 – Strontium geochemistry of modern and ancient calcium sulphate minerals [in:] Purser B.H. (ed.) The Persian Gulf. Springer, New York, Heidelberg, Berlin, 423-470.

Calaforra J.M., Forti P. & Fernandez-Cortes A. 2008 – Speleothems in gypsum caves and their paleoclimatological significance. Environ. Geol., 53: 1099-1105.

Chaudhuri S. & Clauer N. 1992 – History of marines evaporites: constraints from radiogenic isotopes. Lecture Notes in Earth Sciences, 43: 177-198.

Claypool G.E., Holster W.T., Kaplan I.R., Sakai H. an& Zak I. 1980 – The age curves of sulfur and oxygen isotopes in marine sulfate and their mutual interpretation. Chem. Geol., 28: 199-260.

Conley R.F. & Bundy W.M. 1958 - Mechanism of gypsification. Geochim. Cosmochim. Acta, 15: 57-72.

Dadlez R. & Jaroszewski W. 1994 - *Tektonika*. PWN, Warszawa, 743 pp.

Dickin A.P. 2005 – *Radiogenic isotope geology* (2 ed.). Cambridge University Press, 67 pp.

Farnsworth M. 1924 - Effects of temperature and pressure on gypsum and anhydrite. U. S. Bur. Mines Rept. Inv. 2654

Farnsworth M. 1925 – The hydration of Anhydrite. Industrial and Engineering Chemistry, 17, 9: 967-970.

FitzGerald J.D., Etheridge M.A. & Vernon R.H. 1983 – Dynamic recrystallization in a naturally deformed albite. Text Microstruct, 5: 219-237.

Ford D.C. & Williams P.W. 1989 – *Karst geomorphology and geology*. London, Chapman and Hall., 601 pp.

Ford D.C. & Williams P.W. 2007 - *Karst hydrogeology and geology* (2sec ed.). Wiley, 562 pp.

Gutiérrez F., LuchaP. & Guerrero 2004 – La dolina de colapso de la casa azul de Calatayud (noviembre de 2003). Origen, efectos y pronóstico [in:] Benito G., Díez-Herrero A. (eds.) Riesgos naturales y antrópicos en Deomorfología, VII Reunión Nacional de Geomorfología, Toledo, pp. 477-488.

Gutiérrez F., Cooper A.H. & Johnson K.S. 2008 – Identification, prediction, and mitigation of sinkhole hazards in evaporite karst areas. Environ. Geol., 53: 1007-1022.

Hardie L.A. 1967 - The gypsum-anhydrite equilibrium at one atmosphere pressure. Am. Mineral., 52: 171-200.

Ichikuni M. & Setsuko Musha 1978 - Partition of strontium between gypsum and solution. Chemical Geology, 21 (3-4): 359-363.

Hess J., Bender M.L. & Schilling J.G. 1986 – Evolution of the ratio strontium-87 to strontium-86 in seawater from Cretaceous to present. Science, 231: 979-984.

Holster W.T. 1992 – Stable isotope geochemistry of sulfate and chloride rocks. Lecture Notes in Earth Sciences, 43: 153-176.

Jaworska J. 2010 – An oxygen and sulfur isotopic study of gypsum from the Wapno Salt Dome cap-rock (Poland). Geological Quarterly, 54, 1: 25-32.

Jaworska J. & Ratajczak R. 2008 - Geological structure of the Wapno Salt Dome in Wielkopolska (western Poland). Prace Państwowego Instytutu Geologicznego, Warszawa, 190, 69 pp. (in Polish, with English summary).

Johnson K.S. 2008 – Evaporite-karst problems and studies in the USA. Environ. Geol., 53: 937-943.

Jowett E.C., Cathles III L.M. & Davis B.W. 1993 – Predicting depths of gypsum dehydration in evaporitic sedimentary basin. AAPG Bull., 77, 3: 402-413

Kasprzyk A. 1993 – Prawidłowość występowania strontu w gipsach mioceńskich południowego obrzeżenia Gór Świętokrzyskich. Przegląd Geologiczny, 41, 6: 416-421. (in Polish with English summary).

Kasprzyk A. 1994 – Distribution of strontium in the Badenian (Middle Miocene) gypsum deposits of the Nida area, southern Poland. Geological Quarterly, 38, 3: 497-512.

Kasprzyk A., 2003 - Sedimentological and diagenetic patterns of anhydrite deposits in the Badenian evaporite basin of the Carpathian Foredeep, southern Poland. Sedimentary Geology, 158 (3-4): 167-194.

Klimchouk A.B. & Aksem S.D. 2005 – Hydrochemistry and solution rates in gypsum karst: case study from the Western Ukraine. Environ. Geology, 48: 307-319.

Klimchouk A. & Andrejchuk V. 1996 - Sulphate rocks as an arena for karst development. International Journal of Speleology, 25 (3-4): 9-20.

Kubica B. 1972 - O procesie dehydratacji gipsów w zapadlisku przedkarpackim. Przegląd Geologiczny, 20, 4: 184-188.

Lloyd R.M. 1968 - Oxygen isotope behavior in the sulfate-water system. J. Geophys. Res., 73: 6099-6110.

Longinelli A. & Craig H. 1967 - Oxygen-18 variations in sulfate ions and sea water and saline lakes. Science, 156: 56-59.

Machel H.G. 1987 – Sadle dolomite as a by-product of chemical compaction and thernochemical sulfate reduction. Geology, 15: 936-940.

Means W.D. & Ree J.H. 1988 – Seven types of subgrain boundaries in octachloropropane. J. Struct. Geol., 10: 765-770.

Moiola R.J. & Glover E.D. 1965 – Recent anhydrite from Clayton Playa, Nevada. Am. Mineralogists, 50: 2063-2069.

Murray R.C. 1964 - Origin and diagenesis of gypsum and anhydrite. J. Sed. Petrol., 34, 3: 512-523.

Pasieczna A. 1987 – Mineralogical and geochemical analysis of the Zrchstein sulphate deposits of the Puck Bay region. Archiwu Mineralogiczne, 43, 1: 19-40. (in Polish with English summary).

Passchier C.W. and Trouw R.A.J. 1998 – Microtectonics (2ed ed.). Springer – Verlag, Berlin, Heidelberg.

Peryt T.M. 1995 – Geneza złóż polihalitu w cechsztynie rejonu Zatoki Puckiej w świetle badań sedymentologicznych i geochemicznych. Przegląd Geologiczny, 43, 12: 1041-1044.

Peryt T.M. 1996 - Diageneza ewaporatów. Przegląd Geologiczny, 44, 6: 608-611.

Petrichenko O.I. 1989 - Epigenez evaporitov. Naukova Dumka, Kiev, 62 pp.

Petrichenko O.I., Peryt T.M., Poberezski A.W. & Kasprzyk A. 1995 – Inkluzje mikroorganizmów w kryształach badeńskich gipsów Przedkarpacia. Przegląd Geologiczny, 43, 10: 859-862.

Pettijohn F.J. 1957 - Sedimentary rocks (2ed ed.). Harper & Bros., New York, 718 pp.

Pierre C. 1988 - Applications of stable isotope geochemistry to study of evaporites. In: Schreiber BC (ed). Evaporites and Hydrocarbons. Columbia University Press, New York, pp 300-344.

Polański A. & Smulikowski K. 1969 – Geochemia. Wydawnictwa Geologiczne, Warszawa.

Posnjak E. 1940 - Deposition of calcium sulfate from sea water. Am. J. Sci., 238: 559-568.

Raab M. & Spiro B. 1991 - Sulfur isotopic variation during seawater evaporation with fractional crystallization. Chemical Geology, 86: 323-333.

Rosell L., Ortí F., Kasprzyk A., Playà E. & Peryt T.M. 1998 - Strontium geochemistry of Miocene primary gypsum: Messinian of southeastern Spain and Sicily and Badenian of Poland. Journal of Sedimentary Research, 68: 63–79.

Saunders J.A. 1988 – Pb-Zn-Sr mineralization in limestone caprock, Tatum salt dome, Mississippi. Trans. Gulf Coast Assoc. Geol. Soc., 38: 569-576.

Scholle P.A., Ulmer D.S. & Melim L.A. 1992 – Late-stage calcites in the Permian Capitan Formation and its equivalents, Delaware Basin margin, west Texas and New Mexico: evidence for replacement of precursor evaporates. Sedimentology, 39: 207-234.

Shahid S.A., Abdelfattah M.A. & Wilson A. 2007 - A Unique Anhydrite Soil in the Coasta Sabkha of Abu Dhabi Emirate. Soil Surv. Horiz., 48: 75-79.

Slevert T., Wolter A. & Singh N.B. 2005 - Hydratation of anhydrite of gypsum (CaSO$_4$.II) in ball mill. Cement and Concerete Research, 35: 623-630.

Sonnenfeld P. 1984 - Brines and evaporates. Academic Press, London, 613 pp.

Stańczyk I. 1970 – Polihalit w kopalniach soli regionu kujawskiego. Acta Geologica Polonica, 10, 4: 305-820.

Stewart F.H., 1968 – Geochemistry of marine evaporate deposits. Geological Society America Special Paper, 88: 539-540.

Thode H.G. & Monster J. 1965 - Sulfur isotope geochemistry of petroleum evaporites in ancient seas. AAPG Mem., 4: 367-377.

Usdowsky E. 1973 – Das geochemische Verhalten des Strontiums bei der Genese und Diagenese von Ca-Karbonat-und Ca-Sulfat-Mineralen. Contrib. Minerl. Petrol., 38: 177-195.

Veizer J. 1989 – Strontium isotopes in seawater through time. Ann. Rev. Earth Plan. Sci., 17: 141-167.

Warren 1999 – *Evaporites – their Evolution and Economy*. Blackwell Science, 438 pp.

Warren 2006 – *Evaporites: Sediments, Resources and Hydrocarbons*. Springer, 1035 pp.

White J.C. & Mawer C.K. 1988 – Dynamic recrystallisation and associated exsolution in perthites: evidence of deep crystal thrusting. J. Geophys. Res., 93: 325-337.

Wilson A.H. & Versfeld J.A. 1994 – The Elary Archaean Nondweni greenstone belt, southern Kaapvaal Craton, South Africa; Part I, Stratigraphy, sedimentology, mineralization and depositional environment: Precambrian Research, 67: 243-276.
http://minerals.usgs.gov/minerals/pubs/commodity/

2

Separation of Uranyl Nitrate Hexahydrate Crystal from Dissolver Solution of Irradiated Fast Neutron Reactor Fuel

Masaumi Nakahara

Japan Atomic Energy Agency, Nuclear Fuel Cycle Engineering Laboratories
Japan

1. Introduction

Batch crystallization is widely used for the separation and high purification of organic and inorganic materials in the fine chemical, food, pharmaceutical and biochemical industries. In the atomic power industry, application of crystallization to U purification of the Plutonium Uranium Reduction Extraction (PUREX) first cycle product was attempted in Kernforschungszentrum Karlsruhe (KfK), Germany (Ebert et al., 1989). The feed solution had 240−480 g/dm^3 U concentration and 0.1 g/dm^3 fission products (FPs) concentration in 5−6 mol/dm^3 HNO$_3$ solution. Reducing conditions were achieved with 2.4 g/dm^3 of U(IV) which was added to change the Pu valence to Pu(IV) which was required for good separation of Pu from U. In a six-stage cascade crystallizer, the feed solution was cooled down in steps from 30 to −30°C in the course of about 30 min. More than 90% of U was recovered in form of uranyl nitrate hexahydrate (UNH) crystals with an average diameter of 0.2 mm, while a much greater proportion of the transuranium (TRU) elements and FPs remained in the mother liquor. The decontamination factors (DFs) of several of the FPs were determined for one crystal step plus several crystal washing operations. The measured DFs of Pu and Cs were 10^2 and 10^3, respectively.

An advanced aqueous reprocessing for a fast neutron reactor fuel cycle named "New Extraction System for TRU Recovery (NEXT)" has been proposed as one fast neutron reactor fuel reprocessing method (Koyama et al., 2009) and is being developed in Japan Atomic Energy Agency (JAEA). On the advanced aqueous reprocessing for fast neutron reactor fuel cycle, it is supposed to recover not only U and Pu but also minor actinides (MAs; Np, Am and Cm) for the efficient utilization of resources. It will be also effective in decreasing the environmental impact because of their long half-life and high radiotoxicity. These elements are loaded in a fast neutron reactor and are burned as core fuel. Figure 1 shows schematic diagram of NEXT process for fast neutron fuel reprocessing. The NEXT consists of highly efficient dissolution of fuel with HNO$_3$ solution (Katsurai et al., 2009), U crystallization for partial U recovery (Shibata et al., 2009), simplified solvent extraction for U, Pu and Np co-recovery using tri-n-butyl phosphate (TBP) as an extractant (Sano et al., 2009), and extraction chromatography for mutual separation of actinide elements and lanthanide elements from a raffinate (Koma et al., 2009). The powdered fuel was dissolved by the highly efficient dissolution process and the dissolver solution was adjusted to high heavy metal concentration. Then, U is recovered as UNH crystals from dissolver solution derived from fast neutron reactor fuel. Since the amount throughput will be reduced in the simplified solvent extraction process, the adoption of the TRU crystallization process is expected to reduce the radioactive waste, equipment, and hot cell volume. In addition, U/Pu ratio in the dissolver solution is adjusted in the crystallization process to be a suitable Pu content for core fuel fabrication. In the NEXT, Np is changed to Np(VI) in the high HNO_3 concentration feed solution and is co-extracted with U and Pu in the simplified solvent extraction system. The FPs in the raffinate obtained from the simplified solvent extraction process is removed using N,N,N',N'-tetraoctyl-3-oxapentane-1,5-diamide (TODGA) absorbent in the extraction chromatography I. The actinide elements such as Am and Cm is recovered from the solution containing actinide and lanthanide elements by chromatography with 2,6-bis-(5,6-dialkyl-1,2,4-triazine-3-yl)pyridine (R-BTP) absorbent in the extraction chromatography II.

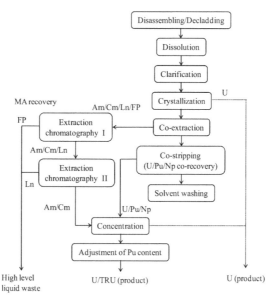

Fig. 1. Schematic diagram of the NEXT process

A dissolver solution of irradiated fast neutron reactor mixed oxide (MOX) fuel in JAEA contains a number of TRU elements and FPs than in KfK. Since U is used as blanket fuel and TRU elements are supposed to recover by other chemical process, it is need to remove TRU elements and FPs from UNH crystals in the U crystallization process. It would be also bring about reduction in the cost for the recovered U storage and the blanket fuel fabrication due to decreased radiation shielding. Therefore, the behavior of TRU elements and FPs in the U crystallization process must be confirmed experimentally.

Since U is recovered as UNH crystal for a blanket fuel fabrication in the U crystallization process, the crystal ratio of U should be evaluated with a dissolver solution of irradiated fast neutron reactor. The crystal ratio of UNH affects HNO_3 concentration in the feed solution. In this study, the feed solution was changed in HNO_3 concentration and the influence on the UNH crystal ratio was examined in the cooling batch crystallization. Two experiments, crystal ratio and the co-existing element behavior, were carried out with a dissolver solution derived from irradiated fast neutron reactor "JOYO" core fuel in a hot cell of the Chemical Processing Facility (CPF), JAEA. Additionally, current status of crystallization apparatus and crystal purification method for the NEXT is described in this paper.

2. Principal of uranium crystallization

In a HNO_3 solution, U ions are crystallized as UNH by the following reaction.

$$UO_2^{2+} + 2NO_3^- + 6H_2O \leftrightarrow UO_2(NO_3)_2 \cdot 6H_2O \tag{1}$$

Figure 2 shows the solubility curves of U in HNO_3 solution (Hart & Morris, 1958). The results represent the mean of two temperatures observed for the first formation and final disappearance of crystal on, respectively, slowly cooling and warming solutions with vigorous agitation. The U ions concentration decreases with decreasing temperature in the solution before reaching the eutectic point, where H_2O and HNO_3 start to crystallize. Thus, U crystallization process should be performed in the right region of the minimum point in this figure. A high HNO_3 solution is desirable for achieving a low U concentration in the solution, therefore yielding more UNH crystals because the eutectic point shifts from right to left as the HNO_3 concentration increases.

In $Pu(NO_3)_4$-HNO_3-H_2O system, the crystallization behavior of plutonyl nitrate hexahydrate (PuNH) was examined. Figure 3 shows the solubility curves of Pu in HNO_3 solution (Yano et al., 2004). The Pu solution was prepared by dissolving PuO_2 powder with 4 mol/dm³ HNO_3 solution containing 0.05 mol/dm³ $AgNO_3$ electrochemically. In the experiments, the Pu valence was adjusted as following methods. The valence of Pu was changed to Pu(IV) with a few drops of 100% H_2O_2. On the other hand, the Pu solution was oxidized to Pu(VI) by Ag^{2+} ion and Ag in the solution was separated by ion exchange. The Pu solution was cooled quickly to −20°C and then cooled at −1 °C/min to −55°C. In the Pu(IV) solution appeared to be a green quasi-liquid (crystals in liquid). In all runs, PuNH was not crystallized in the experimental conditions but crystals of H_2O and $HNO_3 \cdot 3H_2O$ were observed. In the NEXT, PuNH would not precipitate solely in the U crystallization process.

The influence of Pu valence in the feed solution was examined in the U crystallization process (Yano et al., 2004). When Pu(IV) existed in the feed solution, the yellow crystal was observed. On the other hand, the appearance of the crystal was orange in the feed solution adjusted so that Pu valence was Pu(VI), this color likely resulting from the mixture of the yellow crystal of UNH and the red crystal of PuNH. Plutonium(VI) in the feed solution was co-crystallized with U(VI) in the course of U crystallization. The crystal yields of Pu were smaller than those of U (Ohyama et al., 2005). The fact that the crystal ratio of Pu is smaller than that of U suggests a mechanism of U-Pu co-crystallization in which U begins to crystallize when the saturation point of U is reached by cooling the feed solution, and then Pu is crystallized on the UNH crystal.

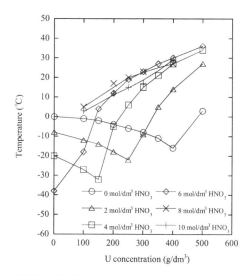

Fig. 2. Solubility of U in HNO₃ solution

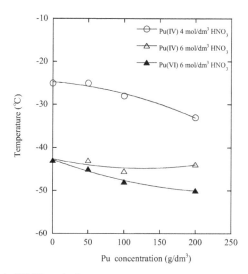

Fig. 3. Solubility of Pu in HNO₃ solution

3. Batch crystallization with dissolver solution of irradiated fast neutron reactor fuel

3.1 Experimental procedure

In the experiments, HNO₃ from Wako Pure Chemical Industries, Ltd., was used without further purification. Irradiated core fuel of the fast neutron reactor "JOYO" Mk-III with an averaged burnup 53 GWd/t and cooling time of 2 y was used for the U crystallization experiments. The sheared pieces of core fuel comprising 166 g of heavy metal were dissolved with 325 cm³ of 8 mol/dm³ HNO₃ solution at 95°C. The valence of Pu in the dissolver solution was changed to Pu(IV) by NO_x gas bubbling. In the crystal ratio experiments, the U and Pu concentrations in the feed solution were approximately 450 and 50 g/dm³, respectively. In co-existing element behavior experiment, the HNO₃ concentration in the feed solution was 4.5 mol/dm³ in the U crystallization process. The CsNO₃ solution was prepared by dissolving CsNO₃ (Wako Pure Chemical Industries, Ltd.) powder in 2 mol/dm³ HNO₃ solution, and added to the dissolver solution. The Cs concentration was 4.0 g/dm³ in the feed solution.

A schematic diagram of the batch cooling crystallizer is shown in Figure 4. The crystallizer made from Pyrex glass was used for cooling the solution volume capacity was 200 cm³, and it had a cooling jacket for cooling and heating media whose temperature was controlled by a thermostat. The feed solution was placed in the crystal vessel and was initially maintained at about 50°C. The feed solution was cooled from 50 to 4°C while being stirred. The spontaneously nucleated and grown crystalline particles were quickly centrifuged from the mother liquor at 3000 rpm for 20 min. After solid-liquid separation, the UNH crystals were washed using 8 mol/dm³ HNO₃ solution at 4°C and then centrifuged at 3000 rpm for 20 min.

The acidity of the solution was determined by acid-base titration (COM-2500, Hiranuma Sangyo Co., Ltd.) and the Pu valence in the feed solution was confirmed as Pu(IV) by optical spectrometry (V-570DS, JASCO Corporation) of the ultraviolet (UV)-visible region. The U and Pu concentrations were measured by colorimetry. The concentrations of Np, Am and Cm were measured by α-ray spectrometry (CU017-450-100: detector and NS920-8MCA: pulse height analyzer, ORTEC). The FPs concentrations were analyzed by γ-ray spectrometry (GEN10: detector and 92XMCA: pulse height analyzer, ORTEC) and inductively coupled plasma atomic emission spectrometry (ICP-AES; ICPS-7500, Shimadzu Corporation).

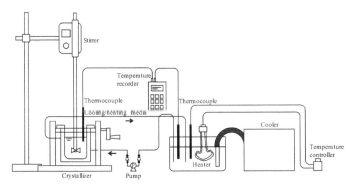

Fig. 4. Schematic diagram of the batch cooling crystallizer

3.2 Crystal ratio of uranyl nitrate hexahydrate

The cooling curve in the U crystallization is shown in Figure 5. The feed solution was placed in the crystallizer and cooled 45.0 to 3.3°C over 150 min. When the temperature of the feed solution reached 23.8°C at 74 min, a small increase in temperature was observed in the feed solution. This indicates the start of crystallization, where heat is released by nucleation.

Figure 6 shows the appearance of UNH crystal recovered the dissolver solution of irradiated fast reactor "JOYO" Mk-III core fuel. After crystal washing, lemon yellow crystals were obtained on a filter.

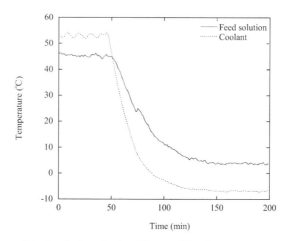

Fig. 5. Cooling curve of feed solution during U crystallization

Fig. 6. Appearance of UNH crystal after crystal washing

The crystal ratio of UNH in the dissolver solution of irradiated fast neutron reactor fuel was examined by changing in HNO_3 solution in the feed solution. The crystal ratio, $R_{c,j}$, is calculated by the following equation.

$$R_{c,j} = 1 - \frac{C_{F,H^+} C_{M,j}}{C_{M,H^+} C_{F,j}} \qquad (2)$$

where C_{F,H^+} and C_{M,H^+} are H^+ concentration in the feed solution and mother liquor, respectively, and $C_{F,j}$ and $C_{M,j}$ are metal j concentration in the feed solution and mother liquor, respectively. Figure 7 shows the relationship between HNO_3 concentration in the feed solution and UNH crystal ratio in the batch cooling crystallization process. These experimental results show high HNO_3 concentration in the feed solution increased with increasing the UNH crystal ratio in the batch cooling crystallization process. These results were in agreement with the reported experimental data (Hart & Morris, 1958).

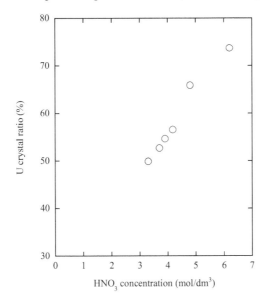

Fig. 7. Relationship between HNO_3 concentration in the feed solution and U crystal ratio

Element	Feed solution (g/dm³)	Decontamination factor (−)	
		Before washing	After washing
U	4.18×10^2	−	−
Pu	4.00×10^1	8.5	24
Ba	3.70×10^{-2}	3.9	4.7
Nuclide	Feed solution (Bq/cm³)	Decontamination factor (−)	
		Before washing	After washing
^{95}Zr	1.84×10^7	7.5	46
^{95}Nb	3.32×10^7	8.2	55
^{106}Ru	3.18×10^8	13	79
^{125}Sb	6.56×10^7	12	129
^{137}Cs	1.04×10^9	2.2	3.4
^{144}Ce	2.42×10^9	15	164
^{144}Pr	2.42×10^9	15	164
^{155}Eu	7.72×10^7	14	118
^{237}Np	2.38×10^4	3.3	6.7
^{241}Am	9.02×10^8	9.8	109
^{242}Cm	2.20×10^7	8.2	115

Table 1. Composition of the feed solution and DFs of metals for UNH crystal

3.3 Behavior of transuranium elements and fission products in uranium crystallization process

Table 1 summarizes the composition of feed solution and the DFs of metals in the U crystallization process. The DFs of metals, $\beta_{c,j}$, are calculated by the following equation.

$$\beta_{c,j} = \frac{\dfrac{C_{F,j}}{C_{F,U}}}{\dfrac{C_{P,j}}{C_{P,U}}} \tag{3}$$

where $C_{F,U}$ is U concentration in the feed solution, and $C_{P,U}$ and $C_{P,j}$ are U and metal j concentrations in the UNH crystal, respectively.

Plutonium behavior in the U crystallization process depends on the Pu valence in the feed solution. In this study, the Pu valence in the feed solution was changed to Pu(IV) by NO_x gas bubbling. After the crystallization, almost all the Pu remained in the mother liquor and attached to the surface of the UNH crystal. The mother liquor on the surface of crystal was efficiently removed after the UNH crystal was washed.

The DF of Np was 3.3 and 6.7 before and after washing, respectively. These experimental results implied Np was present in the form of solid impurities in the mother liquor because its DF was not improved by the crystal washing. Generally, Np can exist simultaneously in three stable oxidation state; Np(IV), Np(V) and Np(VI), in a HNO_3 solution. The oxidation states of Np are interconvertible in HNO_3 medium and exhibit different behavior in the reprocessing. Its valence is strongly affected by oxidation and reduction reactions with agents used in the reprocessing and other co-existing ions. All the kinetics of the oxidation and reduction reactions is not elucidated. One of these reactions is the oxidation of Np(V) by HNO_3, which is the principal influential reaction as follows.

$$2NpO_2^+ + 3H^+ + NO_3^- \leftrightarrow 2NpO_2^{2+} + HNO_2 + H_2O \tag{4}$$

In this reaction, HNO_2 plays the important role of oxidation and reduction between Np(V) and Np(VI). This reaction shows that higher HNO_3 and lower HNO_2 concentrations bring about more oxidation of Np(V) to Np(VI). When the U ions crystallize as UNH in HNO_3 solution, it requires a certain amount of H_2O. As a result, the HNO_3 concentration of the mother liquor is higher than that of the feed solution. It brings about more oxidation of Np(V) to Np(VI) in the mother liquor. When the Pu valence is changed to Pu(VI), Pu(VI) is co-crystallize with U(VI). The chemical behavior of Np(VI) is similar to that of Pu(VI), and it is likely to co-crystallize with U(VI) in the course of U crystallization. Since the Np was incorporated into the UNH crystal, it is difficult to remove from the UNH by the crystal washing operation. If the Np valence is adjusted to Np(IV) or Np(V), the behavior of Np would be different from that of Np(VI). The addition of reductant agent, e.g., U(IV), is effective for preventing from the Np oxidation to Np(VI). Thereby, the Np might remain in the mother liquor after cooling the feed solution.

Americium and Cm are supposed to recover by an extraction chromatography process in the NEXT, and are desired to remain in the mother liquor in the U crystallization process.

The experimental results indicated the DFs of Am and Cm for the UNH crystal were 9.8 and 8.2 before washing, respectively. These elements remained in the mother liquor and attached on the surface of the UNH crystal. The adhesion of liquid impurities was washed away with HNO_3 solution. After washing, the DFs of Am and Cm for the UNH crystal increased by a factor of 109 and 115, respectively.

In the experiments, the behavior of Cs was evaluated in the U crystallization process. The DF of Cs showed 2.2 and 3.4 before and after washing. It is reported that alkali metals react with tetravalent actinide elements and form a double salt in a HNO_3 solution (Staritzky & Truitt, 1949). The reaction of Cs and Pu(IV) is expressed by the following equation.

$$2Cs^+ + Pu(NO_3)_6^{2-} \leftrightarrow Cs_2Pu(NO_3)_6 \tag{5}$$

This reaction indicates that an abundance of $Pu(NO_3)_6^{2-}$ is advantageous for forming of $Cs_2Pu(NO_3)_6$. Figure 8 show the abundance ratio of $Pu(NO_3)_6^{2-}$ in HNO_3 solution (Ryan, 1960). The abundance ratio of $Pu(NO_3)_6^{2-}$ increases with an increase in HNO_3 concentration. In the crystal growth of UNH, a certain amount of H_2O molecules is needed in the feed solution. The mother liquor of HNO_3 concentration is higher than that of feed solution after the U crystallization. Therefore, the formation of $Cs_2Pu(NO_3)_6$ is easy in the course of U crystallization. Anderson reported that double salt of Pu nitrate, $(C_9H_7NH)_2Pu(NO_3)_6$, $Rb_2Pu(NO_3)_6$, $Tl_2Pu(NO_3)_6$, $K_2Pu(NO_3)_6$, $(C_5H_5NH)_2Pu(NO_3)_6$ in addition to $Cs_2Pu(NO_3)_6$ (Anderson, 1949). These materials are less in a dissolver solution of irradiated fast neutron reactor fuel. However, further investigation concerning the double salt of Pu nitrate for the U crystallization process will required experimentally.

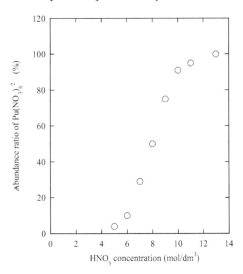

Fig. 8. Abundance ratio of $Pu(NO_3)_6^{2-}$ in HNO_3 solution

Among alkali earth metals, Ba behavior is examined in the cooling batch crystallization. The DFs of Ba was 4.7 after the crystal washing. In the precipitates formation experiments, $Ba_{0.5}Sr_{0.5}(NO_3)_2$ was observed using simulated high level liquid waste solution (Izumida &

Kawamura, 1990). On the other hand, the DF of Sr was decontaminated with the uranyl nitrate solution containing Sr and Ba after the UNH crystal was washed (Kusama et al., 2005). The solubility of Ba is 0.4 g/dm³ in 400 g/dm³ of uranyl nitrate solution with 5 mol/dm³ HNO_3. Therefore, Ba is assumed to precipitate as $Ba(NO_3)_2$ in the U crystallization process. The solid impurities are not removed by the crystal washing with a HNO_3 solution.

Insoluble residues consist of Zr and the elements of platinum group such as Ru in a nuclear fuel reprocessing. Zirconium and Mo precipitate in the form of zirconium molybdate in the dissolution process (Adachi et al., 1990; Lausch et al., 1994; Usami et al., 2010). This compound tends to form at high temperature and with low HNO_3 concentration in the solution. The DF of Zr was high after the crystal washing and Zr remained in the mother liquor. The feed solution was cooled and the acidity in the feed solution increases in the course of U crystallization. Therefore, zirconium molybdate would be difficult to crystallize at low temperature and with high HNO_3 concentration in the mother liquor.

The behavior of Ce, Pr and Eu in the rare earth elements was evaluated in the U crystallization experiments. The DFs of these elements achieved to approximately 10^2 after the crystal washing. Their solubility in the HNO_3 solution was so high that there was no precipitation as solid impurities. Therefore, they remained in the mother liquor during the U crystallization and these elements in the mother liquor that was attached to the surface of the UNH crystal were washed away with the HNO_3 solution.

4. Crystallization apparatus for continuous operation

4.1 Concept of crystallization apparatus

The crystallizer is designed for continuous operation adoption high throughput and equipment scale-up and is developed for the U crystallization in the NEXT (Washiya et al., 2010). Figure 9 shows a schematic diagram of annular type continuous crystallizer. A rotary-driven cylinder has a screw blade to transfer UNH crystal slurry and annular shaped space is formed as crystallization section in between the rotary cylinder and outer cylinder. A dissolver solution of irradiated fast neutron reactor fuel is fed into the annular section from the lower part of the equipment, and the coolant is supplied into the cooling jacket located on the outside cylinder. The dissolver solution is cooled down gradually and is transferred to the outlet in the upper side of the equipment. The UNH slurry is obtained in the annular section and is discharged by the guide blade attached to the rotary cylinder. The mother liquor is separated from the UNH crystal and is discharged from the nozzle located in the more upper side than solution level. The discharged UNH slurry is still accompanied with a little solution. Hence, it needs to be dried by the crystal separator as centrifugal dewatering process.

4.2 Continuous operation with uranyl nitrate solution using annular type continuous crystallizer

The continuous operation experiments were evaluated with a uranyl nitrate solution (Washiya et al., 2010). The feed solution of 450 g/dm³ U concentration in 5 mol/dm³ HNO_3 was cooled at 0°C at 20 rph. Afterwards, amount of the crystal stay was increased gradually, and it reached to steady state in 1–2 h. The moisture content in the UNH slurry obtained from the outlet of the slurry was about 40%. The UNH crystal size was approximately 600 μm and is considered to appropriate size for solid-liquid separation. The U concentration in the mother liquor was reached to steady state within 2 h, and crystal ratio of U was about 84%.

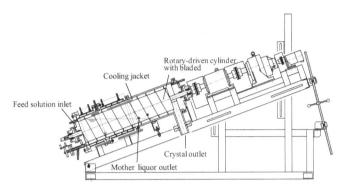

Fig. 9. Schematic diagram of annular type continuous crystallizer

To extract operational failure events comprehensively concerning to the U crystallizer and to clarify their importance, failure mode analysis was carried out by applying Failure Mode and Effects Analysis (FMEA). Significant failure events were identified with failure causes, their effects and probability of these failures were predicted by making use of operation experience. All failure events were evaluated by cause, primary and secondary effects and scored them. As results, crystal accumulation, blockage of mother liquor discharge nozzle, blockage of crystal discharge nozzle were selected as important specific failure events. To investigate how to detect non-steady condition, these three experiments were carried out with screw rotation speed decline, crystal outlet blockage and mother liquor outlet blockage. Also, the resume procedure after non-steady state was examined sequentially to consider countermeasures for each non-steady event (Shibata et al., 2009). The accumulation of UNH crystals can be detected by the torque of the cylinder screw, the liquid level in the annular section and other instruments. These experimental results show that it is possible to recover from non-steady state when the cause of the phenomena such as blockage of crystal outlet is removed by an appropriate operation. The fundamental performance of crystallizer annular type was investigated with uranyl nitrate solution. The experiment will be carried out to confirm the system performance on integrated crystallization system consisting of the engineering-scale crystallizer, crystal separator and related systems.

5. Purification of uranyl nitrate hexahydrate crystal product

5.1 Principal of crystal purification

Generally, crystalline particles produced in crystallizers are often contaminated by the mother liquor which appears on the surface or inside the bodies of the crystals. The UNH crystal recovered from a dissolver solution of irradiated fast neutron reactor fuel is washed away with a HNO_3 solution. Although the TRU elements and FPs on the surface of the UNH crystal are decontaminated by the crystal washing, the inclusions within the crystal and the solid impurities are not removed from the UNH crystal. Therefore, crystal purification method is studied for the purpose of further increasing decontamination performance. One crystal purification method, the grown crystalline particles are purified by heating up to as high as the melting point of the crystal and introducing the mother liquor to the outside of the crystal, which is exhaled along defects and grain boundaries (Zief & Wilcox, 1967; Matsuoka & Sumitani, 1988). This phenomenon is called "sweating" and is applied to organics and metals. The mother liquor and melt in the crystal are discharged by Ostwald ripening and increase in the internal pressure (Matsuoka et al., 1986). The incorporated liquid is expelled from grooves along defects and grain boundaries. It was reported that countless grooves were observed in the organic crystal after sweating (Matsuoka & Sumitani, 1988). The purification of the p-dichlorobenzene (p-DCB) and m-chloronitrobenzene (m-CNB) crystalline particles by sweating was experimentally investigated (Matsuoka et al., 1986). The purity of 99.99% was obtained by a single sweating stage at temperatures about 1°C below the melting points of the pure crystals after the duration of 90 or 120 min of sweating. In the batch operation, the UNH crystal purification experiments were carried out with the dissolver solution of MOX fuel containing simulated FPs (Nakahara et al., 2011). Although the DFs of solid impurities such as $Ba(NO_3)_2$ and $Cs_2Pu(NO_3)_6$ did not change in the sweating process, that of Eu increased with increases in temperature and time. In the batch experiments, the DF of Eu increased to approximately 2.4 times after 30 min at 60°C. There results indicated that liquid impurities such as Eu were effectively removed by the sweating method, but solid impurities such as Pu, Cs and Ba were minimally affected in the batch experiments.

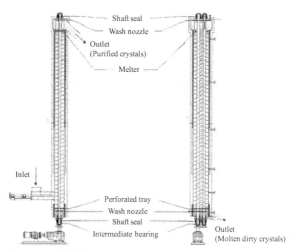

Fig. 10. Schematic diagram of KCP

5.2 Concept of crystal purification apparatus

The crystal purification apparatus, Kureha Crystal Purifier (KCP), has been applied in industrial plants using organic matter (Otawara & Matsuoka, 2002). The schematic diagram of KCP is shown in Figure 10. The apparatus has been developed in the following fashion: feed stock is charged as solids at the bottom of the column, the heating unit is set at the top of the column, and it is possible to contact the melt with crude crystal countercurrently. The KCP features high purity, high yield, energy savings, little maintenance, and a long, stable operation (Otawara & Matsuoka, 2002). The crude crystal is supplied at the bottom of the column and then is carried to the upper side of a column by a double screw conveyor, and then part of the crystal is molten by a heating unit at the top of the column and the melt trickles downward among the crude crystal. The apparatus performs countercurrent contact between the crystal and reflux melts in the course of being conveyed upward, and the crude crystal is washed by a portion of the melt. Therefore, higher DFs of liquid impurities will be obtained by the KCP, because the liquid impurities were washed with melt in addition to sweating effect. The pure crystal product exits from the top of the column.

5.3 Continuous operation with uranyl nitrate solution using kureha crystal purifier

The crystal purification experiments with the KCP were carried out using the UNH crystal recovered from uranyl nitrate solution containing Sr of SUS304L (Yano et al., 2009). Although the DF of liquid impurities, Eu, in the static system was approximately 2.4 at 60°C for 30 min, the DF of Sr was 50 by the KCP. The liquid impurities such as Sr was removed from the UNH crystal not only by the sweating phenomenon but also by washing with U reflux melt, which was produced by the melter at the top of the column of the KCP. On the other hand, the DF of solid impurities, SUS304L, achieved a value of 100 with the KCP. In the static system, the solid impurities remained in the UNH crystal after the sweating operation. In the KCP, the solid impurities were removed from the UNH crystals due to upward movement of the crude crystals from the double screw conveyors; the UNH crystals and solid impurities, which have different densities and particle sizes, separate from each other by gravity and mixing.

6. Conclusion

Experimental studies on the behavior of TRU elements and FPs in the dissolver solution of irradiated fast neutron reactor core fuel were carried out to develop a crystallization method as a part of an advanced aqueous reprocessing. The experimental results show high HNO_3 concentration in the feed solution increased with increasing the UNH crystal ratio in the U crystallization process. Among coexistent elements, Zr, Nb, Ru Sb, Ce, Pr, Eu, Am and Cm remained in the mother liquor at the time of U crystallization. Therefore, portions of these elements in the mother liquor that was attached to the surface of the UNH crystal were washed away with HNO_3 solution. Cesium exhibited different behavior depending on whether Pu was present. Although a high DF of Cs was obtained in the case of uranyl nitrate solution without Pu(IV), Cs was hardly separated at all from the UNH crystal formed from the dissolver solution of irradiated fast neutron reactor core fuel in the case of high Cs concentration in the feed solution. It is likely that a double salt of Pu(IV) and Cs, $Cs_2Pu(NO_3)_6$ precipitated in the course of U crystallization process. Since Ba precipitated as $Ba(NO_3)_2$, its DF was low after the UNH crystal was washed. Neptunium was not removed from the UNH crystal because Np was oxidized to Np(VI) in the feed solution and thus co-crystallized with U(VI). The experimental data on the behavior of TRU elements and FPs will be actually utilized in fast neutron reactor fuel reprocessing. The continuous crystallizer and the KCP were developed, and the apparatus performance was examined with the uranyl nitrate solution containing simulated FPs. In the future, the integrated crystallization system performance will be confirmed for part of U recovery in the NEXT process.

7. Acknowledgment

The author gratefully acknowledges Mr. A. Shibata and Mr. K. Nomura of JAEA for fruitful discussions.

8. References

Adachi, T.; Ohnuki, M.; Yoshida, N.; Sonobe, T.; Kawamura, W.; Takeishi, H.; Gunji, K.; Kimura, T.; Suzuki, T.; Nakahara, Y.; Muromura, T.; Kobayashi, Y.; Okashita, H. & Yamamoto, T. (1990). Dissolution Study of Spend PWR Fuel: Dissolution Behavior and Chemical Properties of Insoluble Residues. *Journal of Nuclear Materials,* Vol. 174, No. 1, (November 1990), pp. 60-71, ISSN 0022-3115

Anderson, H. H. (1949). Alkali Plutonium(IV) Nitrates, In: *The Transuranium Elements, National Nuclear Energy Series IV, Vol. 14B,* G. T. Seaborg, J. J. Katz, W. M. Manning, (Eds.), pp. 964-967, McGraw-Hill Book Co. Inc., New York, USA

Ebert, K.; Henrich, E.; Stahl, R. & Bauder, U. (1989). A Continuous Crystallization Process for Uranium and Plutonium Refinement, *Proceedings of 2nd International Conference on Separation Science & Technology,* pp. 346-352, Paper No. S5b, Hamilton, Ontario, Canada, October 1-4, 1989

Hart, R. G. & Morris, G. O. (1958). Crystallization Temperatures of Uranyl Nitrate-Nitric Acid Solutions, In: *Progress in Nuclear Energy, Series 3, Process Chemistry, Vol. 2,* F. Bruce, (Ed.), pp. 544-545, Pergamon Press, ISSN 0079-6514, New York, USA

Izumida, T. & Kawamura, F. (1990). Precipitates Formation Behavior in Simulated High Level Liquid Waste of Fuel Reprocessing. *Journal of Nuclear Science and Technology,* Vol. 27, No. 3, (March 1990), pp. 267-274, ISSN 0022-3131

Katsurai, K.; Ohyama, K.; Kondo, Y.; Nomura, K.; Takeuchi, M.; Washiya, T. & Myochin, M. (2009). Development of Highly Effective Dissolution Technology for FBR MOX Fuels, *Proceedings of International Conference on The Nuclear Fuel Cycle: Sustainable Options & Industrial Perspectives (GLOBAL 2009),* pp. 108-112, Paper 9219, Paris, France, September 6-11, 2009

Koma, Y.; Sano, Y.; Morita, Y. & Asakura, T. (2009). Adsorbents Development for Extraction Chromatography on Am and Cm Separation, *Proceedings of International Conference on The Nuclear Fuel Cycle: Sustainable Options & Industrial Perspectives (GLOBAL 2009),* pp. 1056-1060, Paper 9325, Paris, France, September 6-11, 2009

Koyama, T.; Washiya, T.; Nakabayashi, H. & Funasaka, H. (2009). Current Status on Reprocessing Technology of Fast Reactor Fuel Cycle Technology Development (FaCT) Project in Japan : Overwiew of Reprocessing Technology Development, *Proceedings of International Conference on The Nuclear Fuel Cycle: Sustainable Options & Industrial Perspectives (GLOBAL 2009),* pp. 46-52, Paper 9100, Paris, France, September 6-11, 2009

Kusama, M.; Chikazawa, T. & Tamaki, Y. (2005). *Estimation Tests for Effecting Factor on Decontamination Property in Crystallization Process,* JNC TJ8400 2005-006, Japan Nuclear Cycle Development Institute, Tokai, Ibaraki, Japan

Lausch, J.; Berg, R.; Koch, L.; Coquerelle, M.; Glatz, J. P.; Walker, C. T. & Mayer, K. (1994). Dissolution Residues of Highly Burnt Nuclear Fuels. *Journal of Nuclear Materials,* Vol. 208, No. 1-2, (January 1994), pp. 73-80, ISSN 0022-3115

Matsuoka, M.; Ohishi, M. & Kasama, S. (1986). Purification of p-Dichlorobenzene and m-Chloronitrobenzene Crystalline Particles by Sweating. *Journal of Chemical Engineering of Japan*, Vol. 19, No. 3, (March 1986), pp. 181-185, ISSN 0021-9592

Matsuoka, M. & Sumitani, A. (1988). Rate of Composition Changes of Organic Soild Solution Crystals in Sweating Operations. *Journal of Chemical Engineering of Japan*, Vol. 21, No. 1, (February 1988), pp. 6-10, ISSN 0021-9592

Nakahara, M.; Nomura, K.; Washiya, T.; Chikazawa, T. & Hirasawa, I. (2011). Removal of Liquid and Solid Impurities from Uranyl Nitrate Hexahydrate Crystalline Particles in Crystal Purification Process. *Journal of Nuclear Science and Technology*, Vol. 48, No. 3, (February 2011), pp. 322-329, ISSN 0022-3131

Ohyama, K.; Yano, K.; Shibata, A.; Miyachi, S.; Koizumi, T.; Koyama, T.; Nakamura, K.; Kikuchi, T. & Homma, S. (2005). Experimental Study on U-Pu Co-crystallization for New Reprocessing Process, *Proceedings of International Conference on Nuclear Energy Systems for Future Generation and Global Sustainability (GLOBAL 2005)*, Paper No. 452, Tsukuba, Japan, October 9-13, 2005

Otawara, K. & Matsuoka, T. (2002). Axial Dispersion in a Kureha Crystal Purifier (KCP). *Journal of Crystal Growth*, Vol. 237-239, No. 3, (April 2002), pp. 2246-2250, ISSN 0022-0248

Ryan, J. L. (1960). Species Involved in the Anion-exchange Absorption of Quadrivalent Actinide Nitrates. *The Journal of Physical Chemistry*, Vol. 64, No. 10, (October 1960), pp. 1375-1385, ISSN 0022-3654

Sano, Y.; Ogino, H.; Washiya, T. & Myochin, M. (2009). Development of the Solvent Extraction Technique for U-Pu-Np Co-Recovery in the NEXT Process, *Proceedings of International Conference on The Nuclear Fuel Cycle: Sustainable Options & Industrial Perspectives (GLOBAL 2009)*, pp. 158-165, Paper 9222, Paris, France, September 6-11, 2009

Shibata, A.; Kaji, N.; Nakahara, M.; Yano, K.; Tayama, T.; Nakamura, K.; Washiya, T.; Myochin, M.; Chikazawa, T. & Kikuchi, T. (2009). Current Status on Research and Development of Uranium Crystallization System in Advanced Aqueous Reprocessing of FaCT Project, *Proceedings of International Conference on The Nuclear Fuel Cycle: Sustainable Options & Industrial Perspectives (GLOBAL 2009)*, pp. 151-157, Paper 9154, Paris, France, September 6-11, 2009

Staritzky, E. & Truitt, A. L. (1949). *Optical and Norphological Crystallography of Plutonium Compounds*, LA-745, Los Almos National Laboratory, Los Alamos, New Mexico, USA

Usami, T.; Tsukada, T.; Inoue, T.; Moriya, N.; Hamada, T.; Purroy, D. S.; Malmbeck, R. & Glatz, J. P. (2010). Formation of Zirconium Molybdate Sludge from an Irradiated Fuel and Its Dissolution into Mixture of Nitric Acid and Hydrogen Peroxide. *Journal of Nuclear Materials*, Vol. 402, No. 2-3, (July 2010), pp. 130-135, ISSN 0022-3115

Washiya, T.; Tayama, T.; Nakamura, K.; Yano, K.; Shibata, A.; Nomura, K.; Chikazawa, T.; Nagata, M. & Kikuchi, T. (2010). Continuous Operation Test at Engineering Scale Uranium Crystallizer. *Journal of Power and Energy Systems*, Vol. 4, No. 1, (March 2010), pp. 191-201, ISSN 1881-3062

Yano, K.; Shibata, A.; Nomura, K.; Koizumi, T.; Koyama, T. & Miyake, C. (2004). Plutonium Behavior Under the Condition of Uranium Crystallization from Dissolver Solution,

Proceedings of International Conference on Advances for Future Nuclear Fuel Cycles (ATALANTE 2004), pp. 1-4, P1-66, Nîmes, France, June 21-25, 2004

Yano, K.; Nakahara, M.; Nakamura, M.; Shibata, A.; Nomura, K.; Nakamura, K.; Tayama, T.; Washiya, T.; Chikazawa, T. & Hirasawa, I. (2009). Reaserch and Development of Crystal Purification for Product of Uranium Crystallization Process, *Proceedings of International Conference on The Nuclear Fuel Cycle: Sustainable Options & Industrial Perspectives (GLOBAL 2009)*, pp. 143-150, Paper 9093, Paris, France, September 6-11, 2009

Zief, M. & Wilcox, W. R. (1967). *Fractional Solidification*, Marcel Dekker Inc., ISBN 978-0824718206, New York, USA

3

Stable and Metastable Phase Equilibria in the Salt-Water Systems

Tianlong Deng
Tianjin Key Laboratory of Marine Resources and Chemistry
College of Marine Science and Engineering
Tianjin University of Science and Technology
TEDA, Tianjin,
CAS Key Laboratory of Salt Lake Resources and Chemistry
Institute of Salt Lakes
Chinese Academy of Sciences
Xining, Qinghai,
People Republic of China

1. Introduction

1.1 Salt lakes and the classification of hydrochemistry

Salt lakes are widely distributed in the world, and some famous salt lake resources are shown in Tables 1 and 2. In China, salt lakes are mainly located in the area of the Qinghai-Xizang (Tibet) Plateau, and the Autonomous Regions of Xinjiang and Inner Mongolia (M.P. Zheng et al., 1989). The composition of salt lake brines can be summarized to the complex salt-water multi-component system (Li - Na – K – Ca - Mg - H - Cl – SO_4 – B_4O_7 – OH- HCO_3 – CO_3 - H_2O).

According to the chemical type of salt lake brines, it can be divided into five types, i.e. chloride type, sulphate type, carbonate type, nitrite type, and borate type among those salt lake resources in the world (Gao et al., 2007).

Chloride type: the component of brines in Death Sea, Mideast and Caerhan Salt Lake in China belongs to the system of chloride type (Na – K – Mg - Cl - H_2O), and the main precipitation of salts are halite (NaCl), sylvite (KCl), carnallite (KCl•$MgCl_2$•7H_2O), and bischofite ($MgCl_2$•6H_2O).

Sulphate type: this kind of salt lake resources is similar with the sea water system (Na – K – Mg – Cl – SO_4 – H_2O), and it can be divided into two kinds of hypotypes i.e. sodium sulphate and magnesium sulphate. As to sodium sulphate hypotype, the Great Salt Lake in America, the gulf of Kara-Bogaz-Golin Urkmenistan, and Da-Xiao Qaidan in China belong to this hypotype with the main deposit of glauberite (Na_2SO_4•$CaSO_4$), glauber salt (Na_2SO_4·10H_2O), halite, galserite (Na_2SO_4•3K_2SO_4), schonenite (K_2SO_4•$MgSO_4$•6H_2O), and so on. As to magnesium sulphate hypotype, there are Yunchen Salt Lake in Shanxi Provine and Chaka Salt Lakes in Qinghai Province, China, especially Salt Lakes of the Qaidam Basin in Qinghai Province are a sub-type of magnesium sulphate brines famous for their abundance of lithium, potassium, magnesium and boron resources (Zheng et al., 1989). The main precipitation of salts are halite, glauber salt, blodite ($Na_2SO_4 \bullet MgSO_4 \bullet 7H_2O$), and epsom salt ($MgSO_4 \bullet 7H_2O$).

Carbonate type: this type belongs to the system (Na – K – Cl – CO_3 - SO_4 – H_2O), and Atacama Salt Lake in Chile and Zabuye Salt Lake in Tibet are the famous carbonate type of salt lake. The main precipitated minerals are thermonatrite ($Na_2CO_3 \bullet 10H_2O$), baking soda ($NaHCO_3$), natron ($Na_2CO_3 \bullet 10H_2O$), glauber salt, and halite.

Nitrite type: the brine composition of this type salt lake can be summarized as the system (Na – K – Mg - Cl – NO_3 - SO_4 – H_2O). The type salt lake main locates in the salt lake area in the northern of Chile among the salt lake group of Andes in the South-America, semi and dry salts in Luobubo and Wuzunbulake Lakes in Xinjiang, the northern of China. There are natratime saltier ($NaNO_3$), niter (KNO_3), darapskite ($NaNO_3 \bullet Na_2SO_4 \cdot H_2O$), POTASSIUM-darapskite ($KNO_3 \bullet K_2SO_4 \bullet H_2O$), humberstonite ($NaNO_3 \bullet Na_2SO_4 \bullet 2MgSO_4 \bullet 6H_2O$).

Borate type: it can be divided into carbonate-borate hypotype and sulphate-borate hypotype. Searles Salt Lake in America, Banguo Lake and Zabuye Salt Lakes in Tibet, China belong to the former, and the brines mostly belong to the system (Na - K - Cl - B_4O_7 - CO_3 - HCO_3 - SO_4 - H_2O). In order to prove the industrial development of Searles Lake brines, Teeple (1929) published a monograph after a series of salt-water equilibrium data on Searles lake brine containing carbonate and borate systems. The latter includes Dong-xi-tai Lake, Da-xiao-chaidan Lake and Yiliping Lake in Qinghai Province, Zhachangchaka Lake in Tibet, China. In those lake area, the natural borate minerals of raphite ($NaO \bullet CaO \bullet 3B_2O_3 \bullet 16H_2O$), pinnoite ($MgO \bullet B_2O_3 \bullet 3H_2O$), chloropinnoite ($2MgO \bullet 2B_2O_3 \bullet MgCl_2 \bullet 14H_2O$), inderite ($2MgO \bullet 3B_2O_3 \bullet 15H_2O$), hungchanoite ($MgO \bullet 2B_2O_3 \bullet 9H_2O$), mcallisterite ($MgO \bullet 3B_2O_3 \bullet 7.5H_2O$), kurnakovite ($2MgO \bullet 3B_2O_3 \bullet 15H_2O$) and hydroborate were precipitated (Zheng et al., 1988; Gao et al., 2007). In addition, the concentration of lithium ion exists in the surface brine of salt lakes.

Salt lakes	Death Sea	Great lake, US	Searles lake, US	Atacama, Chile	Caerhan, China	Zabuye, Tibet
Altitude, /m	-400	1280	512	2300	2900	2677
Area, /km²	1000	3600	1000	1400	5882	120
Dept, m	329	~5	Intragranular brine	Intragranular brine	Intragranular brine	~3
KCl	2×10^9	1×10^8	2.8×10^6	1.1×10^8	3×10^8	6.6×10^7
NaCl	1.2×10^{10}	3.2×10^9	—	—	4.3×10^{10}	2×10^8
MgCl₂	2.2×10^{10}	1.2×10^9	—	1.2×10^8	2.7×10^9	5.7×10^8
MgSO₄	—	1.7×10^7	—	—	—	—
LiCl	1.7×10^7	3.2×10^6	2.7×10^6	2.8×10^6	—	—
CaCl₂	6×10^9	—	—	—	—	—
CaSO₄	1×10^8	—	—	—	—	—
MgBr₂	1×10^9	—	—	—	3.4×10^5	—
B₂O₃	—	1.9×10^6	3×10^7	1.6×10^7	5.5×10^6	1.8×10^6
WO₃	—	—	7.5×10^4	—	—	—

Table 1. Basic data of salt lakes and their salt reserves in the world. unit, /t (Song, 2000)

Type of lithium resources	Country and section	Lithium storage capacity, (Li$_2$O)
Salt lakes	Uyuni, Bolivia	More than 19 million tons
	Silver and Searles, US	More than 10 million tons
	Caerhan and Caida, China	10 million tons
	Kata Baca, Argentina	Sever million tons
Type of crystalline rocks	America	6.34 million tons
	Chile	4.3 million tons
	Canada	6.6 million tons
	Greenbusbse, Australian	6 million tons

Table 2. Statistical distribution of the lithium reserves in the world (Song, 2000; Zhao, 2003)

1.2 Phase equilibria of salt-water systems

It is essential to study the stable and metastable phase equilibria in multi-component systems at different temperatures for its application in the fields of chemical, chemical engineering such as dissolution, crystallization, distillation, extraction and separation.

1.2.1 The stable phase equilibria of salt-water systems

The research method for the stable phase equilibria of salt-water system is isothermal dissolution method. It is worthy of pointing out that the status of the stable phase equilibrium of salt-water system is the in a sealed condition under stirring sufficiently, and the speed of dissolution and crystallization of equilibrium solid phase is completely equal with the marker of no change for the liquid phase composition. As to the thermodynamic stable equilibrium studies aiming at sea water system (Na – K – Mg – Cl – SO$_4$ – H$_2$O), J.H. Vant'hollf (1912) was in the earliest to report the stable phase diagram at 293.15 K with isothermal dissolution method.

In order to accelerate the exploiting of Qaidam Basin, China, a number of the stable phase equilibria of salt-water systems were published at recent decades (Li et al., 2006; Song, 1998, 2000; Song & Du, 1986; Song & Yao, 2001, 2003).

1.2.2 The metastable phase equilibria of salt-water systems

However, the phenomena of super-saturation of brines containing magnesium sulfate, borate is often found both in natural salt lakes and solar ponds around the world. Especially for salt lake brine and seawater systems, the natural evaporation is in a autogenetic process with the exchange of energy and substances in the open-ended system , and it is controlled by the radiant supply of solar energy with temperature difference, relative humidity, and air current, etc. In other word, it is impossible to reach the thermodynamic stable equilibrium, and it is in the status of thermodynamic non-equilibrium.

For the thermodynamic non-equilibrium phase diagram of the sea water system as called "solar phase diagram" in the first, N.S. Kurnakov (1938) was in the first to report the experimental diagrams based on the natural brine evaporation, and further called "metastable phase diagram" for the same system (Na – K – Mg – Cl – SO$_4$ – H$_2$O) at (288.15, 298.15, and 308.15) K was reported on the basis of isothermal evaporation method (Jin, et al., 1980, 2001, 2002; Sun, 1992). Therefore, the metastable phase equilibria research is essential to predict the crystallized path of evaporation of the salt lake brine.

The isothermal evaporation phase diagrams of the sea water system at different temperature show a large difference with Vant'hoff stable phase diagram. The crystallization fields of leonite (MgSO$_4$.K$_2$SO$_4$.4H$_2$O), and kainite (KCl.MgSO$_4$.3H$_2$O) are all disappear whereas the crystallization field of picromerite (MgSO$_4$.K$_2$SO$_4$.6H$_2$O) increases by 20-fold, which is of great importance for producing potassium sulfate or potassium-magnesium fertilizer.

Therefore, in order to separate and utilize the mixture salts effectively by salt-field engineering or solar ponds in Qaidam Basin, studies on the phase equilibria of salt-water systems are focused on the metastable phase equilibria and phase diagrams at present years (Deng et al., 2011; Deng et al., 2008a-g; Deng, et al., 2009a-c; Wang & Deng, 2008, 2010; Li & Deng, 2009; Li et al., 2010; Liu et al., 2011; Meng & Deng, 2011; Guo et al., 2010; Gao & Deng, 2011a-b; Wang et al., 2011a-b).

1.2.3 Solubility prediction for the phase equilibria of salt-water systems

Pitzer and co-workers have developed an ion interaction model and published a series of papers (Pitzer, 1973a-b, 1974a-b, 1975, 1977, 1995, 2000; Pabalan & Pitzer, 1987) which gave a set of expressions for osmotic coefficients of the solution and mean activity coefficient of electrolytes in the solution. Expressions of the chemical equilibrium model for conventional single ion activity coefficients derived are more convenient to use in solubility calculations (Harvie & Weare, 1980; Harvie et al.1984; Felmy & Weare, 1986; Donad & Kean, 1985).

In this chapter, as an example, the stable and metastable phase equilibria in the salt-water system (NaCl - KCl – Na$_2$B$_4$O$_7$ - K$_2$B$_4$O$_7$ - H$_2$O), which is of great importance to describe the metastable behavior in order to separate and purify the mixture salts of borax and halo-sylvite were introduced in detail. The stable phase diagrams of the sub-ternary systems (NaCl - Na$_2$B$_4$O$_7$ - H$_2$O), (KCl –K$_2$B$_4$O$_7$ - H$_2$O), (Na$_2$B$_4$O$_7$ - K$_2$B$_4$O$_7$ - H$_2$O) at 298.15 K and the metastable phase diagrams of the sub-ternary systems (NaCl - Na$_2$B$_4$O$_7$ - H$_2$O) at 308.15 K for the mentioned reciprocal quaternary system were systematically studied on our previous researches under several scientific funding supports. The theoretical prediction for the stable solubility of this reciprocal quaternary system was also briefly introduced based on the ion-interaction model.

2. Apparatus

2.1 Apparatus for the stable phase equilibria in the salt-water system

Stable phase equilibria are the thermodynamic equilibria. In order to reach the isothermal dissolve equilibrium, the apparatus mainly contains two parts i.e. constant temperature installing and equilibrator. Therefore, experimental apparatus depends on the target of temperature. Generally, thermostatic water-circulator bath is used under normal atmospheric temperature, and thermostatic oil-circulator bath is chosen at higher level temperature. Under low temperature, the refrigerator or freezing saline bath is commonly used. Figure 1 shows the common used equalizer pipe with a stirrer. The artificial synthesis complex put in the pipe to gradually reach equilibria under vigorous stirring. In order to avoid the evaporation of water, the fluid seal installing is needed, and the sampling branch pipe is also needed to seal. Usually, for aqueous quaternary system study, a series of artificial synthesis complex, normally no less than 30, was needed to be done one by one the experimental time consume is equivalence large. At present, a thermostatic shaker whose temperature could controlled with temperature precision of ± 0.1 K can be used for the measurement of stable phase equilibrium (Deng et al., 2002; Deng, 2004). The advantage is that a series artificial synthesis complexes which is loading in each sealed bottle can be put in and vigorous shaking together.

In this study, the stable phase equilibria system (NaCl - KCl – $Na_2B_4O_7$ - $K_2B_4O_7$ - H_2O) at 298.15 K, a thermostatic shaker (model HZQ-C) whose temperature was controlled within 0.1 K was used for the measurement of phase equilibrium.

2.2 Apparatus for the metastable phase equilibria in the salt-water system

The isothermal evaporation method was commonly used, and Figure 2 is our designed isothermal evaporation device in our laboratory (Guo et al., 2010). The isothermal evaporation chamber was consisted of evaporating container, precise thermometer to keep the evaporating temperature as a constant and electric fan to simulate the wind in situ, and the solar energy simulating system with electrical contact thermograph, electric relay and heating lamp. The temperature controlling apparatus is made up of an electric relay, an electrical contact thermograph and heating lamps.

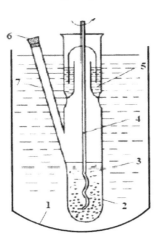

Fig. 1. Apparatus of equalizer pipe. 1, thermostatic water-circulator bath; 2, pipe body; 3, assay; 4, stirrer; 5, fluid seal; 6, rubber seal lock; 7, sampling branch pipe.

In this example of the metastable phase equilibria system (NaCl - KCl – $Na_2B_4O_7$ - $K_2B_4O_7$ - H_2O) at 308.15 K, the isothermal evaporation box was used. In an air-conditioned laboratory, a thermal insulation material box (70 cm long, 65 cm wide, 60 cm high) with an apparatus to control the temperature was installed. When the solution temperature in the container was under (308.15 ± 0.2) K, the apparatus for controlling the temperature formed a circuit and the heating lamp began to heat. Conversely, the circuit was broken and the heating lamp stopped working when the temperature exceeded 308.15 K. Therefore, the temperature in the box could always be kept to (308.15 ± 0.2) K. An electric fan installed on the box always worked to accelerate the evaporation of water from the solutions.

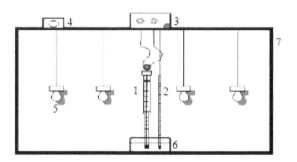

Fig. 2. The schematic diagram of the isothermal evaporation chamber. 1, electrical contact thermograph; 2, precise thermometer; 3, electric relay; 4, electric fan; 5, heating lamp; 6, evaporating container; 7, isothermal container.

Of course, the experimental conditions of an air flow velocity, a relative humidity, and an evaporation rate were controlled as similar as to those of the climate of reaching area in a simulative device.

3. Experimental methods

3.1 Reagents

For phase equilibrium study, reagents used should be high-purity grade otherwise the re-crystallized step was needed. For the stable and metastable phase equilibria in the salt-water system (NaCl - KCl – $Na_2B_4O_7$ - $K_2B_4O_7$ - H_2O), the chemicals used were of analytical grade, except borax which was a guaranteed reagent (GR), and were obtained from either the Tianjin Kermel Chemical Reagent Ltd. or the Shanghai Guoyao Chemical Reagent Co. Ltd: sodium chloride (NaCl, ≥0.995 in mass fraction), potassium chloride (KCl, ≥ 0.995 in mass fraction), borax ($Na_2B_4O_7 \cdot 10H_2O$, ≥ 0.995 in mass fraction), potassium borate tetrahydrate ($K_2B_4O_7 \cdot 4H_2O$, ≥ 0.995 in mass fraction), and were re-crystallized before use. Doubly deionized water (DDW) with conductivity less than 1.2×10^{-4} S·m^{-1} and pH 6.60 at 298.15 K was used to prepare the series of the artificial synthesized brines and chemical analysis.

3.2 Analytical methods

3.2.1 The chemical analysis of the components in the liquids

For phase equilibrium study in this phase equilibrium system (NaCl - KCl – $Na_2B_4O_7$ - $K_2B_4O_7$ - H_2O), the composition of the potassium ion in liquids and their corresponding wet solid phases was analyzed by gravimetric methods of sodium tetraphenyl borate with an uncertainty of $\leq \pm 0.0005$ in mass fraction; Both with an uncertainty of $\leq \pm 0.003$ in mass fraction, the concentrations of chloride and borate were determined by titration with mercury nitrate standard solution in the presence of mixed indicator of diphenylcarbazone and bromphenol blue, and by basic titration in the presence of mannitol, respectively (Analytical Laboratory of Institute of Salt Lakes at CAS, 1982). The concentration of sodium ion was calculated by subtraction via charge balance.

3.2.2 The measurements of the physicochemical properties

For the physicochemical properties determinations, a PHS-3C precision pH meter supplied by the Shanghai Precision & Scientific Instrument Co. Ltd was used to measure the pH of the equilibrium aqueous solutions (uncertainty of ± 0.01). The pH meter was calibrated with standard buffer solutions of a mixed phosphate of potassium dihydrogen phosphate and sodium dihydrogen phosphate (pH 6.84) as well as borax (pH 9.18); the densities (ρ) were measured with a density bottle method with an uncertainty of ± 0.2 mg.cm^{-3}. The viscosities (η) were determined using an Ubbelohde capillary viscometer, which was placed in a thermostat at (308.15 ± 0.1) K. No fewer than five flow times for each equilibrium liquid phase were measured with a stopwatch with an uncertainty of 0.1 s to record the flowing time, and the results calculated were the average. An Abbe refractometer (model WZS-1) was used for measuring the refractive index (n_D) with an uncertainty of ± 0.0001. The physicochemical parameters of density, refractive index and pH were also all placed in a thermostat that electronically controlled the set temperature at (308.15 ± 0.1) K.

3.3 Experimental methods of phase equilibria

3.3.1 Stable phase equilibria

For the stable equilibrium study, the isothermal dissolution method was used in this study. The series of complexes of the quaternary system were loaded into clean polyethylene bottles and capped tightly. The bottles were placed in the thermostatic rotary shaker, whose temperature was controlled to (298.15 ± 0.1) K, and rotated at 120 rpm to accelerate the equilibrium of those complexes. A 5.0 cm^3 sample of the clarified solution was taken from the liquid phase of each polyethylene bottle with a pipet at regular intervals and diluted to 50.0 cm^3 final volumes in a volumetric flask filled with DDW. If the compositions of the liquid phase in the bottle became constant, then equilibrium was achieved. Generally, it takes about 50 days to come to equilibrium.

3.3.2 Metastable phase equilibria

The isothermal evaporation method was used in metastable phase equilibria study. According to phase equilibrium composition, the appropriate quantity of salts and DDW calculated were mixed together as a series of artificial synthesized brines and loaded into clean polyethylene containers (15 cm in diameter, 6 cm high), then the containers were put into the box for the isothermal evaporation at (308.15 ± 0.2) K. The experimental conditions with air flowing velocity of 3.5-4.0 m/s, relative humidity of 20-30%, and evaporation rate of 4-6 mm/d are presented, just like the climate of the Qaidam Basin. For metastable evaporation, the solutions were not stirred, and the crystal behavior of solid phase was observed periodically. When enough new solid phase appeared, the wet residue mixtures were taken from the solution. The solids were then approximately evaluated by the combined chemical analysis, of XP-300D Digital Polarizing Microscopy (Shanghai Caikon Optical Instrument Co,. Ltd., China) using an oil immersion, and further identification with X-ray diffraction (X'pert PRO, Spectris. Pte. Ltd., The Netherlands). Meanwhile, a 5.0 cm³ sample of the clarified solution was taken from the liquid phase of each polyethylene container through a filter pipette, and then diluted to a 250.0 cm³ final volume in a volumetric flask filled with DDW for the quantitative analysis of the compositions of the liquid phase. Some other filtrates were used to measure the relative physicochemical properties individually according to the analytical method. The remainder of the solution continued to be evaporated and reached a new metastable equilibrium.

4. Experimental results

4.1 Mineral identification for the solid phase

For mineral identification when enough new solid phase appeared either in the stable equilibrium system or in the metastable equilibrium system, the wet residue mixtures were taken from the solution according to the experimental method. Firstly, as to the minerals of $Na_2B_4O_7 \cdot 10H_2O$ and $K_2B_4O_7 \cdot 4H_2O$, the former belongs to monoclinic system, and the dual optical negative crystal i.e. 2v(-) whereas the later belongs to trimetric system, and the dual optical positive crystal i.e. 2v(+). Secondly, to the minerals NaCl and KCl, they can be identified through the property of refractive index. The refractive index of NaCl is higher than that of KCl. Observed with a XP-300D Digital Polarizing Microscopy using an oil immersion method, the crystal photos of the single and orthogonal polarized light on representative solid phases in the invariant points (NaCl + KCl + $Na_2B_4O_7 \cdot 10H_2O$) and ($Na_2B_4O_7 \cdot 10H_2O$ + $K_2B_4O_7 \cdot 4H_2O$ + KCl) are presented in Figure 3.

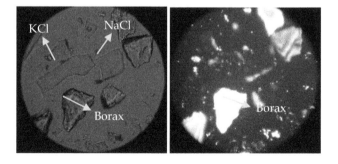

single polarized light (10×10)   orthogonal polarized light (10×10)

(a) Invariant point (NaCl + KCl + Borax)

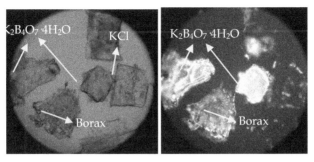

single polarized light (10×10)   orthogonal polarized light (10×10)

(b) Invariant point (KCl + Borax + $K_2B_4O_7 \cdot 4H_2O$)

Fig. 3. Identification of the invariant points for the solid phase in the reciprocal system ($NaCl - KCl – Na_2B_4O_7 - K_2B_4O_7 - H_2O$) with a polarized microscopy using an oil-immersion method. (a), the invariant point ($NaCl + KCl + Na_2B_4O_7 \cdot 10H_2O$); (b), the invariant point ($KCl + Na_2B_4O_7 \cdot 10H_2O + K_2B_4O_7 \cdot 4H_2O$).

The metastable equilibria solid phases in the two invariant points are further confirmed with X-ray diffraction analysis, and listed in Figure 4, except in the invariant points ($NaCl + KCl + Borax$) in Figure 4a which shows that the minerals KCl, NaCl, $Na_2B_4O_7 \cdot 10H_2O$ and a minor $Na_2B_4O_7 \cdot 5H_2O$ are existed. The minor of $Na_2B_4O_7 \cdot 5H_2O$ maybe is formed due to the dehydration of $Na_2B_4O_7 \cdot 10H_2O$ in the processes of transfer operation and/or grinding.

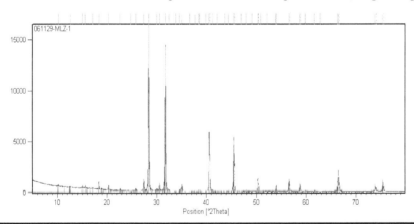

No.	Visible	Ref. Code	Chemical Formula	Score	Scale Factor	Semi-Quant/%
1	True	01-075-0296	KCl	49	0.369	22
2	True	01-075-0296	NaCl	45	0.732	57
3	True	01-075-0296	$B_4O_5(OH)_4(Na_2(H_2O)_8$	39	0.030	20
4	True	01-075-0296	$Na_2B_4O_7(H_2O)_5$	13	0.007	1

(a), the X-ray diffraction photograph and the analytical data for the invariant point ($NaCl + KCl + Borax$)

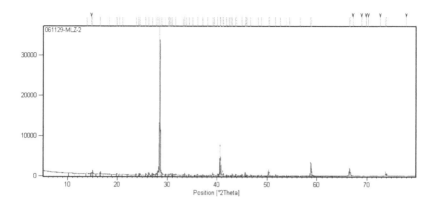

No.	Visible	Ref. Code	Chemical Formula	Score	Scale Factor	Semi-Quant/%
1	True	01-072-1540	KCl	39	0.650	72
2	True	01-076-0753	$K_2(B_4O_5(OH)_4)(H_2O)_2$	37	0.011	22
3	True	01-074-0339	$B_4O_5(OH)_4(Na_2(H_2O)_8$	10	0.004	5

(b), the X-ray diffraction photograph and the analytical data for the invariant point
(KCl + Borax + $K_2B_4O_7 \cdot 4H_2O$)

Fig. 4. The X-ray diffraction data of the invariant points. (a), the invariant point
($NaCl + KCl + Na_2B_4O_7 \cdot 10H_2O$); (b), the invariant point ($Na_2B_4O_7 \cdot 10H_2O + K_2B_4O_7 \cdot 4H_2O +$
KCl).

4.2 Stable phase equilibrium of the quaternary system
($NaCl$ - KCl – $Na_2B_4O_7$ - $K_2B_4O_7$ - H_2O) at 298.15 K

The stable phase equilibrium experimental results of solubilities of the quaternary system
($NaCl$ - KCl – $Na_2B_4O_7$ - $K_2B_4O_7$ - H_2O) at 298.15 K were determined, and are listed in Table
3, respectively. On the basis of the Jänecke index (J_B, $J_B/[mol/100 \, mol(2Na^+ + 2K^+)]$) in Table
3, the stable equilibrium phase diagram of the system at 298.15 K was plotted and shown in
Figure 5.

4.3 Metastable phase equilibrium of the quaternary system
($NaCl$ - KCl – $Na_2B_4O_7$ - $K_2B_4O_7$ - H_2O) at 308.15 K

The experimental results of the metastable solubilities and the physicochemical properties of
the quaternary system ($NaCl$ - KCl – $Na_2B_4O_7$ - $K_2B_4O_7$ - H_2O) at 308.15 K were determined,
and are listed in Tables 4 and 5, respectively. On the basis of the Jänecke index (J_B,
$J_B/[mol/100 \, mol(2Na^+ + 2K^+)]$) in Table 4, the metastable equilibrium phase diagram of the
system at 308.15 K was plotted (Figure 6).

No.	Composition of the solution $100\,w_B^*$				Jänecke index $J_B, /[mol/100mol(2Na^++2K^+)]$			Equilirium solid phase**
	Na^+	K^+	Cl^-	$B_4O_7^{2-}$	$J(2Cl^-)$	$J(2K^+)$	$J(H_2O)$	
1	10.25	0.00	15.80	0.00	100.00	0.00	1807.5	NaCl
2	0.00	13.87	12.58	0.00	100.00	100.00	2310.8	KCl
3	0.72	0.00	0.00	2.41	0.00	0.00	34623.3	N10
4	0.00	4.74	0.00	9.42	0.00	100.00	7864.1	K4
5, E_1	8.02	5.85	17.68	0.00	100.00	30.03	1526.2	NaCl+KCl
6, E_2	9.60	0.00	14.33	1.04	96.79	0.00	1997.5	NaCl+N10
7, E_3	0.00	15.87	12.35	4.48	85.85	100.00	1807.8	KCl +K4
8, E_4	1.15	4.36	0.00	12.54	0.00	69.24	5649.7	N10 + K4
9	1.38	6.11	0.92	14.77	11.98	72.33	3947.5	N10 + K4
10	1.57	11.74	12.35	1.59	24.64	72.41	3496.9	N10 + K4
11	1.59	7.55	3.83	11.97	41.25	73.74	3181.6	N10 + K4
12	3.50	11.39	12.24	7.62	96.67	45.60	1427.4	N10 + K4
13, E	4.72	9.62	14.55	3.17	90.05	54.55	1672.8	KCl + N10 + K4
14	1.13	13.44	12.65	2.78	90.59	88.46	2123.7	KCl+K4
15	1.57	11.74	12.35	1.59	94.45	81.45	2191.0	KCl+K4
16	4.18	10.56	14.96	2.33	93.36	59.79	1670.9	KCl+K4
17	4.15	9.65	13.96	2.61	92.13	57.77	181.0	KCl+K4
18	5.27	9.43	15.50	2.58	93.83	51.84	1602.8	KCl + N10
19	6.11	9.83	17.52	1.79	95.55	48.48	1386.6	KCl + N10
20	6.40	9.10	17.52	1.32	96.67	45.60	1427.4	NaCl + KCl
21	7.50	7.13	17.73	0.67	98.30	35.89	1462.7	NaCl + KCl
22, F	6.16	9.64	17.55	1.52	96.19	47.94	1406.5	NaCl+KCl+N10
23	9.73	1.05	15.35	1.35	96.05	6.12	1830.4	NaCl+N10

* w_B is in mass fraction; ** K4, $K_2B_4O_7 \cdot 4H_2O$; N10, $Na_2B_4O_7 \cdot 10H_2O$.

Table 3. Stable solubilities of the system ($NaCl$ - KCl – $Na_2B_4O_7$ - $K_2B_4O_7$ - H_2O) at 298.15 K

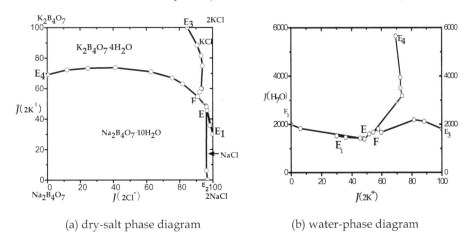

(a) dry-salt phase diagram   (b) water-phase diagram

Fig. 5. Stable phase diagram of the quaternary system ($NaCl$ - KCl – $Na_2B_4O_7$ - $K_2B_4O_7$ - H_2O) at 298.15 K. (a), dry-salt phase diagram; (b), water phase diagram.

No.	Composition of the solution $100w_B$				Jänecke index J_B, /[mol/100mol($2Na^+ + 2K^+$)]			Equilibrium solid phase*
	Na^+	K^+	Cl^-	$B_4O_7^{2-}$	$J(2Cl^-)$	$J(2K^+)$	$J(H_2O)$	
A	1.30	0.00	0.00	4.40	0.00	0.00	2090.17	N10
B	10.47	0.00	16.15	0.00	100.00	0.00	1788.31	NaCl
C	0.00	14.87	13.49	0.00	100	100	18468.05	KCl
D	0.00	3.87	0.00	19.26	0.00	100	4311.59	K4
1, E′₁	7.85	6.56	18.06	0.00	100	32.94	1471.69	NaCl+KCl
2	7.84	6.54	17.76	0.58	98.54	32.89	1469.10	NaCl+KCl
3	7.85	6.60	17.66	0.95	97.6	33.09	1456.10	NaCl+KCl
4, E′	7.87	6.56	17.17	1.99	94.98	32.89	1445.77	NaCl+KCl+N10
5, E′₂	10.62	0.00	15.80	1.28	96.47	0.00	1736.76	NaCl +N10
6	9.88	1.51	15.89	1.59	95.63	8.25	1684.84	NaCl +N10
7	9.36	2.87	16.24	1.74	95.33	15.27	1612.49	NaCl +N10
8	9.10	3.62	16.50	1.81	95.22	18.97	1566.69	NaCl +N10
9	8.15	5.38	16.57	1.94	94.92	27.96	1532.30	NaCl+N10
10, E′₃	0.00	15.22	11.83	4.34	85.66	100	1955.76	KCl+K4
11	0.49	14.75	12.16	4.34	85.99	94.61	1899.70	KCl+K4
12	0.54	14.73	12.29	4.15	86.66	94.16	1894.74	KCl+K4
13	1.22	13.68	12.51	3.88	87.60	86.86	1893.49	KCl+K4
14	1.86	12.88	12.45	4.60	85.56	80.30	1844.98	KCl+K4
15	2.65	12.41	12.51	6.19	81.57	73.39	1699.85	KCl+K4
16, F′	2.84	12.36	12.06	7.74	77.34	71.90	1641.11	KCl+K4+N10
17, E′₄	2.01	8.65	0.00	24.00	0.00	71.66	2346.01	N10+K4
18	1.89	9.86	1.78	22.09	15.03	75.41	2134.79	N10+K4
19	1.74	9.55	4.49	15.01	39.58	76.36	2000.89	N10+K4
20	1.66	10.78	6.11	13.66	49.49	79.22	2160.56	N10+K4
21	1.79	12.05	8.55	11.27	62.43	79.81	1906.21	N10+K4
22	1.87	12.73	9.48	10.86	65.66	79.97	1773.28	N10+K4
23	2.28	12.65	10.89	8.97	72.68	76.57	1712.53	N10+K4
24	2.78	12.61	8.30	11.95	75.94	72.72	1838.60	N10+K4
25	3.11	11.86	12.46	6.78	80.09	69.15	1664.48	N10+KCl
26	3.26	11.41	13.13	4.90	85.45	67.31	1723.59	N10+KCl
27	3.64	10.91	13.59	4.20	87.64	63.8	1717.14	N10+KCl
28	4.06	10.43	14.15	3.45	89.98	60.17	1699.65	N10+KCl
29	4.45	10.17	14.57	3.30	90.62	57.36	1652.70	N10+KCl
30	5.27	9.25	15.16	2.97	91.80	50.79	1604.93	N10+KCl
31	6.17	8.23	15.94	2.27	93.91	43.95	1562.35	N10+KCl

* K4, $K_2B_4O_7 \cdot 4H_2O$; N10, $Na_2B_4O_7 \cdot 10H_2O$; w_B, in mass fraction.

Table 4. Metastable solubilities of the quaternary system ($NaCl - KCl - Na_2B_4O_7 - K_2B_4O_7 - H_2O$) at 308.15 K

No.*	Density $\rho,/(g \cdot cm^{-3})$	pH	Refractive index	Viscosity $\eta/(mPa \cdot s)$
A	1.0441	$-$**	1.3405	$-$
B	1.1935	$-$	1.3800	$-$
C	1.1857	$-$	1.3742	$-$
D	1.2003	$-$	1.3678	$-$
1, E'_1	1.2300	5.63	1.3869	1.1241
2	1.2414	7.29	1.3872	1.1452
3	1.2433	7.72	1.3880	1.1930
4, 'E	1.2524	7.81	1.3890	1.2620
5, E'_2	1.2060	$-$	1.3802	$-$
6	1.2172	8.22	1.3836	1.3023
7	1.2274	8.66	1.3860	1.2865
8	1.2313	9.21	1.3862	1.2829
9	1.2398	8.51	1.3879	1.2729
10, E'_3	1.2270	9.55	1.3782	0.8499
11	1.2347	9.43	1.3798	0.8443
12	$-$	$-$	$-$	$-$
13	$-$	$-$	1.3792	0.8844
14	1.2433	9.02	1.3814	1.0625
15	1.2536	9.10	1.3840	1.0876
16, F'	1.2700	9.30	1.3863	1.1963
17, E'_4	1.3040	10.36	1.3868	2.8279
18	1.3206	10.12	1.3868	2.8032
19	1.2533	9.94	1.3803	1.5661
20	1.2600	9.62	1.3823	1.4548
21	1.2636	9.58	1.3840	$-$
22	1.2764	9.53	$-$	1.3900
23	1.2784	9.47	1.3856	$-$
24	$-$	$-$	$-$	$-$
25	1.2555	9.08	1.3857	1.1682
26	1.2484	9.51	1.3842	1.0812
27	1.2409	9.48	1.3828	1.0480
28	1.2348	9.02	1.3827	1.0252
29	1.2366	9.04	1.3835	1.0421
30	1.2404	8.65	1.3841	1.0809
31	1.2381	8.38	1.3849	1.0130

* Corresponding to the no. column in Table 4; ** not determined.

Table 5. Physicochemical properties of the metastable reciprocal quaternary system (NaCl - KCl – $Na_2B_4O_7$ - $K_2B_4O_7$ - H_2O) at 308.15 K

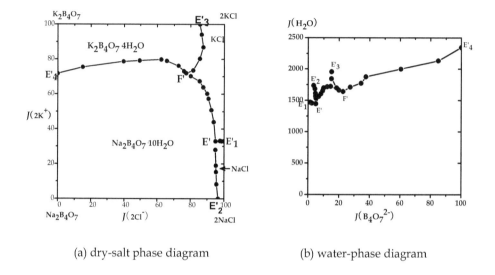

(a) dry-salt phase diagram   (b) water-phase diagram

Fig. 6. Metastable phase diagram of the quaternary system (NaCl - KCl – Na$_2$B$_4$O$_7$ - K$_2$B$_4$O$_7$ - H$_2$O) at 308.15 K. (a), dry-salt phase diagram; (b), water-phase diagram.

On the basis of physicochemical property data of the metastable system (NaCl - KCl – Na$_2$B$_4$O$_7$ - K$_2$B$_4$O$_7$ - H$_2$O) at 308.15 K in Table 5, the diagram of physicochemical properties versus composition was drawn and shows in Figure 7. The physicochemical properties of the metastable equilibrium solution vary regularly with the composition of borate mass fraction. The singular point on every curve of the composition versus property diagram corresponds to the same invariant point and on the metastable solubility.

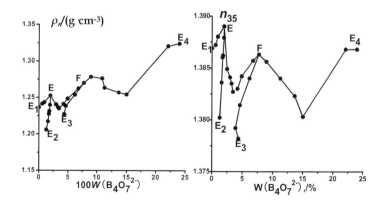

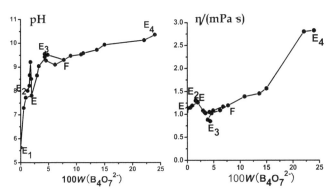

Fig. 7. Diagram of physicochemical properties versus composition for the metastable quaternary system (NaCl - KCl – $Na_2B_4O_7$ - $K_2B_4O_7$ - H_2O) at 308.15 K

4.3 Comparison of the stable and metastable phase diagram of the quaternary system (NaCl - KCl – $Na_2B_4O_7$ - $K_2B_4O_7$ - H_2O)

A comparison of the dry-salt diagrams of the metastable phase equilibrium at 308.15 K and the stable phase equilibrium at 298.15 K for the same system is shown in Figure 8. The metastable crystallization regions of borax and potassium chloride are both enlarged while the crystallized area of other minerals existed is decreased. When compared with the stable system, the solubility of borax in water in the metastable system is increased from 3.13 % to 5.70 %. The metastable phenomenon of borax is obvious in this reciprocal quaternary system.

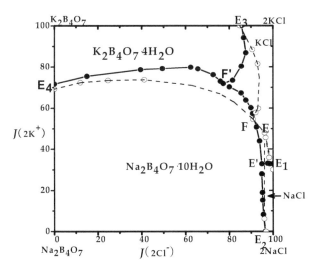

Fig. 8. Comparison of the metastable phase diagram at 308.15 K in solid line and the stable phase diagram at 298.15 K in dashed line for the quaternary system (NaCl - KCl – $Na_2B_4O_7$ - $K_2B_4O_7$ - H_2O). - ● -, metastable experimental points; -○-, stable experimental points.

5. Solubility theoretic prediction of salt-water system

5.1 Ion-interaction model

As to any electrolyte, its thermodynamic prosperity varied from weak solution to high concentration could be calculated through 3 or 4 Pitzer parameters. Pitzer ion-interaction model and its extended HW model of aqueous electrolyte solution can be briefly introduced in the following (Pitzer, 1975, 1977, 2000; Harvie & Wear, 1980; Harvie et al., 1984; Kim & Frederich, 1988a-b).

As to the ion-interaction model, it is a semiempirical statistical thermodynamics model. In this model, the Pitzer approach begins with a virial expansion of the excess free energy of the form to consider the three kinds of existed potential energies on the ion-interaction potential energy in solution.

$$G^{ex} / (n_w RT) = f(I) + \sum_i \sum_j \lambda_{ij}(I) m_i m_j + \sum_i \sum_j \sum_k \mu_{ijk} m_i m_j m_k + ... \tag{1}$$

Where n_w is kilograms of solvent (usually in water), and m_i is the molality of species i (species may be chosen to be ions); i, j, and k express the solute ions of all cations or anions; I is ion strength and given by

$$I = \frac{1}{2} \sum m_i z_i^2 \text{, here } z_i \text{ is the number of charges on the } i\text{-th solute.}$$

The first term on the right in equation (1) is the first virial coefficients. The first virial coefficients i.e. the Debye-Hückel limiting law, $f(I)$, is a function only of ionic strength to express the long-range ion-interaction potential energy of one pair of ions in solution and not on individual ionic molalities or other solute properties.

Short-range potential effects are accounted for by the parameterization and functionality of the second virial coefficients, λ_{ij}, and the third virial coefficients, μ_{ijk}. The quantity λ_{ij} represents the short-range interaction in the presence of the solvent between solute particles i and j. This binary interaction parameter of the second virial coefficient does not itself have any composition dependence for neutral species, but for ions it is dependent it is ionic strength.

The quantity μ_{ijk} represents short-range interaction of ion triplets and are important only at high concentration. The parameters μ_{ijk} are assumed to by independent of ionic strength and are taken to be zero when the ions i, j and k are all cations or all anions.

Taking the derivatives of equation ? with respect to the number of moles of each components yields expressions for the osmotic and activity coefficients.

5.1.1 For pure electrolytes

For the pure single-electrolyte MX, the osmotic coefficient defined by Pitzer (2000):

$$\phi - 1 = |z_M z_X| f^\phi + m \frac{2v_M v_X}{v} B_{MX}^\phi + m^2 \frac{2(v_M v_X)^{3/2}}{v} C_{MX}^\phi \tag{2}$$

φ is the osmotic coefficient; Z_M and Z_X are the charges of anions and cautions in the solution. m is the molality of solute; υ_M, υ_X, and υ ($\upsilon = \upsilon_M + \upsilon_X$) represent the stiochiometric coefficients of the anion, cation, and the total ions on the electrolyte MX.

In equation (1), f^φ, B^φ_{MX} and C^φ_{MX} are defined as following equations. In equation (1a), here b is a universal empirical constant to be equal 1.2 $kg^{1/2}\,mol^{-1/2}$.

$$f^\phi = -\frac{A^\phi I^{1/2}}{1+bI^{1/2}} \tag{2a}$$

For non 2-2 type of electrolytes, such as several 1-1-,2-1-, and 1-2-type pure salts, the best form of B^φ_{MX} is following (Pitzer, 1973):

$$B^\phi_{MX} = \beta^{(0)}_{MX} + \beta^{(1)}_{MX} e^{-\alpha\sqrt{I}} \tag{2b}$$

For 2-2 type of electrolytes, such as several 3-1- and even 4-1-type pure salts, an additional term is added (Pizter, 1977):

$$B^\phi_{MX} = \beta^{(0)}_{MX} + \beta^{(1)}_{MX} e^{-\alpha_1\sqrt{I}} + \beta^{(2)}_{MX} e^{-\alpha_2\sqrt{I}} \tag{2c}$$

$$A^\phi = \frac{1}{3}\left(\frac{2\pi N_0 \rho_W}{1000}\right)^{1/2}\left(\frac{e^2}{DkT}\right)^{3/2} \tag{2d}$$

A^φ is the Debye-Hückel coefficient for the osmotic coefficient and equal to 0.3915 at 298.15 K. Where, N_0 is Avogadro's number, d_w and D are the density and static dielectric constant of the solvent (water in this case) at temperature and e is the electronic charge. k is Boltzmann's constant. In equation (1b), $\beta^{(0)}{}_{MX}$, $\beta^{(1)}{}_{MX}$, C^φ_{MX} are specific to the salt MX, and are the single-electrolyte parameters of MX. The universal parameters $\alpha = 2.0$ $kg^{1/2}\,mol^{-1/2}$ and omit $\beta^{(2)}_{MX}$ for several 1-1-,2-1-, and 1-2-type salts at 298.15 K. As salts of other valence types, the values $\alpha_1 = 1.4$ $kg^{1/2}\,mol^{-1/2}$, and $\alpha_2 = 12$ $kg^{1/2}\,mol^{-1/2}$ were satisfactory for all 2-2 or higher valence pairs electrolytes at 298.15 K. The parameter $\beta^{(2)}_{MX}$ is negative and is related to the association equilibrium constant.

The mean activity coefficient $\gamma_\pm$ is defined as:

$$\ln\gamma_\pm = |z_M z_X|f^\gamma + m\frac{2\upsilon_M\upsilon_X}{\upsilon}B^\gamma_{MX} + m^2\frac{2(\upsilon_M\upsilon_X)^{3/2}}{\upsilon}C^\gamma_{MX} \tag{3}$$

$$f^\gamma = -A^\phi[I^{1/2}/(1+bI^{1/2})+(2/b)\ln(1+bI^{1/2})] \tag{3a}$$

$$B^\gamma_{MX} = B_{MX} + B^\phi_{MX} \tag{3b}$$

$$B_{MX} = \beta^{(0)}_{MX} + \beta^{(1)}_{MX}g(\alpha_1 I^{1/2}) + \beta^{(2)}_{MX}g(\alpha_2 I^{1/2})] \tag{3c}$$

$$g(x) = 2[1-(1+x)\exp(-x)/x^2] \tag{3d}$$

$$C_{MX}^{\gamma} = 3C_{MX}^{\phi} / 2 \tag{3e}$$

5.1.2 For mixture electrolytes

In order to treat mixed electrolytes, the following sets of equations are identical with the form used by Harvie & Weare (1984) for modeling the osmotic coefficient and the activity coefficient of a neutral electrolyte based on Pitzer Equations.

$$\sum_i m_i(\phi-1) = 2(-A^{\phi}I^{3/2} / (1+1.2I^{1/2}) + \sum_{c=1}^{N_c}\sum_{a=1}^{N_a} m_c m_a (B_{ca}^{\phi} + ZC_{ca})$$

$$+ \sum_{c=1}^{N_c-1}\sum_{c'=c+1}^{N_c} m_c m_{c'}(\Phi_{cc'}^{\phi} + \sum_{a=1}^{N_a} m_a \psi_{cc'a}) + \sum_{a=1}^{N_a-1}\sum_{a'=a+1}^{N_a} m_a m_{a'}(\Phi_{aa'}^{\phi}) \tag{4}$$

$$+ \sum_{c=1}^{N_c} m_c \psi_{aa'c}) + \sum_{n=1}^{N_n}\sum_{c=1}^{N_c} m_n m_c \lambda_{nc}$$

$$ln\,\gamma_M = z_M^2 F + \sum_{a=1}^{N_a} m_a(2B_{Ma} + ZC_{Ma}) + \sum_{c=1}^{N_c} m_c(2\Phi_{Mc} + \sum_{a=1}^{N_a} m_a \psi_{Mca}) +$$

$$\sum_{a=1}^{N_a-1}\sum_{a'=a+1}^{N_a} m_a m_{a'} \psi_{aa'M} + |z_M| \sum_{c=1}^{N_c}\sum_{a=1}^{N_a} m_c m_a C_{ca} + \sum_{n=1}^{N_n} m_n(2\lambda_{nM}) \tag{5}$$

$$ln\,\gamma_X = z_X^2 F + \sum_{c=1}^{N_c} m_c(2B_{cX} + ZC_{cX}) + \sum_{a=1}^{N_a} m_a(2\Phi_{Xa} + \sum_{c=1}^{N_c} m_c \psi_{Xac}) +$$

$$\sum_{c=1}^{N_c-1}\sum_{c'=c+1}^{N_c} m_c m_{c'} \psi_{cc'X} + |z_X| \sum_{c=1}^{N_c}\sum_{a=1}^{N_a} m_c m_a C_{ca} + \sum_{n=1}^{N_n} m_n(2\lambda_{nX}) \tag{6}$$

$$ln\,\gamma_N = \sum_{c=1}^{N_c} m_c(2\lambda_{nc}) + \sum_{a=1}^{N_a} m_a(2\lambda_{na}) \tag{7}$$

In equations (3), (4), (5) and (6), the subscripts M, c, and c' present cations different cations; X, a, and a' express anions in mixture solution. N_c, N_a and N_n express the numbers of cations, anions, and neutral molecules; r_M, Z_M, m_C and r_X, Z_X, m_a, Φ present the ion activity coefficient, ion valence number, ion morality, and the permeability coefficient; γ_n, m_n, λ_{nc}, and λ_{na} express activity coefficient of neutral molecule, morality of neutral molecule interaction coefficient between neutral molecules with cations c and anion a.

In equations from (3) to (6), the function symbols of $F, C, Z, A^{\phi}, \psi, \Phi, B^{\phi}, B$ are as following, respectively:

1. The term of F in equations (4) to (5) depends only on ionic strength and temperature. The defining equation of F is given by equation (7).

$$F = -A^{\phi}[I^{1/2} / (1+1.2I^{1/2}) + 2/1.2\,ln(1+1.2I^{1/2})] + \sum_{c=1}^{N_c}\sum_{a=1}^{N_a} m_c m_a B_{ca}'$$

$$+ \sum_{c=1}^{N_c-1}\sum_{c'=c+1}^{N_c} m_c m_{c'}\Phi_{cc'}' + \sum_{a=1}^{N_a-1}\sum_{a'=a+1}^{N_a} m_a m_{a'}\Phi_{aa'}' \tag{8}$$

2. The single-electrolyte third virial coefficient, C_{MX}, account for short-range interaction of ion triplets and are important only at high concentration. These terms are independent of ionic strength. The parameters C_{MX} and C^{ϕ}_{MX}, the corresponding coefficients for calculating the osmotic coefficient, are related by the equation (1-6) (Pitzer & Mayorga, 1973):

$$C_{MX} = C^{\phi}_{MX} / (2|Z_M Z_X|^{1/2})$$ (9)

3. The function Z in the equation (8) is defined by:

$$Z = \sum_i |z_i| m_i$$ (10)

Where, m is the molality of species i, and z is its charge.

4. A^{φ} is the Debye-Hückel coefficient for the osmotic coefficient and equal to 0.3915 at 298.15 K, and it is decided by solvent and temperature as equation (1d).

5. The third virial coefficients, $\psi_{i,j,k}$ in equations (3) to (5) are mixed electrolyte parameters for each cation-cation-anion and anion-anion-cation triplet in mixed electrolyte solutions.

6. The parameters $B^{\phi}_{CA}, B, B'_{CA}$ which describe the interaction of pair of oppositely charged ions represent measurable combinations of the second virial coefficients. They are defined as explicit functions of ionic strength by the following equations (Kim & Frederick, 1988).

$$B^{\phi}_{CA} = \beta^{(0)}_{CA} + \beta^{(1)}_{CA} exp(-\alpha_1 I^{1/2}) + \beta^{(2)}_{CA} exp(-\alpha_2 I^{1/2})$$ (11)

$$B_{CA} = \beta^{(0)}_{CA} + \beta^{(1)}_{CA} g(\alpha_1 I^{1/2}) + \beta^{(2)}_{CA} g(\alpha_2 I^{1/2})$$ (12)

$$B'_{CA} = [\beta^{(1)}_{CA} g'(\alpha_1 I^{1/2}) + \beta^{(2)}_{CA} g'(\alpha_2 I^{1/2})] / I$$ (13)

Where the functions g and g' in equations (10), (11) and (12) are defined by

$$g(x) = 2[1-(1+x)exp(-x)] / x^2$$ (14)

$$g'(x) - -2[1-(1+x+x^2/2)exp(x)] / x^2$$ (15)

In equations (13) and (14), x = $\alpha_1 I^{1/2}$ or = $\alpha_2 I^{1/2}$.

In Pitzer's model expression in Eqns. (10) to (12), α is a function of electrolyte type and does not vary with concentration or temperature. Following Harvie et al. (1984), when either cation or anion for an electrolyte is univalent, the first two terms in equations (10) to (12) are considered, $\beta^{(2)}_{CA}$ can be neglect and α_1 =2.0 kg$^{1/2}$ mol$^{-1/2}$, α_2 = 0 at 298.15 K. For higher valence type, such as 2-2 electrolytes for these higher valence species accounts for their increased tendency to associate in solution, the full equations from (10) to (12) are used, and α_1=1.4 kg$^{1/2}$ mol$^{-1/2}$ and α_2=12 kg$^{1/2}$ mol$^{-1/2}$ at 298.15 K.

7. $\Phi_{ij}^{\phi}, \Phi_{ij}, \Phi_{ij}'$ which depend upon ionic strength, are the second virial coefficients, and are given the following form (Pitzer, 1973).

$$\Phi_{ij}^{\phi} = \theta_{ij} + {}^{E}\theta_{ij} + I\,{}^{E}\theta_{ij}' \tag{16}$$

$$\Phi_{ij} = \theta_{ij} + {}^{E}\theta_{ij} \tag{17}$$

$$\Phi_{ij}' = {}^{E}\theta_{ij}' \tag{18}$$

In equations (15), (16) and (17), $\theta_{i,j}$ is an adjustable parameter for each pair of anions or cations for each cation-cation and anion-anion pair, called triplet-ion-interaction parameter. The functions, ${}^{E}\theta_{ij}$ and ${}^{E}\theta_{ij}'$ are functions only of ionic strength and the electrolyte pair type. Pitzer (1975) derived equations for calculating these effects, and Harvie and Weare (1981) summarized Pitzer's equations in a convenient form as following:

$$^{E}\theta_{ij} = (Z_i Z_j / 4I)[J(x_{ij}) - J(x_{ii}) / 2 - J(x_{jj}) / 2] \tag{19}$$

$$^{E}\theta_{ij}' = -({}^{E}\theta_{ij} / I) + (Z_i Z_j / 8I^2)[x_{ij}J'(x_{ij}) - x_{ii}J'(x_{ii}) / 2 - x_{jj}J'(x_{jj}) / 2] \tag{20}$$

$$x_{ij} = 6Z_i Z_j A^{\phi} I^{1/2} \tag{21}$$

In equations (18) and (19), $J(x)$ is the group integral of the short-range interaction potential energy. $J'(x)$ is the single-order differential quotient of $J(x)$, and both are independent of ionic strength and ion charges. In order to give the accuracy in computation, $J(x)$ can be fitted as the following function:

$$J(x) = 1 / 4x - 1 + 1 / x \int_0^{\infty} [1 - exp(-x / yeC_1 x^{-C_2} \cdot exp(-C_3 x^{C_4})]^{-1} \tag{22}$$

$$J'(x) = [4 + C_1 x^{-C_2} \cdot exp(-C_3 x^{C_4})]^{-1} \\ + [4 + C_1 x^{-C_2} exp(-C_3 x^{C_4})]^{-2} [C_1 x exp(-C_3 x^{C_4})(C_2 x^{-C_2-1} + C_3 C_4 x^{C_4-1} x^{-C_2})] \tag{23}$$

In equations (21) and (22), $C_1 = 4.581$, $C_2 = 0.7237$, $C_3 = 0.0120$, $C_4 = 0.528$.

Firstly, x_{ij} can be calculated according to equation (20), and $J(x)$ and $J'(x)$ were obtained from equations (21) and (22), and then to obtained ${}^{E}\theta_{ij}$ and ${}^{E}\theta_{ij}'$ from equations (18) and (19); finally, $\Phi_{ij}^{\phi}, \Phi_{ij}, \Phi_{ij}'$ can be got through equations from (15) to (17). Using the values of $\Phi_{ij}^{\phi}, \Phi_{ij}, \Phi_{ij}'$, the osmotic and activity coefficients of electrolytes can be calculated via equations from (3) to (6).

Using the osmotic coefficient, activity coefficient and the solubility products of the equilibrium solid phases allowed us to identify the coexisting solid phases and their compositions at equilibrium.

On Pitzer ion-interaction model and its extended HW model, a numbers of papers were successfully utilized to predict the solubility behaviors of natural water systems, salt-water systems, and even geological fluids (Felmy & Weare, 1986; Kim & Frederich, 1988a, 1988b; Fang et al., 1993; Song, 1998; Song & Yao, 2001, 2003; Yang, 1988, 1989, 1992, 2005).

By the way, additional work has centered on developing variable-temperature models, which will increase the applicability to a number of diverse geochemical systems. The primary focus has been to broaden the models by generating parameters at higher or lower temperatures (Pabalan & Pitzer, 1987; Spencer et al., 1990; Greenberg & Moller, 1989).

5.2 Model parameterization and solubility predictions

As to the borate solution, the crystallized behavior of borate salts is very complex. The coexisted polyanion species of borate in the liquid phase is difference with the differences of boron concentration, pH value, solvent, and the positively charged ions. The ion of $B_4O_7^{2-}$ is the general statistical express for various possible existed borates. Therefore, the structural formulas of $Na_2B_4O_7 \cdot 10H_2O$ and $K_2B_4O_7 \cdot 4H_2O$ in the solid phases of the quaternary system ($NaCl$ - KCl – $Na_2B_4O_7$ - $K_2B_4O_7$ - H_2O) are $Na_2[B_4O_5(OH)_4] \cdot 8H_2O$ and $K_2[B_4O_5(OH)_4] \cdot 2H_2O$, respectively. Borate in the liquid phase corresponding to the equilibrium solid phase maybe coexists as $B_4O_5(OH)_4^{2-}$, $B_3O_3(OH)_4^{-}$, $B(OH)_4^{-}$, and son on due to the reactions of polymerization or depolymerization of boron anion.

Therefore, in this part of predictive solubility of the quaternary system ($NaCl$ - KCl – $Na_2B_4O_7$ - $K_2B_4O_7$ - H_2O), the predictive solubilities of this system were calculated on the basis of two assumptions: Model I: borate in the liquid phase exists all in statistical form of $B_4O_7^{2-}$ i.e. $B_4O_5(OH)_4^{2-}$; Model II: borate in the liquid phase exists as various boron species of $B_4O_5(OH)_4^{2-}$, $B_3O_3(OH)_4^{-}$, $B(OH)_4^{-}$.

The necessary model parameters for the activity coefficients of electrolytes in the system at 298.15 K were fit from obtained osmotic coefficients and the sub-ternary subsystems by the multiple and unary linear regression methods.

5.2.1 Model I for the solubility prediction

Model I: Suppose that borate in solution exists as in the statistical expression form of $B_4O_7^{2-}$ i.e. $B_4O_5(OH)_4^{2-}$, and the dissolved equilibria in the system could be following:

$$Na_2B_4O_7 \cdot 10H_2O = 2Na^+ + B_4O_5(OH)_4^{2-} + 8H_2O$$

$$K_2B_4O_7 \cdot 4H_2O = 2K^+ + B_4O_5(OH)_4^{2-} + 2H_2O$$

$$NaCl = Na^+ + Cl^-$$

$$KCl = K^+ + Cl^-$$

So, the dissolved equilibrium constants can be expressed as:

$$K_{N10} = (m_{Na^+} \cdot \gamma_{Na^+})^2 \cdot (m_{B4} \cdot \gamma_{B4}) \cdot a_w^8 \tag{24}$$

$$K_{K4} = (m_{K^+} \cdot \gamma_{K^+})^2 \cdot (m_{B4} \cdot \gamma_{B4}) \cdot a_w^2 \tag{25}$$

$$K_{NaCl} = (m_{Na^+} \cdot \gamma_{Na^+}) \cdot (m_{Cl^-} \cdot \gamma_{Cl^-}) \tag{26}$$

$$K_{KCl} = (m_{K^+} \cdot \gamma_{K^+}) \cdot (m_{Cl^-} \cdot \gamma_{Cl^-}) \tag{27}$$

And the electric charge balance exists as:

$$m_{Na^+} + m_{K^+} = m_{Cl^-} + 2m_{B_4O_5(OH)_4^{2-}} \tag{28}$$

Where, K, r, m, and a_w express equilibrium constant, activity coefficient, and water activity, and N10, K4 instead of the minerals of $Na_2B_4O_7 \cdot 10H_2O$, $K_2B_4O_7 \cdot 4H_2O$ (the same in the following), respectively. Then, the equilibria constants K are calculated with μ^0/RT and shown in Table 6.

The single salt parameters $\beta^{(0)}$, $\beta^{(1)}$, $C^{(\varphi)}$ of NaCl, KCl, $Na_2[B_4O_5(OH)_4]$, and $K_2[B_4O_5(OH)_4]$, two-ion interaction Pitzer parameters of $\theta_{Na, K}$, $\theta_{Cl, B4O5(OH)4}$ and the triplicate-ion Pitzer parameters of $\Psi_{Cl, B4O5(OH)4, Na}$, $\Psi_{Cl, B4O5(OH)4, K}$, $\Psi_{Na, K, Cl}$, $\Psi_{Na, K, B4O5(OH)4}$ in the reciprocal quaternary system at 298.15 K were chosen from Harvie et al. (1984), Felmy & Weare (1986), Kim & Frederick (1988), and Deng (2001) and summarized in Tables 7 and 8.

According to the equilibria constants and the Pitzer ion-interaction parameters, the solubilities of the quaternary system at 298.15 K have been calculated though the Newton's Iteration Method to solve the non-linearity simultaneous equations system, and shown in Table 9.

5.2.2 Model II for the solubility prediction

Model II: Suppose that borate in solution exists as in various boron species of $B_4O_5(OH)_4^{2-}$, $B_3O_3(OH)_4^-$, $B(OH)_4^-$ to further describe the behaviors of the polymerization and depolymerization of borate anion in solution, and the dissolved equilibria in the system could be following:

$$Na_2B_4O_7 \cdot 10H_2O = 2Na^+ + B_4O_5(OH)_4^{2-} + 8H_2O$$

$$K_2B_4O_7 \cdot 4H_2O = 2K^+ + B_4O_5(OH)_4^{2-} + 2H_2O$$

$$B_4O_5(OH)_4^{2-} + 2H_2O = B_3O_3(OH)_4^- + B(OH)_4^-$$

$$NaCl = Na^+ + Cl^-$$

$$KCl = K^+ + Cl^-$$

So, the dissolved equilibrium constants can be expressed as:

$$K_{N10} = (m_{Na^+} \cdot \gamma_{Na^+})^2 \cdot (m_{B4} \cdot \gamma_{B4}) \cdot a_w^8 \tag{29}$$

$$K_{K4} = (m_{K^+} \cdot \gamma_{K^+})^2 \cdot (m_{B4} \cdot \gamma_{B4}) \cdot a_w^2 \tag{30}$$

$$K_{B4B3B} = \frac{(m_{B3} \cdot \gamma_{B3}) \cdot (m_B \cdot \gamma_B)}{(m_{B4} \cdot \gamma_{B4}) \cdot a_w^2} \tag{31}$$

$$K_{NaCl} = (m_{Na^+} \cdot \gamma_{Na^+}) \cdot (m_{Cl^-} \cdot \gamma_{Cl^-}) \tag{32}$$

$$K_{KCl} = (m_{K^+} \cdot \gamma_{K^+}) \cdot (m_{Cl^-} \cdot \gamma_{Cl^-}) \tag{33}$$

Where, B4, B3 and B to instead of $B_4O_5(OH)_4^{2-}$, $B_3O_3(OH)_4^-$, and $B(OH)_4^-$ for short; K_{B4B3B} expresses the equilibrium constant of the polymerized species reaction of $B_4O_5(OH)_4^{2-}$, $B_3O_3(OH)_4^-$, $B(OH)_4^-$.

And the electric charge balance exists as:

$$m_{Na^+} + m_{K^+} = m_{Cl^-} + 2m_{B_4O_5(OH)_4^{2-}} + m_{B_3O_3(OH)_4^-} + m_{B(OH)_4^-} \tag{34}$$

From this reaction of B4, B3 and B, i.e. $B_4O_5(OH)_4^{2-} + 2H_2O = B_3O_3(OH)_4^- + B(OH)_4^-$, the molalities of B3 and B are in equal. In the meantime, we suppose that two-ion and triplicate-ion interaction of different boron species would be weak, and the mixture ions parameters of different boron species should be ignored.

Similar as in model I, then, the equilibrium constant K existed solid phase is calculated with μ^0/RT, and also shown in Table 6, where another four possible borate salts of $NaB_3O_3(OH)_4$, $NaB(OH)_4$, $KB_3O_3(OH)_4$, $KB(OH)_4$ were also listed. The single salt parameters, binary ion interaction parameters, triplet mixture parameters and more parameters of $\theta_{Cl,B3O3(OH)4}$, $\theta_{Cl,B(OH)4}$, $\Psi_{Cl,B3O3(OH)4,Na}$, and $\Psi_{Cl,B(OH)4,Na}$ were considered, and shown in Table 8. According to the equilibria constants and the Pitzer ion-interaction parameters, the solubilities of the quaternary system at 298.15 K have been calculated though the Newton's Iteration Method to solve the non-linearity simultaneous equations system, and shown in Table 10. In fact, this theoretic calculation for the reciprocal quaternary system is equivalence of the calculated solubilities for the six-component system (Na - K - Cl - $B_4O_5(OH)_4$ - $B_3O_3(OH)_4$ - $B(OH)_4$ – H_2O). It is worthy saying that although the concentrations of Na^+, K^+, Cl^-, $B_4O_5(OH)_4^{2-}$, $B_3O_3(OH)_4^-$, $B(OH)_4^-$ in molalities could be got (Table 10), the concentrations including $B_4O_5(OH)_4^{2-}$, $B_3O_3(OH)_4^-$, $B(OH)_4^-$ should be all inverted into the concentration of $B_4O_7^{2-}$ when the Jänecke index of $B_4O_7^{2-}$ calculation.

Species	μ^0/RT	Refs	Species	μ^0/RT	Refs
H_2O	-95.6635		$B(OH)_4^-$	-465.20	
Na^+	-105.651	Harvie et al., 1984	$Na_2B_4O_5(OH)_4 \cdot 8H_2O$	-2224.16	Felmy & Weare, 1986
K^+	-113.957		$K_2B_4O_5(OH)_4 \cdot 2H_2O$	-1663.47	
Cl^-	-52.955				
$B_4O_5(OH)_4^{2-}$	-1239.10	Felmy & Weare, 1986	$NaCl$	-154.99	Harvie et al., 1984
$B_3O_3(OH)_4^-$	-963.77		KCl	-164.84	

Table 6. μ^0/RT of species in the system ($NaCl$ - KCl – $Na_2B_4O_7$ - $K_2B_4O_7$ - H_2O) at 298.15 K

On the basis of the calculated solubilities, a comparison diagram among model I, model II, experimental values for the reciprocal quaternary system at 298.15 K are shown in Figure 9.

Cation	Anion	$\beta_{MX}^{(0)}$	C_{MX}^{ϕ} $\beta_{MX}^{(1)}$	$C_{MX}^{(\phi)}$	Refs
Na$^+$	Cl$^-$	0.07722	0.25183	0.00106	Kim & Frederick, 1988
Na$^+$	B$_4$O$_5$(OH)$_4$$^{2-}$	-0.11	-0.40	0.0	
Na$^+$	B$_3$O$_3$(OH)$_4$$^-$	-0.056	-0.91	0.0	Felmy & Weare, 1986
Na$^+$	B(OH)$_4$$^-$	-0.0427	0.089	0.0114	
K$^+$	Cl$^-$	0.04835	0.2122	-0.00084	Harvie et al., 1984
K$^+$	B$_4$O$_5$(OH)$_4$$^{2-}$	-0.022	0.0	0.0	
K$^+$	B$_3$O$_3$(OH)$_4$$^-$	-0.13	0.0	0.0	Felmy & Weare, 1986
K$^+$	B(OH)$_4$$^-$	0.035	0.14	0.0	

Table 7. Single-salt Pitzer parameters in the system (NaCl - KCl – Na$_2$B$_4$O$_7$ - K$_2$B$_4$O$_7$ - H$_2$O) at 298.15 K

Parameters	Values	Refs
$\theta_{Na+, K+}$	-0.012	Harvie et al., 1984
$\theta_{Cl^-, B4O5(OH)42-}$	0.074	
$\theta_{Cl^-, B3O3(OH)4-}$	0.12	Felmy & Weare, 1986
$\theta_{Cl^-, B(OH)4-}$	-0.065	
$\theta_{B4O5(OH)42-, B3O3(OH)4-}$	—	—
$\theta_{B4O5(OH)42-, B(OH)4-}$	—	—
$\theta_{B3O3(OH)4-, B(OH)4-}$	—	—
$\Psi_{Cl^-, B4O5(OH)42-, Na+}$	0.025	
$\Psi_{Cl^-, B3O3(OH)4-, Na+}$	-0.024	Felmy & Weare, 1986
$\Psi_{Cl^-, B(OH)4-, Na+}$	-0.0073	
$\Psi_{B4O5(OH)42-, B3O3(OH)4-, Na+}$	—	—
$\Psi_{B4O5(OH)42-, B(OH)4-, Na+}$	—	—
$\Psi_{B3O3(OH)4-, B(OH)4-, Na+}$	—	—
$\Psi_{Cl^-, B4O5(OH)42-, K+}$	0.0185245	Deng, 2004
$\Psi_{Cl^-, B3O3(OH)4-, K+}$	—	—
$\Psi_{Cl^-, B(OH)4-, K+}$	—	—
$\Psi_{B4O5(OH)42-, B3O3(OH)4-, K+}$	—	—
$\Psi_{B4O5(OH)42-, B(OH)4-, K+}$	—	—
$\Psi_{B3O3(OH)4-, B(OH)4-, K+}$	—	—
$\Psi_{Na+, K+, Cl^-}$	-0.0018	Harvie et al., 1984
$\Psi_{Na+, K+, B4O5(OH)42-}$	0.289823	Deng, 2004
$\Psi_{Na+, K+, B3O3(OH)4-}$	—	—
$\Psi_{Na+, K+, B(OH)4-}$	—	—

Table 8. Mixing ion-interaction Pitzer parameters in the system (NaCl - KCl – Na$_2$B$_4$O$_7$ - K$_2$B$_4$O$_7$ - H$_2$O) at 298.15 K

Though the theoretical calculation on the basis of model II, it was found that the boron species are mainly existed B$_3$O$_3$(OH)$_4$$^-$ and B(OH)$_4$$^-$ while the concentration of B$_4$O$_5$(OH)$_4$$^{2-}$ is very low when the total concentration of boron is low in weak solution. This result demonstrated that the polymerization or depolymerization behaviors of borate are complex.

No.	Composition liquid phase molality, $/(mol/kgH_2O)$				Jänecke index, $J/(mol/100mol$ dry salts)		Equilibrium solid phases*
	Na^+	K^+	Cl^-	$B_4O_7^{2-}$	$J(2Na^+)$	$J(B_4O_7^{2-})$	
1	0.2748	1.2843	0.00	0.7796	17.62	100.00	N10+K4
2	0.2768	1.3032	0.2000	0.6900	17.52	87.34	N10+K4
3	0.2804	1.3279	0.3500	0.6292	17.43	78.24	N10+K4
4	0.2863	1.3619	0.5000	0.5741	17.37	69.66	N10+K4
5	0.3014	1.4496	0.7800	0.4855	17.21	55.45	N10+K4
6	0.3177	1.5386	1.0000	0.4282	17.11	46.13	N10+K4
7	0.3603	1.7573	1.4400	0.3388	17.02	32.00	N10+K4
8	0.4297	2.0854	2.0000	0.2575	17.08	20.48	N10+K4
9	0.5028	2.4030	2.5000	0.2029	17.30	13.97	N10+K4
10	0.5858	2.7319	3.0000	0.1589	17.66	9.58	N10+K4
11	0.6774	3.0673	3.5000	0.1224	18.09	6.54	N10+K4
12	0.8903	3.7442	4.5000	0.06725	19.21	2.90	N10+K4
13,A1	1.0298	4.1687	5.1107	0.04392	19.81	1.69	N10+K4+KCl
14	0.00	4.8834	4.7149	0.08423	0.00	3.45	K4+KCl
15	0.1500	4.7688	4.7699	0.07444	3.05	3.03	K4+KCl
16	0.3000	4.6592	4.8262	0.06649	6.05	2.68	K4+KCl
17	0.4500	4.5534	4.8835	0.05996	8.99	2.40	K4+KCl
18	0.6000	4.4509	4.9418	0.05457	11.88	2.16	K4+KCl
19	0.7500	4.3511	5.0009	0.05009	14.70	1.96	K4+KCl
20	0.9000	4.2536	5.0609	0.04636	17.46	1.80	K4+KCl
21	4.8000	2.2523	7.0507	8.13E-4	68.06	0.023	N10+KCl
22	4.3000	2.4641	6.7621	9.67E-4	63.57	0.029	N10+KCl
23	3.6000	2.7821	6.3794	0.00139	56.41	0.044	N10+KCl
24	3.2000	2.9750	6.1713	0.00185	51.82	0.060	N10+KCl
25	2.8000	3.1758	5.9705	0.00263	46.86	0.088	N10+KCl
26	2.4000	3.3843	5.7762	0.00405	41.49	0.14	N10+KCl
27	2.0000	3.6007	5.5869	0.00686	35.71	0.25	N10+KCl
28	1.6000	3.8255	5.3992	0.01316	29.49	0.49	N10+KCl
29	5.2183	1.9000	7.1163	9.75E-4	73.31	0.027	N10+NaCl
30	5.4046	1.5000	6.9014	0.00159	78.28	0.046	N10+NaCl
31	5.6432	1.0000	6.6369	0.00313	84.95	0.094	N10+NaCl
32	5.8894	0.5000	6.3762	0.0066	92.17	0.21	N10+NaCl
33	6.1479	0.00	6.1178	0.01504	100.00	0.49	N10+NaCl
34,B1	5.1148	2.1256	7.2389	7.53E-4	70.64	0.021	N10+NaCl+KCl
35	5.1147	2.1259	7.2394	6.00E-4	70.64	0.017	NaCl+KCl
36	5.1145	2.1264	7.2401	4.00E-4	70.63	0.011	NaCl+KCl
37	5.1143	2.1269	7.2408	2.00E-4	70.63	0.0055	NaCl+KCl
38	5.1142	2.1273	7.2415	0.00	70.62	0.00	NaCl+KCl

Table 9. Calculated solubility data of the system ($NaCl$ - KCl – $Na_2B_4O_7$ - $K_2B_4O_7$ - H_2O) at 298.15 K on the basis of Model I. * N10, $Na_2B_4O_7 \cdot 10H_2O$; K4, $K_2B_4O_7 \cdot 4H_2O$.

No.	Composition liquid phase molality, /(mol/kgH$_2$O)*						Jänecke index, J /(mol/100mol dry-salt)		Equilibrium solid phases
	Na$^+$	K$^+$	Cl$^-$	B4	B3	B	J(2Na$^+$)	J(B$_4$O$_7^{2-}$)	
1	0.3362	1.4923	0.00	0.7084	0.2058	0.2058	18.38	100.00	N10+K4
2	0.3394	1.5158	0.2000	0.6362	0.1914	0.1914	18.28	89.22	N10+K4
3	0.3431	1.5428	0.3500	0.5868	0.1812	0.1812	18.19	81.44	N10+K4
4	0.3485	1.5771	0.5000	0.5413	0.1715	0.1715	18.10	74.03	N10+K4
5	0.3628	1.6608	0.7800	0.4667	0.1551	0.1551	17.93	61.45	N10+K4
6	0.3777	1.7431	1.0000	0.4165	0.1439	0.1439	17.81	52.85	N10+K4
7	0.4168	1.9432	1.4400	0.3347	0.1253	0.1253	17.66	38.98	N10+K4
8	0.4818	2.2464	2.0000	0.2562	0.1079	0.1079	17.66	26.69	N10+K4
9	0.5519	2.5451	2.5000	0.2018	0.0967	0.0967	17.82	19.28	N10+K4
10	0.6326	2.8591	3.0000	0.1574	0.0884	0.0884	18.12	14.08	N10+K4
11	0.7234	3.1822	3.5000	0.1207	0.0821	0.0821	18.52	10.38	N10+K4
12	0.9355	3.8426	4.5000	0.0654	0.0736	0.0736	19.58	5.82	N10+K4
13,A2	1.0770	4.2295	5.0794	0.04305	0.0705	0.0705	20.30	4.28	N10+K4+KCl
14	0.00	4.9588	4.6844	0.08694	0.0502	0.0502	0.00	5.53	K4+KCl
15	0.1500	4.8452	4.7369	0.07661	0.0525	0.0525	3.00	5.17	K4+KCl
16	0.3000	4.7368	4.7905	0.06822	0.0549	0.0549	5.96	4.89	K4+KCl
17	0.4500	4.6326	4.8449	0.06133	0.0576	0.0576	8.85	4.68	K4+KCl
18	0.6000	4.5320	4.8999	0.05563	0.0604	0.0604	11.69	4.52	K4+KCl
19	0.7500	4.4342	4.9558	0.05089	0.0633	0.0633	14.47	4.41	K4+KCl
20	0.9000	4.3390	5.0122	0.04692	0.0665	0.0665	17.18	4.33	K4+KCl
21	4.8000	2.2665	7.0005	8.237E-4	0.0322	0.0322	67.93	0.93	N10+KCl
22	4.3000	2.4815	6.7150	9.779E-4	0.0323	0.0323	63.41	0.98	N10+KCl
23	3.6000	2.8052	6.3356	0.00141	0.0334	0.0334	56.20	1.09	N10+KCl
24	3.2000	3.0021	6.1289	0.00188	0.0347	0.0347	51.60	1.18	N10+KCl
25	2.8000	3.2078	5.9289	0.00268	0.0368	0.0368	46.61	1.31	N10+KCl
26	2.4000	3.4226	5.7346	0.00415	0.0398	0.0398	41.22	1.51	N10+KCl
27	2.0000	3.6474	5.5442	0.0071	0.0445	0.0445	35.41	1.83	N10+KCl
28	1.6000	3.8845	5.3532	0.0138	0.0519	0.0519	29.17	2.40	N10+KCl
29	5.2316	1.9000	7.0663	9.963E-4	0.0316	0.0316	73.36	0.91	N10+NaCl
30	5.4169	1.5000	6.8530	0.00163	0.0303	0.0303	78.31	0.92	N10+NaCl
31	5.6544	1.0000	6.5902	0.00321	0.0289	0.0289	84.97	0.96	N10+NaCl
32	5.8998	0.5000	6.3310	0.00679	0.0276	0.0276	92.19	1.07	N10+NaCl
33	6.1580	0.00	6.0739	0.0155	0.0266	0.0266	100.00	1.37	N10+NaCl
34,B2	5.1668	2.1309	7.2304	7.463E-4	0.0329	0.0329	70.80	0.92	N10+KCl+NaCl
35	5.1612	2.1307	7.2318	6.00E-4	0.0294	0.0294	70.78	0.82	NaCl+KCl
36	5.1523	2.1303	7.2339	4.00E-4	0.0239	0.0239	70.75	0.67	NaCl+KCl
37	5.1410	2.1296	7.2364	2.00E-4	0.0168	0.0168	70.71	0.47	NaCl+KCl
38	5.1142	2.1273	7.2415	0.00	0.00	0.00	70.62	0.00	NaCl+KCl

Table 10. Calculated solubility data of the system (NaCl - KCl – Na$_2$B$_4$O$_7$ - K$_2$B$_4$O$_7$ - H$_2$O) at 298.15 K on the basis of Model II. * B4, B3, B express for B$_4$O$_5$(OH)$_4^{2-}$, B$_3$O$_3$(OH)$_4^-$, B(OH)$_4^-$; N10, Na$_2$B$_4$O$_7$·10H$_2$O; K4, K$_2$B$_4$O$_7$·4H$_2$O.

In Figure 9, compared with Models I and II, the calculated values in the boundary points and the cosaturated point of ($Na_2B_4O_7 \cdot 10H_2O$ + KCl + NaCl) based on model II were in good agreement with the experimental data. However, in the cosaturated point of ($Na_2B_4O_7 \cdot 10H_2O$ + $K_2B_4O_7 \cdot 4H_2O$ + KCl), a large difference on the solubility curve still existed. Reversely, the predictive result based on model II closed to the experimental curve. There were two possible reasons: one is that the structure of borate in solution is very complex, an the Pitzer's parameters of borate salts is scarce; the other one is the high saturation degree of borate, the difference between the experimental equilibrium constant and the theoretic calculated equilibrium constant was large enough.

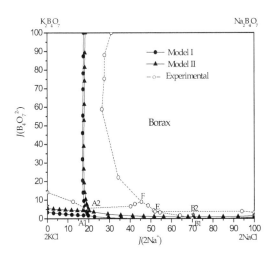

Fig. 9. Comparison of the experimental and calculated phase diagram of the quaternary system (NaCl - KCl – $Na_2B_4O_7$ - $K_2B_4O_7$ - H_2O) at 298.15 K. -●-, Calculated based on Model I; -▲-, Calculated based on Model II; -○-, Experimental.

6. Acknowledgments

Financial support from the State Key Program of NNSFC (Grant.20836009), the NNSFC (Grant. 40773045), the "A Hundred Talents Program" of the Chinese Academy of Sciences (Grant. 0560051057), the Specialized Research Fund for the Doctoral Program of Chinese Higher Education (Grant 20101208110003), The Key Pillar Program in the Tianjin Municipal Science and Technology (11ZCKFGX2800) and Senior Professor Program of Tianjin for TUST (20100405) is acknowledged. Author also hopes to thank all members in my research group and my Ph.D. students Y.H. Liu, S.Q. Wang, Y.F. Guo, X.P. YU, J. Gao, L.Z. Meng, DC Li, Y. Wu, and D.M. Lai for their active contributions on our scientific projects.

7. References

Analytical Laboratory of Institute of Salt Lakes at CAS. (1988). *The analytical methods of brines and salts*, 2nd ed., Chin. Sci. Press, ISBN 7-03-000637-2, pp. 35-41 & 64-66, Beijing

Deng, T.L. (2004). Phase equilibrium on the aqueous five-component system of lithium, sodium, potassium, chloride, borate at 298.15 K, *J. Chem. Eng. Data*, Vol. 47, No. 9, pp. 1295-1299, ISSN 0021-9568

Deng, T.L.; Li, D. (2008a). Solid-liquid metastable equilibria in the quaternary system (NaCl + LiCl + CaCl$_2$ + H$_2$O) at 288.15 K, *J. Chem. Eng. Data*, Vol. 53, No. 11, pp. 2488–2492, ISSN 0021-9568

Deng, T.L.; Li, D.C. (2008b). Solid-liquid metastable equilibria in the quaternary system (NaCl + KCl + CaCl$_2$ + H$_2$O) at 288.15 K, *Fluid Phase Equilibria*, Vol. 269, No. 1-2, pp. 98-103, ISSN 0378-3812

Deng, T.L.; Li, D.C.; Wang, S.Q. (2008c). Metastable phase equilibrium on the aqueous quaternary system (KCl – CaCl$_2$ – H$_2$O) at (288.15 and 308.15) K, *J. Chem. Eng. Data*, Vol. 53, No. 4, pp. 1007-1011, ISSN 0021-9568

Deng, T.L.; Meng, L.Z.; Sun, B. (2008d). Metastable phase equilibria of the reciprocal quaternary system containing sodium, potassium, chloride and borate ions at 308.15 K, *J. Chem. Eng. Data*, Vol. 53, No. 3, pp. 704-709, ISSN 0021-9568

Deng, T.L.; Wang, S.Q. (2008e). Metastable phase equilibrium in the reciprocal quaternary system (NaCl + MgCl$_2$ + Na$_2$SO$_4$ + MgSO$_4$ + H$_2$O) at 273.15 K, *J. Chem. Eng. Data*, Vol. 53, No. 12, pp. 2723-2727, ISSN 0021-9568

Deng, T.L.; Wang, S.Q.; Sun, B. (2008f). Metastable phase equilibrium on the aqueous quaternary system (KCl + K$_2$SO$_4$ + K$_2$B$_4$O$_7$ + H$_2$O) at 308.15 K, *J. Chem. Eng. Data*, Vol. 53, No. 2, pp. 411-414, ISSN 0021-9568

Deng, T.L.; Yin, H.A.; Tang, M.L. (2002). Experimental and predictive phase equilibrium of the quary System at 298K, *J. Chem. Eng. Data*, Vol. 47, pp. 26-29, ISSN 0021-9568

Deng, T.L.; Yin, H.J.; Guo, Y.F. (2011). Metastable phase equilibrium in the aqueous ternary system (Li$_2$SO$_4$ + MgSO$_4$+ H$_2$O) at 323.15 K, *J. Chem. Eng. Data*, Vol. 56, No. 9, pp. 3585-3588, ISSN 0021-9568

Deng, T.L.; Yin, H.J.; Li, D.C. (2009a). Metastable phase equilibrium on the aqueous ternary system (K$_2$SO$_4$ – MgSO$_4$ – H$_2$O) at 348.15 K, *J. Chem. Eng. Data*, Vol. 54, No. 2, pp. 498-501, ISSN 0021-9568

Deng, T.L.; Yu, X.; Li, D.C. (2009b). Metastable phase equilibrium on the aqueous ternary system (K$_2$SO$_4$ – MgSO$_4$ – H$_2$O) at (288.15 and 308.15) K, *J. Solution Chem*, Vol. 38, No. 1, pp. 27-34, ISSN 0095-9782

Deng, T.L.; Yu, X.; Sun, B. (2008g). Metastable phase equilibrium in the aqueous quaternary system (Li$_2$SO$_4$ – K$_2$SO$_4$ – MgSO$_4$ – H$_2$O) at 288.15 K, *J. Chem. Eng. Data*, Vol. 53, No. 11, pp. 2496–2500, ISSN 0021-9568

Deng, T.L.; Zhang, B.J.; Li, D.C.; Guo, Y.F. (2009c). Simulation studies on the metastable phase equilibria in the aqueous ternary systems (NaCl – MgCl$_2$ – H$_2$O) and (KCl – MgCl$_2$ – H$_2$O) at 308.15 K, *Front. J. Chem. Eng.*, Vol. 3, No. 2 , pp. 172-175, ISSN 2095-0179

Donad, S.P.; Kean, H.K. (1985). The application of the Pitzer equations to 1-1 electrolytes in mixed solvents, *J. Solution Chem*, Vol. 14, No. 4, pp. 635-651, ISSN 0095-9782

Fang, C.H.; Song, P.S.; Chen, J.Q. (1993). Theoretical calculation of the metastable phase diagram of the quinary system(Na + K + Cl + SO$_4$ + CO$_3$ + H$_2$O) at 25°C, *J. Salt Lakes Res.*, Vol. 1, No. 2, pp. 16-22, ISSN1008-858X

Felmy, A.R.; Weare, J.H. (1986). The prediction of borate mineral equilibria in nature waters: Application to Searles Lake, California, *Geochim. Cosmochim. Acta*, Vol. 50, No. 10, pp. 2771-2783, ISSN 0016-7037

Gao, J.; Deng, T.L. (2011a). Metastable phase equilibrium in the aqueous quaternary system (LiCl + MgCl$_2$ + Li$_2$SO$_4$ + MgSO$_4$+ H$_2$O) at 308.15 K, *J. Chem. Eng. Data*, Vol. 56, No. 4, pp. 1452-1458, ISSN 0021-9568

Gao, J.; Deng, T.L. (2011b). Metastable phase equilibrium in the aqueous quaternary system (MgCl$_2$ + MgSO$_4$+ H$_2$O) at 308.15 K, *J. Chem. Eng. Data*, Vol. 56, No. 5, pp. 1847-1851, ISSN 0021-9568

Gao, S.Y.; Song, P.S.; Xia, S.P.; Zheng, M.P. (2007). *Salt Lake Chemistry – A new type of lithium and boron salt lake*, Chinese Science Press, ISBN 978-7-03-016972-3, Beijing

Greenberg, J.P.; Moller, N. (1989). The prediction of mineral solubilities in natural waters: A chemical equilibrium model for the Na-K-Ca-Cl- SO$_4$-H$_2$O system to high concentration from 0°C to 250°C, *Geochim. Cosmochim. Acta*, Vol. 53, No. 9, pp. 2503-2518, ISSN 0016-7037

Guo, Y.F.; Yin, H.J.; Wu, X.H.; Deng, T.L. (2010). Metastable phase equilibrium in the aqueous quaternary system (NaCl + MgCl$_2$ + Na$_2$SO$_4$ + MgSO$_4$ + H$_2$O) at 323.15 K, *J. Chem. Eng. Data*, Vol. 55, No. 10, pp. 4215-4220, ISSN 0021-9568

Harvie, C.E.; Moller, N.; Weare, J.H. (1984). The prediction of mineral solubilities in natural waters: The Na-K-Mg-Ca-H-C1-SO$_4$-OH-HCO$_3$-CO$_3$-CO$_2$-H$_2$O system to high ionic strength at 25°C, *Geochim.Cosmochim. Acta*, Vol. 48, No. 4, pp. 723-751, ISSN 0016-7037

Harvie, C.E.; Weare, J.H. (1980). The prediction of mineral solubilities in natural waters: the Na-K-Mg-Ca-Cl-SO$_4$-H$_2$O system from zero to high concentration at 25°C, *Geochim. Cosmochim. Acta*, Vol. 44, No. 7, pp. 981-997, ISSN 0016-7037

Jin, Z. M.; Xiao, X. Z.; Liang, S. M. (1980). Study of the metastable equilibrium for pentanary system of (Na$^+$, K$^+$, Mg^{2+}),(Cl$^-$, SO$_4^{2-}$), H$_2$O. *Acta Chim. Sinica*, Vol. 38, No. 2, pp. 313-321, ISSN 0567-7351

Jin, Z. M.; Zhou, H. N.; Wang, L. S. (2001). Study on the metastable phase equilibrium of Na$^+$, K$^+$, Mg^{2+}//Cl$^-$, SO$_4^{2-}$ -H$_2$O quinary system at 35℃. *Chem. J. Chin. Univ.* Vol. 22, No. 3, pp. 634-638, ISSN 0251-0790

Jin, Z. M.; Zhou, H. N.; Wang, L. S. (2002). Studies on the metastable phase equilibrium of Na$^+$, K$^+$, Mg^{2+}//Cl$^-$, SO$_4^{2-}$ -H$_2$O quinary system at 15℃. *Chem. J. Chin. Univ.* Vol. 23, No. 3, pp. 690-694, ISSN 0251-0790

Kim, H-T.; Frederick, W.J. (1988a). Evaluation of Pitzer ion interaction parameters of aqueous electrolytes at 25 ℃.1.Single salt parameters, *J. Chem. Eng. Data*, Vol. 33, No. 2, pp. 177-184, ISSN 0021-9568

Kim, H-T.; Frederick, W.J. (1988b). Evaluation of Pitzer ion interaction parameters of aqueous electrolytes at 25℃.2.Ternary mixing parameters, *J. Chem. Eng. Data*, Vol. 33, No. 3, pp. 278-283, ISSN 0021-9568

Kurnakov, N. S. (1938). Solar phase diagram in the regions of potassium salts for the five-component system of (NaCl - MgSO$_4$ – MgCl$_2$ – KCl - H$_2$O at (288.15, 298.15, and 308.15) K, *Eehv. Fieh. Khim. Analieha*, Vol.10, No. 5, pp. 333-366.

Li, D.C.; Deng, T.L. (2009). Solid-liquid metastable equilibria in the quaternary system (NaCl + KCl + CaCl$_2$ + H$_2$O) at 308.15 K, *J. Therm. Anal. Calorim*, Vol. 95, No. 2, pp. 361-367, ISSN 0021-9568

Li, Y.H.; Song, P.S.; Xia, S.P. (2006). Solubility prediction for the HCl–MgCl$_2$ –H$_2$O system at 40℃ and solubility equilibrium constant calculation for HCl ·MgCl$_2$ ·7H$_2$O at 40℃, *CALPHAD*, Vol. 30, No. 1, pp. 61-64, ISSN 0364-5916

Li, Z.Y.; Deng, T.L.; Liao, M.X. (2010). Solid-liquid metastable equilibria in the quaternary system Li$_2$SO$_4$ + MgSO$_4$ + Na$_2$SO$_4$ + H$_2$O at T = 263.15 K, *Fluid Phase Equilibria*, Vol. 293, No.1, pp. 42-46, ISSN 0378-3812

Liu, Y.H.; Deng, T.L.; Song, P.S. (2011).Metastable phase equilibrium of the reciprocal quaternary system LiCl + KCl + Li$_2$SO4 + K$_2$SO$_4$ + H$_2$O at 308.15 K, *J. Chem. Eng. Data*, Vol. 56, No. 4, pp. 1139-1147, ISSN 0021-9568

Meng, L.Z.; Deng, T.L. (2011). Solubility prediction for the system of (MgCl$_2$–MgSO$_4$–MgB$_4$O$_7$–H$_2$O) at 298.15 K using the ion-interaction model, *Russ. J. Inorg. Chem*, Vol. 56, No. 8, pp. 1-4, ISSN 0036-0236

Pabalan, R.T.; Pitzer, K.S. (1987). Thermodynamics of concentrated electrolyte mixtures and the prediction of mineral solubilities to high temperatures for mixtures in the system Na-K-Mg-Cl-SO$_4$-OH-H$_2$O, *Geochim. Cosmochim. Acta*, Vol. 51, No. 9, pp. 2429-2443, ISSN 0016-7037

Pitzer, K.S. (1973a). Thermodynamics of electrolytes I: theoretical basis and general equation. *J. Phys. Chem*, Vol. 77, No. 2, pp. 268-277, ISSN 1932-7447

Pitzer, K.S. (1973b). Thermodynamics of electrolytes. II. Activity and osmotic coefficients for strong electrolytes with one or both ions univalent, *J. Phys. Chem*, Vol. 77, No. 19, pp. 2300-2308, ISSN 1932-7447

Pitzer, K.S. (1974a). Thermodynamics of electrolytes. III. Activity and osmotic coefficients for 2-2 electrolytes, *J. Solution Chem*, Vol. 3, No. 7, pp. 539-546, ISSN 0095-9782

Pitzer, K.S. (1974b). Thermodynamics of electrolytes. IV. Activity and osmotic coefficients for mixed electrolytes, *J. Am. Chem. Soc*, Vol. 96, No. 18, pp. 5701-5707, ISSN 0002-7863

Pitzer, K.S. (1975). Thermodynamics of electrolytes. V. Effects of higher-order electrostatic terms, *J. Solution Chem*, Vol. 4, No. 3, pp. 249-265, ISSN 0095-9782

Pitzer, K.S. (1977). Electrolytes theory-improvements since Debye-Hückel, *Account Chem. Res*, Vol. 10, No. 10, pp. 371-377, ISSN 0013-4686

Pitzer, K.S. (1995). *Semiempirical equations for pure and mixed electrolytes. Thermodynamics*, 3rd ed.; McGraw-Hill Press, ISBN 0-07-050221-8, New York, Sydney, Tokyo, Toronto

Pitzer, K.S. (2000). *Activity coefficients in electrolyte solutions*, 2rd ed.; CRC Press, ISBN 0-8493-5415-3, Boca Raton, Ann Arbor, Boston, London

Song, P.S. (1998). Calculation of the metastable phase diagram for sea water system, *J. Salt Lakes Res.*, Vol. 6, No. 2-3, pp. 17-26, ISSN1008-858X

Song, P.S. (2000). Salt lakes and their exploiting progresses on the relatively resources. *J. Salt Lakes Res.*. Vol. 8, No. 1, pp. 1-16, ISSN1008-858X

Song, P.S.; Du, X.H. (1986). Phase equilibrium and properties of the saturated solution in the quaternary system Li$_2$B$_4$O$_7$-Li$_2$SO$_4$-LiCl-H$_2$O at 25°C, *Chinese Science Bulletin*, Vol. 31, No. 19, pp. 1338-1343, ISSN 0567-7351

Song, P.S.; Yao, Y. (2001). Thermodynamics and phase diagram of the salt lake brine system at 25°C I. Li$^+$, K$^+$, Mg^{2+}//Cl$^-$, SO$_4$$^{2-}$–H$_2$O system, *CALPHAD*, Vol. 25, No. 3, pp. 329-341, ISSN 0364-5916

Song, P.S.; Yao, Y. (2003). Thermodynamic model for the salt lake brine system and its applications–I. Applications in physical chemistry for the system Li$^+$, Na$^+$, K$^+$, Mg^{2+}/Cl$^-$, SO$_4^{2-}$ – H$_2$O, *J. Salt Lakes Res.*, Vol. 11, No. 3, pp. 1-8, ISSN1008-858X

Spencer, R.J.; Moller, N.; Weare, J. (1990). The prediction of mineral solubilities in natural waters: A chemical equilibrium model for the Na-K-Ca-Mg-Cl-SO$_4$-H$_2$O system at temperatures blow 25°C, *Geochim. Cosmochim. Acta*, Vol. 54, No. 4, pp. 575-590, ISSN 0016-7037

Su, Y. G.; Li, J.; Jiang, C. F. (1992). Metastable phase equilibrium of K$^+$, Na$^+$, Mg^{2+}//Cl$^-$, SO$_4^{2-}$ - H$_2$O quinary system at 15°C. *J. Chem. Ind. Eng.* Vol. 43, No. 3, pp. 549-555, ISSN 0438-1157

Teeple, J.E. (1929). *The development of Searles Lake brines: with equilibrium data*, American Chemical Society Monograph Series, Chemical Catalog Company Press, ISBN 0-07-571934-6, New York.

Vant'hoff, J.H. (1912). Solubilities of the five-component system of (NaCl - MgSO$_4$ – MgCl2 – KCl - H$_2$O at 298.15 K, in: Howard L. Silcock. (1979). Solubilities of inorganic and organic compounds, Vol.3 Part 2: Ternary and multi-component systems of inorganic substances, Pergamon Press, ISBN 0-08-023570-0, Oxford, New York, Toronto, Sydney, Paris, Frankfurt.

Wang, S.Q.; Deng, T.L. (2008). Solid-liquid isothermal evaporation phase equilibria in the aqueous ternary system (Li$_2$SO$_4$+ MgSO$_4$+ H$_2$O) at T = 308.15 K, *Journal of Chemistry Thermodynamics*, Vol. 40, No. 6, pp. 1007-1011, ISSN 0021-9614

Wang, S.Q.; Deng, T.L. (2010). Metastable phase equilibria of the reciprocal quaternary system containing lithium, sodium, chloride, and sulfate ions at 273.15 K, *J. Chem. Eng. Data*, Vol. 55, No. 10, pp. 4211-4215, ISSN 0021-9568

Wang, S.Q.; Guo, Y.F.; Deng, T.L. (2011a). Solid-liquid metastable equilibria in the aqueous ternary system containing lithium, magnesium, and sulfate ions at 273.15 K, *Proceedings of CECNet*, Vol. 6, pp. 5243-5246, XianNing, China, ISBN 978-1-61284-470-1, April 16-18, 2011

Wang, S.Q.; Guo, Y.F.; Deng, T.L. (2011b). Solubility predictions for the reciprocal quaternary system (NaCl + MgCl2 + Na2SO4 + MgSO4 + H2O) at 283.15 K using Pitzer ion-interaction model, *Proceedings of CECNet*, Vol. 2, pp. 1661-1664, XianNing, China, ISBN 978-1-61284-470-1, April 16-18, 2011

Yang, J.Z. (1988). Thermodynamics of electrolyte mixtures. Activity and osmotic coefficient with higher-order limiting law for symmetrical mixing, *J. Solution Chem*, Vol., No. 7, pp. 909-924, ISSN 0095-9782

Yang, J.Z. (1989). The application of the ion-interaction model to multi-component 1-1 type electrolytes in mixed solvents, *J. Solution Chem*, Vol. 3, No. 2, pp. 201-210, ISSN 0095-9782

Yang, J.Z. (1992). Thermodynamics of amino acid dissociation in mixed solvents. 3: Glycine in aqueous glucose solutions from 5 to 45°C, *J. Solution Chem*, Vol. 11, No. 9, pp. 1131-1143, ISSN 0095-9782

Yang, J.Z. (2005). Medium effect of an organic solvent on the activity coefficients of HCl consistent with Pitzer's electrolyte solution theory, *J. Solution Chem*, Vol. 1, No. 1, pp. 71-76, ISSN 0095-9782

Zhao, Y.Y. (2003). Chinese salt lake resources and the exploiting progresses. *Mineral Deposits*, Vol. 22, No. 1, pp. 99-106, ISSN 0258-7106

Zheng, M.P.; Xiang, J.; Wei, X.J.; Zheng, Y. (1989). *Saline lakes on the Qinghai-Xizang (Tibet) Plateau*, Beijing Science and Technology Press, ISBN 7-5304-0519-5, Beijing

Zheng, X. Y.; Tang, Y.; Xu, C.; Li, B.X.; Zhang, B.Z.; Yu, S.S. (1988). *Tibet saline lake*; Chinese Science Press, ISBN 7-03-000333-0, Beijing

4

"Salt Weathering" Distress on Concrete by Sulfates?

Zanqun Liu[1,2,3], Geert De Schutter[2], Dehua Deng[1,3] and Zhiwu Yu[1,3]

[1]School of Civil Engineering, Central South University, Changsha, Hunan,
[2]Magnel Laboratory for Concrete Research, Department of Structural Engineering,
Ghent University, Ghent,
[3]National Engineering Laboratory for High Speed Railway Construction,
Changsha, Hunan,
[1,3]P.R China
[2]Belgium

1. Introduction

Salt weathering, also called salt crystallization or physical salt attack, is defined as the basic degradation mechanism that a porous material, such as stone and masonry, undergoes at and near the Earth's surface [1]. The parts of porous materials in contact with relatively dry air near the Earth's surface will be severely deteriorated but the parts buried in salts environment look sound.

Generally, the idea of sulfate attack on concrete means that a complex physiochemical process including several harmful productions formation through chemical reaction, such as ettringite and gypsum, following the crystal growth of these productions in cracks or pores resulting in concrete damage. However, another concept was given more and more attention that "salt weathering/physical salt attack" on concrete partially exposed to environment specially containing Na_2SO_4 or $MgSO_4$. ACI (American Concrete Institute) created a new subcommittee, ACI 201-E (Salt Weathering/Physical Salt Attack) in 2009. In 2011, an ballot was performed to discuss if it is necessary to separate the "physical salt attack" from chapter 6 "sulfate attack" as chapter 8 for ACI 201.2R. There were also more and more reports discussing this topic [2-9]. It seems that this topic will be high interest and relevance for the concrete community.

Certainly, concrete is also a kind of porous material. When partially exposed to an environment containing salts (especially sodium sulfate), such as in the case of a foundation, dam, column, flatwork and tunnel, a large amount of efflorescence will appear on the surface of the concrete accompanied with a similar scaling manner as salt weathering distress on masonry, showing a freezing-and-thawing-like deterioration on the surface of concrete [2] (Fig. 1). Therefore, concrete technologists logically and involuntarily define this phenomenon as salt weathering distress on concrete or physical attack on concrete.

Apparently, it seems reasonable to attribute salt weathering to the decay of concrete partially exposed to sulfate environment. Concrete technologists subjectively accepted that salt weathering or salt crystallization cannot be avoided in concrete, because concrete is also a kind of porous material similar to stone. However, in effect, some field and indoor research results of "salt weathering" distress on concrete have shown a number of appearances opposite to the basic principles of salt weathering on porous materials. Therefore, it is necessary and imperative to present this problem to avoid further confusion.

Fig. 1. Deterioration of railway tunnel (Southwestern Region, China)

This review paper includes three parts. First, the basic principles of salt weathering on porous materials are reviewed. Second, some field and indoor tests of "salt weathering" on concrete by sulfates are presented. Some appearances, which were generated by "salt weathering" on concrete but were opposite to the basic principles of salt weathering on porous materials, are analyzed in detail. Several points that need further study are presented in the third part.

2. Salt weathering distress on porous materials

2.1 Salt crystallization in pore

The work of Carl W. Correns on crystallization pressure is undoubtedly a milestone in the field of durability of porous materials [10], and the equation (Eq. (1)) exhibited in his paper written in 1949 for crystallization is broadly used and quoted.

$$P = \frac{RT}{v}\ln(\frac{C}{Cs})$$ (1)

Where R is the ideal gas constant, T is the absolute temperature, v is the molar volume, C is the concentration of solution, and Cs is the concentration of saturated solution. C/Cs is the supersaturation.

The above equation indicates that supersaturation is the key factor for crystallization. The supersaturation should be maintained during the process of salt crystallization. The crystal will grow until the supersaturation is consumed. He also pointed out that a thin layer/film of aqueous solution always remains between the crystal and the internal solid walls of the porous network. The thin layer allows the solute to diffuse from the pore solution to the crystal surface that is growing against the pore wall. If this thin layer did not exist, the crystal would go into contact with the pore wall, the growth would stop and no crystallization pressure would form [11]. Diffusion through this thin layer will equalize the concentration at the tip of the crystal and in the gap between the side of the crystal and the pore wall [12] [13]. The concentration and mobility of ions within this gap have a profound impact on the crystallization stress [14].

On the other hand, for a crystal, when the equilibrium is established between solution and crystal, the solubility product will satisfy:

$$\gamma_{cl}\kappa_{cl} = \frac{RT}{v}\ln(\frac{C}{Cs}) \qquad (2)$$

Where, γ_{cl} is the crystal /liquid interfacial energy; κ_{cl} is the surface curvature of crystal. Eq. (2) means two facts: a smaller spherical crystal is in equilibrium with a higher concentration than a larger flat crystal (equilibrium growth). The larger crystal (a relatively flat crystal) will grow and consume the supersaturation. Consequently, the smaller crystal will dissolve and the liberated solution will diffuse to the larger crystal (non-equilibrium growth) [14].

For equilibrium growth, a confined crystal can only exert stress if it is in contact with a pore solution that is supersaturated with respect to the unloaded face of the crystal [15]. The stress can be obtained by Eq. (3) [11]:

$$\sigma_W = \gamma_{cl}(\kappa_{cl}^C - \kappa_{cl}^E) \qquad (3)$$

Where, κ_{cl}^E is the curvature of the pore entrance (labeled point E), and κ_{cl}^C is the curvature of other internal points (labeled point C) (Fig.2).

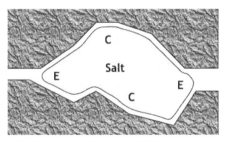

Fig. 2. Schematic of crystal of salt growing in a pore [14]

Because κ_{cl}^E is less convex (positive) than κ_{cl}^C, the compressive strength is negative, but it creates a tensile stress in the hoop direction around the pore. This tensile stress is the destructive "crystallization pressure" A high equilibrium crystallization pressure requires a confined crystal in a pore of any geometry with a very small pore entrance [16]. Therefore, the stones with a bimodal pore size distribution are extremely susceptible to salt attack [17-19].

For non-equilibrium growth, all of the crystals in internal pores of a matrix with a distribution of pore sizes are unstable with respect to macroscopic crystals that nucleate in large voids. During the drying (evaporation) or in the presence of a sharp temperature gradient, the smaller crystals will dissolve and feed the growth of the larger one, reaching another equilibrium. During this equilibrium, a high transient stress can be produced (Eq. (4)) [14].

$$\sigma_W = \frac{\gamma_{CL}}{r_s} - \frac{R_g T}{V_m} \ln(\frac{C}{Cs}) \tag{4}$$

Where, r_s is the radius of the small pore entrance.

The duration and intensity of the transient crystallization pressure depend on three factors [14]: (1) the rate of supply of solute; (2) the rate of growth of crystal; (3) the rate of diffusion of solute to macro-pores. High evaporation can result in high supersaturation, and increase the growth of crystal and result in a high transient stress [14] [20], leading to severe damage by salt crystallization.

The supersaturation can be produced by cooling, evaporation and drying and wetting cycle. If the temperature dependence of the solubility of a salt is high, a drop of temperature can result in supersaturation. Supersaturation caused by evaporation always occurs when one face of the porous material is in contact with the solution and the other face is exposed to relatively dry conditions, i.e., the salt weathering process.

As to the relationship between strength and durability of porous materials, it always shows positive correlation [21-23]: porous materials with higher strength can suffer stronger salt crystallization distress.

2.2 Characteristics of salt weathering distress

In the process of salt weathering, efflorescence and sub-efflorescence will occur. Efflorescence always occurs on the surface of the material, and shows little or no damage. On the contrary, sub-efflorescence forms under the material surface and results in significant damage [24-26]. Some interesting studies showed that addition of ferrocianides ($[Fe(CN)_6]^{4-}$) can promote NaCl efflorescence growth as opposed to sub-efflorescence growth in porous stones, and minimize salt damage [27, 28].

Wick action is the transport of water (and any species it may contain) through a concrete (porous material) element face in contact with water to a drying face with less than 100% relative humidity of air [29]. The mechanism involves capillary sorption and evaporation.

During the process of wick action, if there is no evaporation, the solution level can increase through capillary rise in the concrete according to Eq. 5: [30]

$$h = \frac{2\gamma_{LV} \cdot \cos\theta}{rg\rho} \tag{5}$$

where h is the height of capillary rise, γ_{LV} is the liquid/vapor interfacial energy, θ is the contact angle, r is the pore radius, g is the gravitational acceleration, and ρ is the density of the solution. In the case of water in concrete $\cos\theta \approx 1$, γ_{LV} is ~ 400 mJ·m^{-2}, and r is the typically 10~100nm. Therefore, h is about 1-10m. However, the pores will easily lose water due to evaporation. The pores of 10 um will start to empty when the relative humidity is lower than 95%. So, when the relative humidity is lower, h will decrease. After some time, a state of equilibrium (wet-drying interface) may be reached. Then the rate of water entering the concrete by capillary sorption matches the rate of water leaving the opposite face of the concrete element by water vapour diffusion.

If the water is containing salts, these salts cannot be carried by the vapour and therefore build up at this position. This concentration effect causes back-diffusion of salt away from the wet-dry interface. If the salts concentration near the wet-dry interface ever exceeds the solubility of the salt compounds present, precipitation is likely to occur [31-34]. The absorption-diffusion relationship can be described by the definition of the Peclet number [34]:

$$Pe \equiv \frac{hL}{\theta_m D_c} \qquad (6)$$

Where, h (m³ m⁻² s⁻¹) is the drying rate, L (m) the length of the sample, and θ_m (m³ m⁻³) the maximum fluid content by capillary saturation. D_c (m² s⁻¹) is the diffusion coefficient of the ions in the moisture in the porous medium. For Pe<<1 diffusion dominates and the ion-profiles will be uniform, whereas for Pe>>1 absorption dominates and ions will be accumulated at the drying surface.

Y. T. Puyate et al discussed the chloride transport due to wick action in the concrete in detail [31-34]. One vital conclusion is that it was the vapour pressure of the solution and the relative humidity of air which control the position of the wet-dry interface [33]. The position of dry-wet interface locates in the inner of the concrete faced to a low relative humidity situation (0%) (Pe>>1), and a sharp peak of chloride concentration exceeding the saturation value occurs [34]. In contrast, in a high relative humidity condition (78%) (Pe<<1), the location of the interface is close to the concrete surface [31, 32]. Therefore, high evaporation can induce severe crystallization distress.

Nuclear magnetic resonance (NMR) is used to study the crystallization of sodium chloride due to wick action. Measuring the moisture and ion profile in a fired-clay brick cylinder (Ø 20×45mm), an efflorescence pathway diagram is plotted [35, 36] as shown in Fig. 3.

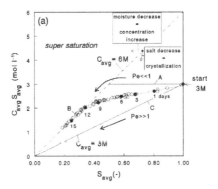

Fig. 3. Efflorescence pathway diagram [35]

According to this diagram (Fig.3), when Pe<<1, i.e. very slow drying or high relative humidity, the ion profile remains homogeneous and for some time no crystallization will occur. The average NaCl concentration slowly increases until the complete sample has reached saturation, forming a high concentration pore solution zone. When Pe>>1, i.e. very fast drying or low relative humidity, ions are directly advected with the moisture to the top of the sample and a saturation peak will build up with a very small width. If the rate of crystallization is high enough, i.e. if there are enough nucleation sites at the top to form crystals, the average NaCl concentration of the solution in the sample itself will remain constant at nearly the initial concentration.

The mechanism of efflorescence is the crystals growing at a free surface: the crystals in the pores cannot be stable and will dissolve and diffuse towards the atmosphere (an infinite pore) (Eq.2). Because the crystals are in contact with the solution only in their bases, they cannot grow laterally but form long needles like whisker [14]. This is the reason why efflorescence is un-harmful for the porous materials. Sub-efflorescence precipitates when the evaporative flux is greater than the capillary flux in the porous materials where the solution is supplied by the capillary suction and evaporation [37].

In summary, the efflorescence and sub-efflorescence of salt weathering distress on the porous material can be schematically shown in Fig. 4.

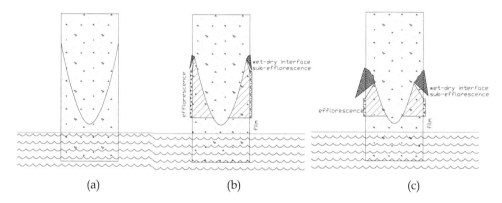

(a)   (b)   (c)

Fig. 4. Schematic of salt weathering distress on porous material.
(a) no evaporation condition, (b) low evaporation condition, (c) high evaporation condition

As we know, the capillary absorption just occurs in the interconnected pores between air and water. When a porous element is partially subjected to the sulfate solution under no evaporation condition, a pore solution zone will be formed as shown in Fig. 4(a). The solution cannot rise from the solution surface to the top of the element by capillary absorption due to few or no interconnected pores from the bottom to the top in a relatively long distance. The interconnected pores can form from the solution surface to the side surface of the element in a relatively short distance, resulting in the generation of capillary absorption.

Under a low evaporation condition, the wet-dry interface can occur in the tip of the pore solution zone, where the rate of evaporation is fast compared with the rate of solution rise, because solution rises into the bulk at a rate that decreases with height. At the same time the sulfate concentration of pore solution will slowly increase until the complete sample has reached saturation, forming a high concentration pore solution zone, where the efflorescence occurs. Near the solution a liquid film occurs on the surface of the element, where the rate of rise if fast compared with the evaporation and the sulfate concentration is close to the exposure solution [20]. In this case the deterioration due to salt crystallization is minor (as shown in Fig. 4 (b)).

Under a low relative humidity condition, due to the high evaporation rate, the position of the wet-dry interface will move to the inner part of the element and closer to the bottom of the element, where a saturation peak will build up with a very small width, forming supersaturation, and resulting in sub-efflorescence and more severe deterioration. The breadth of efflorescence zone decreases and the average concentration of the pore solution will remain constant at nearly the initial concentration of exposure solution (as shown in Fig. 4 (c)).

2.3 Crystallization of Na_2SO_4 and $MgSO_4$

Sodium sulfate is known to be a salt that causes the worst crystallization decay on porous materials and has become widely used in accelerated durability testing [38]. However, the sodium sulfate system is complicated, because under different conditions (temperature and relative humidity), it will form two stable phases (thenardite, Na_2SO_4 and mirabilite, $Na_2SO_4 \cdot 10H_2O$) or one metastable phase (heptahydrate, $Na_2SO_4 \cdot 7H_2O$) [12, 13] [39]. The metastable phase ($Na_2SO_4 \cdot 7H_2O$) is formed during the rehydration of the anhydrous sodium sulfate phase (Na_2SO_4) to the nucleation of mirabilite. Prior to mirabilite [12] [39], the crystallization pressure exerted by heptahydrate does not cause damage under the condition of the cooling experiments [36], and it can not be observed in building stone [13].

Regarding the damage caused by the crystallization of sodium sulfate, there are two views. One school thinks that the crystallization of thenardite is more destructive [40], because the crystallization of thenardite can generate higher pressure than mirabilite at the same supersaturation [41]. However, more and more experimental results support another school that the dissolution of thenardite producing a solution highly supersaturated with respect to mirabilite will cause the precipitation of mirabilite and result in the damage of porous materials [13] [38]. I.e. the transformation between thenardite and mirabilite can generate severe large crystallization pressure, resulting in porous materials damage.

The only naturally occurring members of the $MgSO_4$ nH_2O series on Earth are epsomite ($MgSO_4 \cdot 7H_2O$, 51 wt% water), hexahydrite ($MgSO_4$ $6H_2O$, 47 wt% water) and kieserite ($MgSO_4 \cdot H_2O$, 13 wt% water). In aqueous systems, epsomite is stable at T below 48.4°C, hexahydrite is stable in the T range 48.4–68 °C, and kieserite is stable at T > 68 °C [42]. Thus, at the normal temperature, the crystallization of epsomite ($MgSO_4 \cdot 7H_2O$,) is the distress reason.

2.4 Summary

In summary, according to above review, the following basic principles of salt weathering on porous materials can be concluded:

1. Supersaturation is the key factor for salt crystallization. During the process of salt weathering the supersaturation must be maintained at a high level.
2. High evaporation results in the formation of strong sub-efflorescence, causing severe deterioration. Low evaporation results in weak crystallization distress but causes the formation of a pore solution zone with high concentration in the part of porous materials in contact with air.

3. Experimental studies of "salt weathering" on concrete

3.1 Long term field tests

Since 1940, a long-term study of "salt weathering" on concrete was carried out by the Portland Cement Association (PCA) [43-45]. Thousands of concrete beams (152×152×762mm) were laid horizontally to a depth of 75 mm in sulfate rich soils (about 5.6% sulfate ion by weight of soil) basins in Sacramento, California. About 10 to 12 wetting and drying cycles are carried out every year. As the experiment progressed, commercial salts were added into the soils to replenish losses through leakage, overflow, and possibly other undetermined causes. Water was added to the basins just before the soils began to show drying to maintain the soils saturated.

Three reports were published [43-45]: the first [43] provided the experimental results of initial set of beams resistance to sulfate attack between 1940 and 1949. The second [44] described the performance development of concrete beams for 5 years field exposure to soils containing sodium sulfate. The third [45] introduced the experimental results for 16 years exposure.

Some other five years field tests were carried out by Irassar and Di Maio [46]. Concrete cylinders with the size of Ø150 × 300mm were buried at half height in a soil containing approximately 1% sodium sulfate. There are several important common experimental observations of the above field experiences:

1. The parts of the beams above ground, regardless of their cement content, cement composition, mineral additions, surface treatments and type of coarse aggregates, were deteriorated severely. The parts of the beams under ground, however, show little or no deterioration;
2. Pozzolanic additions, such as fly ash, furnace slag or silica fume, play a negative role in the performance of concrete exposed to these conditions;
3. The water-to-cementitious material ratio (W/CM) is the primary factor affecting the durability and performance of concrete in contact with sulfate soils: applying a low W/CM ratio results in a higher resistance to sulfate attack;
4. According to XRD, optical microscopy and SEM analysis, a large amount of chemical sulfate attack products, such as ettringite, gypsum and thaumasite, were identified in the upper part of concrete in contact with air. However, the samples for these tests were drilled with water [45, 46].

Concerning the deterioration mechanism, researchers attributed the failure of concrete to physical attack or salt crystallization. However, in effect the experimental results cannot be explained by salt weathering. Fig. 5 is the evolution of visual rating of the upper part of concrete in contact with air, obtained by Irassar and Di Maio [46].

In the test, the mix proportions of different mixes (the ratio of water : binder : sand : aggregate) were almost the same with different dosages of fly ash, slag and natural pozzolan. From Fig.5, we can deduce some interesting observations:

1. Comparing concrete H1 and H2, the difference between them was the air content, namely 1.3% resp. 4.4%. Thus, if the damage mechanism of the upper part of concrete was caused due to salt weathering, there would be a big difference in visual rating of these two concretes due to the different pore structures. However, after 5 years exposure the visual ratings were almost the same.

2. As to the role of mineral additions, with the increase of dosage of fly ash and natural pozzolan, the concrete cylinders showed worse visual observation. This may be explained by the fact that salt crystallization in smaller pores can form higher crystallization stress due to the refinement of mineral additions. On the other hand, if we compare concrete H3 and H5, H4 and H6, we can find that with the same dosage the natural pozzolan (H5 and H6) played a more negative role in concrete damage than fly ash (H3 and H4). However the compressive strength of reference cylinders of H5 and H6 were higher than H3 and H4 respectively. The tests of PCA also showed similar results. This appearance cannot be explained by the salt weathering.

3. Comparing the normal concrete and blended concrete, Irassar et al [46] attributed an increase in capillary suction height caused by the pore size refinement of mineral addition to the more severe deterioration of blended concrete. However, this is in conflict with the following observations. Following the explanation based on the height of capillary sorption, concrete with a low W/C ratio should be more susceptible to salt weathering than with high W/C ratio. In the paper by Nehdi and Hayek [47], we can find that the sorption height of mortar with W/C of 0.45 is higher than with W/C of 0.6. If the above explanation is right, the mortar with low W/C (0.45) should be more susceptible to damage than the mortar with high W/C (0.6). Obviously, this conclusion is opposite to the result of field tests.

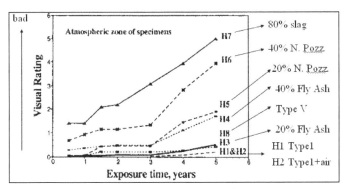

Fig. 5. Evolution of visual rating at the atmosphere part of concrete [46]

Besides, there are also some field cases, in which a wide variety of efflorescence salts (sodium sulfate, sodium chloride and magnesium sulfate) were routinely observed on the evaporating surface of the foundation concrete in South California [48, 49] However, the damage was not caused by salt crystallization. Much of the cement paste had lost its integrity, mainly as a result of the removal of portlandite, de-calcification of the calcium silicate phases and the ultimate replacement of calcium with magnesium in many of the cementitious compounds. Many reaction products, in particular magnesium silicate hydrate, brucite, Friedel's salt ($3CaO \cdot Al_2O_3 \cdot CaCl_2 \cdot 10H_2O$), sodium carbonate and thaumasite were found at some depth within the concrete.

Another interesting case [50] concerns the slabs of Yongan Dam, which is in Keshi, Xinjiang, P.R. China, where the land is arid and rich in various kinds of salts, especially sulfate salts. The slabs were constructed in August – October 2003. The air temperature fluctuates significantly: the highest temperature is up to 37 °C and at night it is only 15-20 °C. Within this temperature range, the transformation between thenardite and mirabilite can occur. However, it was found that the slabs became gray and mushy throughout the thickness where they were in contact with groundwater due to thaumasite sulfate attack.

In summary, according to the above analysis the appearances of long term field tests did not show convincing evidences to support "salt weathering" causing the deterioration of concrete partially exposed to sodium sulfate environment.

3.2 Indoor tests

Rodriguez-Navarro and Doehne [26] studied the effect of evaporation on salt weathering distress on stone. After 30 days of exposure to a saturated sodium sulfate solution at constant 20 ºC, larger amounts of efflorescence and lower weight losses were observed when the crystallization took place at a relative humidity of 60% instead of at 30% RH. H. Haynes and his coworkers [9] carried out some tests partially exposing the same concrete cylinders to 5% NaCl and Na_2CO_3 solutions. Severe damage was observed for concrete cylinders, which were placed in constant environment at 20 ºC and 54% relative humidity from day 28 to day 530, and then at 20 ºC and 32% RH from day 530 to day 1132. The specimens kept in a constant environment at 20 ºC and 82% relative humidity from day 28 to day 1132 looked sound. These observations can be explained by the fact that a low relative humidity results in more evaporation, leading to sub-efflorescence [20] that forms deep in the material and results in significant damage [24,25] as explained in the section 2.2.

However, contradictory observations can be found with respect to concrete exposed to sodium sulfate solution [51]. Two concrete specimens with the same mixture proportions were partially immersed in 10% Na_2SO_4, solution. One specimen was placed at 80%±5% RH and the other was placed at 30% ±5%. After 75 days of exposure to a constant temperature of 25 ºC, the specimens at 80% RH showed signs of deterioration first over a very large area, starting from above the solution level. On the contrary, at 30% RH, the zone of deterioration was narrower and was situated at a certain distance above the solution level. In this case, the first sign of deterioration was a crack, not spalling (shown in Fig. 6).

Fig. 6. Concrete cylinders exposed to sodium sulfate solution for 75 days [51]

Similar results were also observed in the tests performed by H. Haynes and his coworkers [2]. Narrower spalling zone was found in case of concrete specimens exposed to constant environment at 20 °C and 54% relative humidity from day 28 to day 530, and then at 20 °C and 32% RH from day 530 to day 1132. Extensive spalling zone was found under constant environment at 20 °C and 82% relative humidity from day 28 to day 1132. These tests will be discussed in detail as follow.

3.2.1 H. Haynes tests [2, 9]

H. Haynes and his coworkers performed very important and systemical tests about the salt weathering distress on concrete. In the two papers [2, 9], different ambient conditions were created within storage cabinets whose temperature and relative humidity were controlled. The concrete cylinders (Ø76 × 145mm) were partially exposed to 5% Na_2SO_4, $NaCO_3$ and NaCl solutions. A partial submergence condition was achieved by wetting the specimen to a height of 25 mm. At the height of 50mm, a plastic cover to the container functioned as a quasi-vapor retarder to minimize evaporation. The plastic cover did not touch the cylinder. Hence, within the region of 25 to 50mm the cylinder was exposed to a moist environment. Above 50mm (2 in.), the concrete was exposed to ambient environmental conditions. In the test program the author said that "the sulfate solution and tap water were replaced on a monthly basis; however, replacements for evaporation loss were provided at 2-week intervals. Much of the solution evaporated in the 40 °C and 31% relative humidity environment where, in general, at the end of 2 weeks, minor amounts of solution remained; and at times, no solution remained".

The tests were divided into two Phases for 3.1 years. The performance of concrete cylinders under five storage conditions was studied in detail, the exposures were:

Condition 1: steady at 20°C and 54% relative humidity from 28 to 530 days (Phase I), and then 20°C and 32% relative humidity from 530 to 1132 days (Phase II),

Condition 2: steady at 20°C and 82% relative humidity from 28 to 530 days (Phase I), and then from 530 to 1132 days (Phase II),

Condition 3: 40°C and 74% relative humidity from 28 to 406 days (Phase I), and then 40°C and 31% relative humidity from 406 to 1132 days (Phase II),

Condition 4: 2-week cycles between 20°C and 54% relative humidity and 20°C and 82% relative humidity from 28 to 530 days (Phase I), and then 2-week cycles between 20°C and 31% relative humidity and 20°C and 82% relative humidity from 530 to 1132 days(Phase II),

Condition 5: exposed to 2-week cycles between 20°C and 82% relative humidity and 40°C and 74% relative humidity from 28 to 406 days(Phase I), and then 2-week cycles between 20°C and 82% relative humidity and 40°C and 31% relative humidity from 406 to 560 (847) days (Phase II).

The effects of Na_2SO_4, Na_2CO_3 and NaCl were compared. The visual observation was photographed, the average mass of scaling materials was collected, the species of concrete were identified by petrographic analysis, and chemical analysis was employed to study the ions distribution. According to the experimental results, they concluded that salt weathering plays the predominant role in concrete damage. However, there are a few questionable points about the relationship between the evidences and the conclusion based on the basic principles.

3.2.1.1 Visual observation of concrete cylinders

According to the photographs of visual observation at the end of exposure, we can find that:

1. When concrete cylinders were exposed to Na_2SO_4 solution, as abovementioned the concrete cylinders exposed to high relative humidity condition were deteriorated more severely than low relative humidity condition (as shown in Fig. 7).

Condition 1   Condition 2

Fig. 7. Visual observation of concrete deterioration. Condition 1: steady at 20°C and 54% relative humidity from 28 to 530 days (Phase I), and then 20°C and 32% relative humidity from 530 to 1132 days; Condition 2: steady at 20°C and 82% relative humidity from 28 to 530 days (Phase I), and then from 530 to 1132 days [2]

Correspondingly, Fig. 8 shows the visual observation of concrete exposed to Na_2CO_3 and NaCl solutions under Condition 1 and Condition 2.

(Na_2CO_3)

Condition 1   Condition 2

(NaCl)

Condition 1	Condition 2

Fig. 8. Visual observation of concrete deterioration exposed to Na_2CO_3 and NaCl solutions [9]

2. Under Condition 3 (40°C /74%RH, 40°C /31%), concrete showed the most significant scaling in the case of NaCl, on the contrary, the concrete showed least scaling in the cases of Na_2CO_3 and Na_2SO_4 solutions. These appearances are contradictory because the concretes should show similar scaling manners due to the salt weathering in case of Na_2CO_3 and NaCl solutions. If the mechanism of concrete damage is also salt weathering in case of Na_2SO_4 solution, the concrete should also show similar scaling manners in NaCl solutions. If the mechanism of concrete damage is the chemical sulfate attack, the scaling manners of concrete cylinders by Na_2CO_3 and Na_2SO_4 solutions should show big difference. The possible reason of the above contradictory appearances may be the fast evaporation rate of Na_2CO_3 and Na_2SO_4 solutions. At high ambient temperature, Na_2CO_3 and Na_2SO_4 solutions would dry up soon, but some NaCl solution would remain. The tests under Condition 3 should be further studied to avoid the effect of evaporation of solution.

3.2.1.2 Mass of scaling materials

During Phase I from 28 days to 406 or 530 days: the worst damage occurred under Condition 5 (cycle 20°C/82% RH and 40°C/70% RH) in the case of Na_2SO_4 solution (the mass of scaled material was about 16g). However, in the case of Na_2CO_3 solution the worst damage appeared under Condition 1 (20°C/54% RH) (the mass of scaled material was just about 2.8g). This means that high RH can accelerate concrete damage by Na_2SO_4.

During Phase II from 406 or 530 days to the end of tests:

1. In the case of Na_2SO_4 solution the mass of scaled material under Condition 1(20°C/32% RH) was less than Condition 2(20°C/82% RH). The opposite appearance was observed in case of Na_2CO_3: the mass of scaled material was about 22g under Condition 1(20°C/32% RH) and about 1g under Condition 2 (20°C/82% RH). Similar appearance also observed in the case of NaCl solution.
2. The mass of scaled material of concrete under Condition 1 (20°C/32% RH) was less than Condition 4(cycle 20°C/82% RH and 20°C/31% RH) in case of Na_2SO_4. However, the mass of scaled material of concrete under Condition 1 (20°C/32% RH) was almost the same as Condition 4(cycle 20°C/82% RH and 20°C/31% RH) in the case of Na_2CO_3.

Corresponding to Phase I, this also means that high RH is in favor of the deterioration effect of Na_2SO_4 on concrete.

3.2.1.3 Petrographic analysis

According to petrographic analysis, abundant gypsum deposits were detected in large and small cracks, microcracks and voids near the surface of concrete. However, the authors presented two points to show that gypsum cannot result in concrete damage:

1. "Although trace amounts of gypsum were found near the outer surfaces, gypsum formation is a one-time occurrence, whereas crystallization of mirabilite and thenardite occurred repeatedly due to the biweekly cyclic changes in environmental conditions. Hence, the cycles of mirabilite and thenardite crystallization appear to be responsible for any significant expansion force". It is not clear why the authors thought that "gypsum formation is a one-time occurrence". According to above review, the solution can be drawn into the concrete continuously during wick action. In the presence of sulfate, gypsum crystals can continuously grow. Moreover, the authors pointed out that the pH value of the pore solution in the concrete should have been reduced due to carbonation, whereas for gypsum formation, the pH value of solution should be less than 11.9 [52].

2. "Despite the extensive alteration of the microstructure and the formation of gypsum, the concrete below the solution line was mostly intact with no mass loss, whereas there was substantial mass loss at the surface of the cylinder above the solution line. If gypsum did not cause scaling below the solution line, there is little reason to suspect that gypsum would cause scaling above the solution line. This indicated that salt crystallization alone, or in conjunction with gypsum, caused the scaling above the solution line. As salt crystallization by itself is known to damage rocks, the presence of gypsum is not necessary". This point seems reasonable, however, the authors did not give the quantitative analysis of gypsum. Because according to abovementioned wick action, a much higher concentration pore solution will be formed in the cylinder above the solution than under the solution, resulting in more severe sulfate attack and forming more gypsum.

3.2.1.4 Chemical analysis

Several cylinders were cut vertically to obtain a 25 mm thick midsection slice. This slice was then cut vertically into two 17 mm exterior sections and one 34 mm interior section. Starting at the bottom, the vertical sections were cut horizontally into six pieces 25 mm each. These pieces were crushed and pulverized to minus No. 50 mesh. The SO_3 contents and Na_2O contents were determined. The distributions of SO_3 contents and Na_2O contents are schematically shown in Fig. 9.

The salt distribution in the concrete cylinder provides a powerful evidence supporting that salt weathering is not the major mechanism causing concrete damage.

Based on the above review on the basic principles of salt weathering, supersaturation is the key factor for the salt crystallization. The salt crystals will deposit from the solution during the process of salt crystallization, however, the salt concentration of pore solution must be maintained high for the formation of supersaturation during the whole process. I.e. the salt contents should be highest where salt crystallization distress occurs in the concrete. This is the reason why the position where Na_2O content is highest corresponds with most severe deterioration of concrete exposed to Na_2CO_3 and $NaCl$ solutions. However the situation is opposite in the case of Na_2SO_4, the position, where SO_3 content is highest, locates on the top portion of cylinder that shows little or no deterioration (as shown the black line).

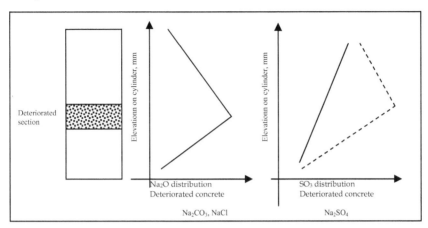

Fig. 9. Schematic of salt distribution in the concrete cylinders

In summary, based on the above analysis of indoor tests, two conclusions can be deduced:

1. Concrete partially exposed to Na_2SO_4 is susceptive to being more severely deteriorated under high RH environment than low RH environment. This appearance is in conflict with the basic principles of salt weathering.
2. The most severe deterioration does not occur in the portion of concrete containing the highest sulfates content. This is also in conflict with the basic principles of salt weathering.

3.3 Our tests [53, 54]

The starting point of our tests is to find a trace of salt crystals in the concrete as a direct evidence by means of XRD, SEM and EDS [53, 54]. Sulfate crystals can be easily identified in stone [23, 55]. However, in case of concrete elements, it is hard to identify them. Concrete technologists always attribute this to the coring and sawing operations when preparing samples for experimental analysis, as lapping water can readily dissolve salts from original and treated surfaces [2, 3, 4]. However this is not the main cause for the problem. Samples also can be taken in a dry manner to avoid the influence of water. Furthermore, In our study, the tests were designed to avoid the influence of water within the detection process of sulfate crystals.

Cement paste and cement – fly ash paste specimens and normal concrete specimens were partially exposed to Na_2SO_4 and $MgSO_4$ solution under constant and fluctuating storage conditions respectively. After a period of exposure, the specimens were moved out from the solution and did not touch solution or water any more. The surface of the specimens was cleared by a thin blade and a soft brush. The samples for XRD and SEM were dried in a vacuum container with silica gel.

3.3.1 Cement paste partially exposed to Na_2SO_4 and $MgSO_4$ solution under a constant storage condition

The test of cement paste specimens (20×20×150mm) partially exposed to Na_2SO_4 and $MgSO_4$ solution under a constant storage condition (20°C and 60%RH) is based on the study performed by Ruiz-Agudo [42] in which limestone specimens were partially submerged in a 19.4 g/100ml sodium sulfate solution and a 33.5 g/100ml magnesium sulfate solution respectively and located in a controlled environment (20°C±2°C, and 45%±5% RH). Results showed that the limestone specimens were severely damaged in both cases. While salt weathering by sodium sulfate consisted of a detachment of successive stone layers, magnesium sulfate induced the formation and propagation of cracks within the bulk stone. Thenardite (Na_2SO_4) and epsomite ($MgSO_4 \cdot 7H_2O$) crystals were identified by ESEM in the pores of limestone.

Before immersion minor shrinkage cracks were observed in the cement paste specimens. These cracks were focused upon in detail because narrow micro-fissures appear to be important in the decay process due to the effectiveness of crystallization pressure generated by salt growth [37]. So, if crystallization is the mechanism of decay of cement paste, salt crystallization should first occur in the shrinkage cracks and sodium sulfate or magnesium sulfate crystals should be identified in these cracks.

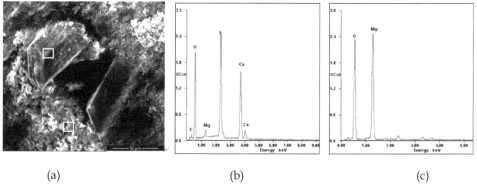

(a)   (b)   (c)

Fig. 10. ESEM and EDS analysis of white substance on the surface of shrinkage crack [54]
(a) ESEM image of white substance (b) EDS analysis of prismatic crystal
(c) EDS analysis of flocculent crystals

However, based on the micro-analysis results the sulfate crystals were not detected in the atmospheric part of the paste partially exposed to Na_2SO_4 solution. On the contrary large amounts of ettringite crystals, the main chemical sulfate attack product, were identified as the reason for paste spalling. Another important observation is that a layer of white substance was formed on the surface of shrinkage cracks in the atmospheric part of the paste partially exposed to $MgSO_4$ solution. Two distinct crystals can be distinguished: prismatic crystals surrounded by flocculent crystals. According to the EDS analysis, the prismatic crystal is gypsum and the flocculent crystal is brucite (shown in Fig.10). The products of this white substance in the shrinkage cracks are the same as the products in the interfacial zone of concrete fully immersed in magnesium sulfate solution [56].

3.3.2 Cement – fly ash paste partially exposed to Na₂SO₄ solution under a constant storage condition

Cement – fly ash paste specimens (20×20×150mm) were partially exposed to Na_2SO_4 solution. After 5 months exposure under the constant storage condition (20°C and 60%RH), some cracks were found near the upper edge above solution level of the cement-fly ash paste specimen. Along the crack, small pieces were carefully broken off using a thin blade. The ESEM image of a small piece is shown in Fig. 11.

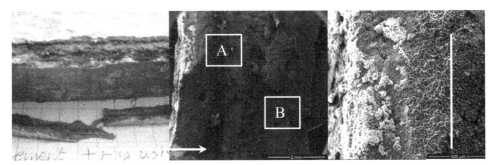

Fig. 11. Cracks in cement – fly ash paste [54]

The middle image is the zoomed surface of a small piece with magnification of 25 times. The left side is the outer surface of paste in contact with air, and zone A is the surface of a crack. Zone B is a small point in the bulk of paste.

Two distinct parts can be observed at the right and left hand side of the white line in Zone A. At the right side a large amount of dense granular crystals cover the surface (Fig. 12), while at the left side porous crystals can be found accompanied with white substance (Fig. 13). It can be found that the crystals at left and right sides are both calcite. However, some calcite crystals at left side are peeled off and crystal caves are left. Some crystals are honeycombed with small pores. According to the EDS analysis, Na and S are also present. Obviously, the crystallization of sodium sulfate results in damage of the calcite crystals. At the right side, the calcite crystals show no damage.

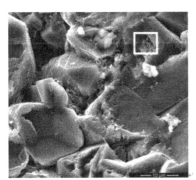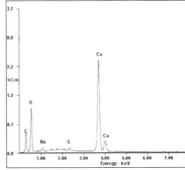

Fig. 12. ESEM image and EDS analysis of f the granular crystal at left side [54]

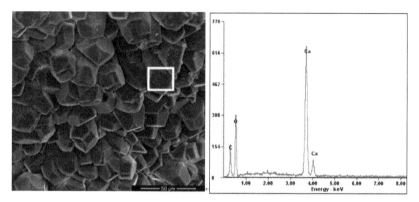

Fig. 13. ESEM image and EDS analysis off the granular crystal at right side [54]

According to the above observation, two conclusions can be drawn:

1. If salt crystallization is causing crack formation, the salt crystals should be identified at the right side in Fig. 11 to form sub-efflorescence instead of in the area in contact with air.
2. Salt crystallization can occur in the calcite crystals.

As we know, the crack formation is attributed to some expansive products present in the paste. When a small piece was broken off along the crack, the inner zone B on the surface of piece was a weak part in the paste and the source of crack initiation. The analysis of the products in this zone can disclose the real reason for the crack formation. Fig. 14 shows the ESEM image of the surface of Zone B.

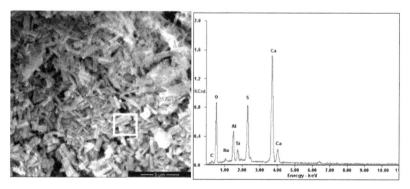

Fig. 14. ESEM and EDS analysis of white square [54]

In Fig. 14, a large amount of short needle crystals are found in this zone. According to the EDS analysis, there are Ca, Al, Si, S, Na, and O elements. Combining the XRD analysis, thenardite was not identified and the crystals were ettringite.

3.3.3 Cement – fly ash paste partially exposed to Na₂SO₄ solution under a fluctuating condition

Specimens (10×40×150 mm) were placed in a fluctuating condition (40±2°C and 35±5% RH for 24 hours, 10±1°C and 85±5% RH for 24 hours). After 3 cycles they were broken into several small pieces along some cracks. We checked the products on the surface of a crack. Fig. 15 shows the SEM image and EDS analysis.

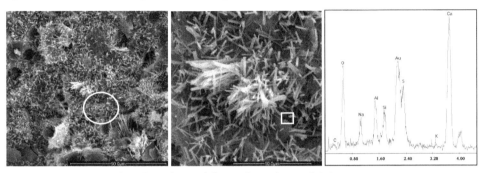

Fig. 15. SEM image and EDS analysis of the surface of a crack [54]

It can be found that a large amount of needle-like crystals grow on the surface like a hedgehog. Some pores are filled with a cluster of the needle-like crystals. Based on the EDS analysis and combining the XRD analysis, the needles are ettringite.

3.3.4 Normal concrete partially exposed to Na₂SO₄ and MgSO₄ solution under a constant storage condition

As we know, in concrete the weak interfacial transition zone (ITZ) plays a particularly important and even determining role in the main characteristic of concrete. A number of full immersion tests already showed that concrete deterioration occurred first in the ITZ by sulfate attack [56-60]. In this test, the concrete was made with just cement and aggregate to emphasize the role of ITZ in concrete deterioration. The concrete specimens (10×40×150 mm) were partially exposed to Na₂SO₄ and MgSO₄ for 8 months. The results showed that: (1) the harmful effect of MgSO₄ is much weaker than Na₂SO₄. This appearance also cannot be explained by the mechanism of salt weathering. This will be discussed in detail in section 4; (2) in the case of Na₂SO₄, damage also initiated in the ITZ. A large amount of gypsum crystals were formed on the surface of cement paste of ITZ in the upper part of concrete above solution (shown in Fig. 16).

Besides, the effect of carbonation on the salt weathering on concrete was studied. Before exposure a group of concrete cylinders were placed in an accelerated carbonation chamber with 10% CO_2 concentration at 20 ± 2 °C and 60% ± 5% RH for 14 days. Then, these cylinders were partially exposed to Na₂SO₄ solution. After 8 months exposure, the carbonated cylinders were deteriorated more severely than normal concrete (shown in Fig. 17).

During the process of cleaning the surface of cylinders, a lot of small mortar pieces could be easily brushed off. According to the XRD analysis (Fig.18) Na₂SO₄ crystals and $CaCO_3$ crystals were present in the mortar. This appearance also means that salt crystallization can occur in the $CaCO_3$ crystals.

Fig. 16. ESEM image and EDS analysis of surface of cement paste [53]

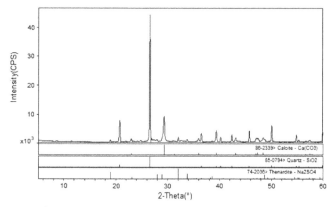

Fig. 17. Visual observation of normal and carbonated concrete specimens exposed to sodium sulfate solution [53]

Fig. 18. XRD pattern of mortar [53]

In summary, according to the above test results, two main conclusions can be deduced:

1. Sulfate crystals cannot be identified in the cement paste or concrete partially exposed to Na_2SO_4 and $MgSO_4$ solutions. The chemical reaction products, ettringite, gypsum and brucite, were the determining factors for material damage.

2. Salt crystallization can occur in the calcite crystals, the carbonated products of concrete.

3.4 Summary

The "salt weathering" on concrete was just received a lot of attention in the recent years. Based on the above analysis of a limited number of research reports, the experimental results already showed convincing appearances that were completely opposite to the basic principles of salt weathering distress on porous materials. On the contrary, the experimental results of long term field tests and indoor tests rather tended to indicate that chemical sulfate attack is the mechanism for the concrete damage.

4. Further research

According to the limited literature review, the conclusion can be made that the so-called "salt weathering" on concrete in effect is rather chemical sulfate attack. In order to systematically disclose the principles of this appearance, some issues should be further studied.

4.1 Study of the reason why salt crystallization cannot occur in concrete

The reason why salt cannot occur in concrete may be explained as follows:

Sulfates likely do not crystallize in a cement paste because in the highly alkaline pore solution other less soluble salts, e.g. ettringite, or gypsum, are preferably precipitated according to chemical equilibria theory. Salt crystallization in porous materials is difficult because salt crystallization occurrence has to reach and even exceed a threshold-supersaturation. However, the chemical reactions in pore solution can occur regardless of the sulfate concentration and decrease the possibility of physical attack due to consuming sulfates and decreasing the sulfate concentration of pore solution, moreover, high concentration solution will increase the rate of chemical reaction. This will make it is very difficult that the pore solution reaches supersaturation.

In Fig. 9, the SO_3 distribution showed some powerful evidence that the sulfates were consumed, resulting in the severest concrete damage. I.e. if there was no chemical reaction and if it were salt weathering causing concrete damage, the ion distribution curves of Na_2SO_4 should show similar features to Na_2O distribution of Na_2CO_3 and $NaCl$. Certainly, this explanation is not convincing enough to disclose the mechanism. Further studies may be performed through thermodynamic calculation to check the negative effect of chemical reaction on the supersaturation formation.

4.2 Study of the concentration of solution on the formation of pore solution zone in concrete

In the previous tests, an opposite appearance to salt weathering was that the concrete was susceptible to be damaged under a higher relative humidity condition. Combining the role of relative humidity in wick action and chemical sulfate attack, it can be explained that a wider sulfate pore solution can be formed in the upper part of concrete in contact with moist air and chemical sulfate attack occurring in the pore solution zone resulted in concrete damage.

In our previous study [61], the pore solution expression test method was used to squeeze the pore solution in the cement paste. Cement paste samples were partially exposed to 10% Na_2SO_4 solution under the constant storage condition (20°C and 60% RH). The sulfate concentrations of pore solution in the lower part under solution (labeled L), film zone (labeled M) and efflorescence zone (labeled U) (shown in Fig. 4) were measured respectively. Fig. 19 gives the results.

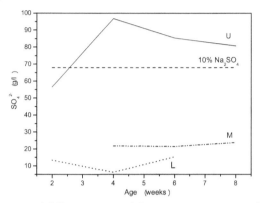

Fig. 19. SO_4^{2-} concentration of different parts of the cement paste partially exposed to the 10% Na_2SO_4 solution [61]

The results confirm that a pore solution zone can be formed in the efflorescence zone in the concrete, and the sulfate concentration was much higher than the lower part under solution and even the exposure solution (10% by mass). The strong chemical reactions occurring in this high concentration pore solution cause severe concrete decay. This also confirms the wick action theory.

Certainly, the ambient temperature and relative humidity of environment are always fluctuating. The boundary, the sulfate concentration and the formation time of pore solution zone were controlled by the evaporation rate due to the interactive effect of temperature and relative humidity. To study the pore solution zone formation needs further study.

4.3 Study of chemical sulfate attack mechanism

As we know, the main hydrated phases of cement paste are calcium silicate hydrate (C-S-H), calcium hydroxide (CH), calcium aluminate hydrate (C-A-H) ettringite (AFt) and mono-sulfoaliuminate (AFm). However, these three hydrated phases are not stable in the external environment containing sulfates. The following reactions can occur [62]:

$$Ca(OH)_2 + C\text{-}S\text{-}H + SO_4^{2-} + H_2O \rightarrow CaSO_4 \cdot 2H_2O \tag{7}$$

$$Ca(OH)_2 + C\text{-}S\text{-}H + MgSO_4 + H_2O \rightarrow CaSO_4 \cdot 2H_2O + Mg(OH)_2 + SiO_2 \cdot xH_2O \tag{8}$$

$$3CaO \cdot Al_2O_3 \cdot Ca(OH)_2 \cdot (12\text{-}18)H_2O + SO_4^{2-} \cdot 2H_2O + H_2O \rightarrow 3CaO \cdot Al_2O_3 \cdot 3CaSO_4 \cdot 32H_2O \tag{9}$$

$$Ca(OH)_2 + C\text{-}S\text{-}H + SO_4^{2-} + CO_3^{2-} + H_2O \rightarrow CaSiO_3 \cdot CaCO_3 \cdot CaSO_4 \cdot H_2O \tag{10}$$

The main reaction products are gypsum, ettringite, thaumasite, brucite and silica gel. Gypsum and ettringite are the common products of sulfate attack. Brucite and silica gel are found in case of magnesium sulfate. Thaumasite is formed when CO_3^{2-} is presented.

However, the product formation depends on the exposure conditions, such as sulfate content and pH value of sulfate environment, temperature and relative humidity.

Concerning gypsum, Bellmann et al have discussed the influence of sulfate concentration and pH value of solution on the gypsum formation in detail [52]. They indicate that portlandite will react to gypsum at a minimal sulfate concentration of approximately 1400 mg/l (pH=12.45). With rising pH, higher concentrations of sulfate ions are needed for the reaction to proceed. Between pH values of 12.45 and 12.7, the sulfate concentration slowly increases, whereas it rises dramatically from that level on. In solutions in which sodium ions are the counterpart of the hydroxide ions, the precipitation of gypsum can take place until pH values of approximately **12.9**. Beyond that mark, a further increase of the sulfate concentration is unable to lead to the formation of gypsum [52].

Concerning ettringite, ettringite is not stable in an environment with pH value below 11.5-12.0. At this low pH range, ettringite decomposes and forms gypsum [63, 64].

Concerning thaumasite, a number of experimental studies show that a high pH value (above 10.5) is in favor of the thaumasite formation [65-67]. If the pH value drops below 10.5 and even further towards 7, thaumasite is unstable, calcite and another calcium-bearing phase will be generated in the field cases [68, 69]. Thaumasite formation needs a relatively cold condition (below 15°C) [70].

In summary, the sulfate concentration, pH value and temperature control the reaction products.

Normally, in the full immersion test, 5% sulfate solutions stored at 23.0 ± 2.0°C are used in laboratories [71]. Compared to ground water in the field, a 5% sulfate solution used in the tests is much more concentrated [72]. Thus, the concrete immersed in the 5% sulfate solution can be regarded as an accelerated test. However, a high sulfate contents pore solution (higher than 5% and 10%) can be formed in the concrete in contact with air (Fig. 19) [61]. Due to concrete carbonation the pH value of pore solution in the concrete will decrease. The ambient temperature during the process of salt weathering in the field is always fluctuating. The exposure conditions of "salt weathering" on concrete are different from the full immersion tests. According to the XRD analysis, the results of long term field tests and indoor tests indicated that gypsum likely was the main reaction products and responsible for the concrete damage [2, 45, 46, 53]. Certainly, the mechanism of chemical sulfate attack should be further and systematically studied.

4.4 Study of the role of mineral additions in "Salt weathering" on concrete

An important result of long term field tests is the negative role of mineral additions in the concrete sulfate resistance. Normally, the indoor tests [74-80] always showed that the mineral additions can improve the sulfate resistance of cementitious materials based on the full immersion in 5% sulfate solutions stored at 23.0 ± 2.0°C. However, the long term field tests showed that the mineral additions accelerated the concrete decay.

As pointed out by Mehta [73], when concrete is fully immersed in the sulfate solution, for the prevention of sulfate attack "control of permeability is more important than control of the chemistry of cement". The pore size refinement due to mineral additions will prevent the sulfates to penetrate into concrete and lighten the negative effect of sulfate attack on concrete. Therefore, a number of previous researches all supported the idea that the mineral additions, such as fly ash, slag powder, silica fume and metakaolin, play a positive role in making sulfate-resisting concrete [74-80], not depending on the exposure conditions (sodium sulfate, magnesium sulfate, or ammonium sulfate).

However, in the case of partial immersion the pore size refinement due to mineral addtions will contribute to an increase in the capillary sorption height following Eq. 5. The pore solution expression tests showed that this process can draw more sulfates into fly ash concrete than into normal concrete, resulting in a pore solution with a higher sulfate concentration as shown in Fig. 20 [61].

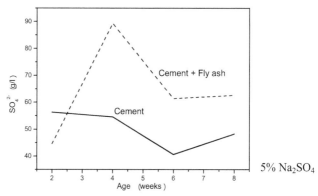

Fig. 20. SO_4^{2-} concentration of the pore solution in efflorescence and film zone of cement paste and cement-FA paste exposed to 5% Na_2SO_4 solution [61]

Another reason for mineral additions to lighten the negative effect of sulfate attack on concrete is the dilution effect induced by the partial cement replacement since it entails a reduction in the C_3A content [81]. Thus, based on laboratory tests mineral additions have always been regarded as an effective constituent to increase concrete's sulfate resistance in the field [82]. However, people maybe just remember the good things and ignore the bad ones. In the 1960s and 1970s extensive studies at the U.S. Bureau of Reclamation had reminded that [83, 84] concretes containing 30 percent low-calcium fly ashes showed greatly improved sulfate resistance to a standard sodium sulfate solution. However, the use of high-calcium fly ashes generally reduced the sulfate resistance. The high-calcium fly ashes containing highly reactive alumina in the form of C_3A or $C_4A_3\hat{S}$ are therefore less suitable than low-calcium fly ashes for improving the sulfate resistance of concrete. Taylor also pointed out that if slag has low alumina content, it improves the sulfate resistance, but with a high content of alumina, the reverse is the case [85,86]. M. Nehdi[47] also pointed out that it should not be overlooked that fly ash contains a large amount of reactive aluminum and that binders with an increased Al_2O_3 content can be more susceptive to the formation of ettringite. P. Nobst and J. Stark [87] carried out a very interesting test. Hardened cement pastes modified by different mineral additions were ground to a fineness of <200 μm and were mixed with stoichiometric parts of high quality gypsum powder (CaSO₄·2H₂O) and chemically produced calcite (CaCO₃) as well as with an excess of 20% of distilled water to investigate the thaumasite formation without the physical obstacle. The products identification showed unexpected results: (1) concerning cement-FA paste, the amount of ettringite increased with an increasing Al₂O₃ content at 20 °C while at 6 °C fly ash promoted a little more thaumasite formation; (2) slag cements which are generally classified as high sulfate resisting cements showed the most intensive thaumasite formation; (3) the use of micro-silica strongly accelerated the thaumasite formation. These findings indicate that mineral additions have a potentially negative effect in the concrete's resistance to sulfate attack depending on the exposure conditions. In the paper [87], the negative effect emerged due to no physical obstacle.

In the fly ash the aluminum phase existing as solid glass spheres is stable, but can be activated in a thermal, mechanical or chemical way [88]. It should be noted that Na₂SO₄ is an effective activator which is often used to activate the pozzolanic fly ash reaction in cement-fly ash pastes [89, 90]. What is worse, the ambient temperature may rise and also play a positive role in activating the aluminum phase, promoting the ettringite formation. According to the thermal analysis results [61], the cement and cement – fly ash (25%) pastes were immersed in the 5% Na₂SO₄ solution at 30 °C for 6 months. The amount of ettringite in the cement and cement-FA pastes amounted to 0.173 mg /mg and 0.217 mg/mg respectively. On the other hand, more gypsum was also detected in the cement – fly ash paste than cement paste. Fig. 21 is the thermal analysis of pastes exposed to 15% Na₂SO₄ solution at 30°C for 6 months. More ettringite and gypsum were generated in the cement – fly ash paste than in cement paste. Moreover, according to the wick action, 15% Na₂SO₄ can be formed in the upper portion of concrete in contact with air during the process of "salt weathering" on field concrete.

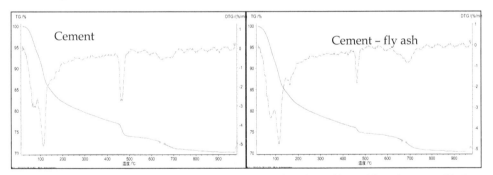

Fig. 21. Thermal analysis of pastes partially immersed in the 15% Na₂SO₄ solution at 30°C for 6 months [61]

In the paper [87], the negative effect of mineral additions on concrete sulfate resistance opposite to the normal results was attributed to no physical obstacle. In the process of "salt weathering" on concrete, a similar no physical obstacle appearance also can be defined. As abovementioned, solution goes into concrete by capillary suction. For porous materials, the capillary suction is a kind of active process, i.e. the solution is invited into the concrete. This can also be regarded as a no physical obstacle process. The sulfates can homogeneously distribute in the cement paste similar to the alkali activated cement in which Na_2SO_4 and powders are mixed before adding water, resulting in Na_2SO_4 homogeneously distributing in the cement paste.

In summary, concerning the role of mineral additions in the sulfate attack on partially exposed concrete, the exposure conditions and the solution transport mechanism are different from the full immersion cases. It needs further research.

4.5 Study of the effect of different kinds of sulfates in "salt weathering" on concrete

As abovementioned, Ruiz-Agudo [42] studied salt weathering distress of limestone specimens submerged in sodium sulfate and magnesium sulfate solutions respectively. The results showed that these two sulfates both severely damaged stone.

In the full immersion attack, because of the simultaneous significant decomposition of the C-S-H gel that accompanies the formation gypsum and ettringite, admittedly, people think the overall corrosive action of magnesium sulfate is greater than that of sodium sulfate [81, 91]

However, during the process of "salt weathering" on concrete, the test results showed the opposite appearance. Nehdi and Hayek [47] observed the appearances of the cement mortar partially exposed to 10% sodium sulfate and magnesium sulfate solution in a RH cycling between 32±3% and >95% condition respectively. The results showed that a large amount of efflorescence covers the surface when mortar is exposed to sodium sulfate solution. On the contrary, the surface of mortar subjected to magnesium sulfate solution is clean. It seems that sodium sulfate performs more corrosive effect on mortar than magnesium sulfate. The tests [53] also showed the same results. The aggregates and cement paste were completely separated in the upper part of concrete after 8 months exposure. However, the samples exposed to magnesium sulfate solution showed little damage. As shown in Fig. 22.

Fig. 22. Visual observation of concrete specimens partially exposed to Na_2SO_4 and $MgSO_4$ solutions [53]

This appearance may indicate that the concrete damage cannot be explained by salt weathering. First, $MgSO_4$ showed a harmful effect on stone due to salt weathering. As a porous material concrete should also show a similar scaling manner. Secondly, Fig. 23 shows the surface tensions of NaCl, Na_2SO_4 and $MgSO_4$ [92]. According to Eq. 5 the equilibrium heights of capillary rise of sodium sulfate and magnesium sulfate should be almost the same, showing similar efflorescence zone due to salt weathering. The reason for the opposite appearance of $MgSO_4$ in concrete may be the insoluble brucite due to chemical reaction that blocks the capillary. The role of sulfates in the "salt weathering" on concrete also needs further research.

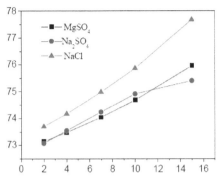

Fig. 23. Surface tensions of NaCl, Na_2SO_4 and $MgSO_4$ [92]

4.6 Study of the role of concrete carbonation in "salt weathering" on concrete

The negative effect of carbonation on corrosion of reinforcing steel in concrete is well known. As to the sulfate attack on concrete, as a result of carbonation, the total porosity would be reduced and the permeability of concrete could be improved [93-94]. So, Gao [95] pointed out that the carbonation layer could mitigate diffusion of sulfate ions to some extent in the full immersion situation.

However, when the concretes are partially exposed to sulfate solutions, the situation may be different. V.T. Ngala [94] studied the effect of carbonation on the ratio of capillary to total porosity of cement paste. The results showed that the capillary pore fraction greatly was improved after carbonation (shown in Fig. 24). This will promote the capillary suction of concrete, forming a more severe sulfate pore solution in the concrete and resulting more severer concrete damage.

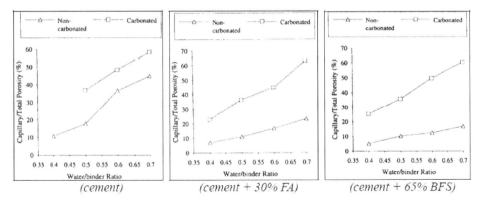

Fig. 24. Ratio of capillary to total porosity of non-carbonated and carbonated paste [94]

From Fig. 24, compared to cement paste, the ratios of capillary pores fraction of cement + 30% FA and cement + 65% BFS were higher than cement paste after carbonation. Some research showed that blended concrete has high carbonation rate [96], high degree of carbonation [97] or large carbonation depth [98] compared to the ordinary cement concrete. The carbonation susceptibility of blended concrete may be another reason for the negative effect of mineral addition on sulfate resistance of partially exposed concrete.

Besides, according to the review of indoor tests of "salt weathering" on concrete, two experimental results were observed showing that sulfate crystallization can be detected in the calcite crystals, the carbonation products of concrete (Fig. 12) and that carbonation could accelerate the concrete damage (Fig. 17). It might be that the efflorescence also occurs after concrete carbonation.

In summary, the effect of carbonation on sulfate resistance of partially exposed concrete is not clear. Further research will contribute to disclose the mechanism of "salt weathering" on concrete.

5. Conclusions

"Salt weathering" on concrete by sulfates is a deceptive and misleading phenomenon. In this paper, according to the comparison between the basic principles of salt weathering on porous materials and the abnormal appearances of "salt weathering" on concrete, the conclusion can be drawn that the salt weathering distress is not the major reason causing concrete damage when partially exposed to the sulfate environment. Chemical sulfate attack occurring in a high concentration pore solution is more likely the degradation mechanism for concrete deterioration similar to the full immersion cases of sulfate attack on concrete.

6. Acknowledgements

This work was financially supported by the National Science Foundation of P. R. China under contract #50378092, the scholarship from CSC (China Scholarship Council) and the co-funding from Ghent University of Belgium. The research was performed under a Bilateral Cooperation Agreement between Ghent University of Belgium and Central South University of P.R. China.

7. References

[1] D. Benavente, M.A. García Del Cura, A. Bernabéu, S. Ordóñez, Quantification of salt weathering in porous stones using an experimental continuous partial immersion method, Eng. Geol 2001; 59(3-4):313-325.

[2] Harvey Haynes, Robert O'Neill, Michael Neff, and P. Kumar Mehta, Salt weathering distress on concrete exposed to sodium sulfate environment, ACI Mater. J. 2008, 105(1):35-43.

[3] K. J.Folliard, P. Sandberg, Mechanisms of Concrete Deterioration by Sodium Sulfate Crystallization, Durability of Concrete, SP-145, American Concrete Institute, Farmington Hills, MI, 1994: 933-945.

[4] Haynes, H.; O'Neill, R.; and Mehta, P. K. Concrete Deterioration from Physical Attack by Salts, Concr. Int. 1996, 18 (1):63-68.

[5] W. G. Hime, R. A. Martinek, L. A. Backus, S. L. Marusin, Salt Hydration Distress, Concr. Int. 2001.23(10): 43-50.

[6] Thaulow, Niels, Sahu, Sadananda. Mechanism of concrete deterioration due to salt crystallization, Mater. Charact.2004, 53(2-4):123-127.

[7] YANG Quanbing，YANG Qianrong. Effects of salt-crystallization of sodium sulfate on deterioration of concrete, J. Chin. Chem. Soc 2007, 35(7): 877-880+885 (Chinese).

[8] MA Kunlin，XIE Youjun，LONG Guangcheng，LIU Yunhua. Deterioration characteristics of cement mortar by physical attack of sodium sulfate, J. Chin. Chem. Soc 2007, 35(10):1376-1381.

[9] Harvey Haynes, Robert O'Neill, Michael Neff, and P. Kumar Mehta, Salt Weathering of Concrete by Sodium Carbonate and Sodium Chloride, ACI Mater. J. 2010, 107(3):256-266

[10] Robert J. Flatt, Michael Steiger, George W. Scherer, A commented translation of the paper by C.W. Correns and W. Steinborn on crystallization pressure, Environ. Geol. 2007, 52(2): 221-237.

[11] George W. Scherter, crystallization in pores, Cem. Concr. Res. 1999, 29(8): 1347-1358.

[12] Rosa M. Espinosa Marzal, George W. Scherer, Crystallization of sodium sulfate salts in limestone, Environ. Geol. 2008, 56(3-4): 605-621.;

[13] Robert J. Flatt, Salt damage in porous materials: How high supersaturations are generated, J. Cryst. Growth 2002, 242(3-4): 435-454.

[14] George W. Scherer, Factors affecting crystallization pressure, International RILEM TC 186-ISA workshop and internal sulfate attack and delayed ettringite formation, 2002, Villars, Switzerland.

[15] Michael Steiger, Crystal growth in porous materials - I: The crystallization pressure of large crystals, J. Cryst. Growth 2005, 282(3-4):455-469.

[16] Michael Steiger, Crystal growth in porous materials - II: Influence of crystal size on the crystallization pressure, J. Cryst. Growth 2005, 282(3-4):470-481.

[17] Joerg Ruedrich, Siegfried Siegesmund, Salt and ice crystallization in porous sandstones, Environ. Geol. 2007, 52(2): 343-367.

[18] R.M. Espinosa, L. Franke, G.. Deckelmann, Model for the mechanical stress due to the salt crystallization in porous materials, Constr. Build. Mater. 2008, 22(7):1350-1367.

[19] G. Cultrone, L.G.Russo, C. Calabrò, M. Uroševic, A. Pezzino, Influence of pore system characteristics on limestone vulnerability: A laboratory study, Environ. Geol. 2008, 54(6):1271-1281.

[20] George W. Scherer, Stress from crystallization of salt, Cem. Concr. Res. 2004, 29(9): 1613-1624.

[21] D.Benavente, N. Cueto, J. Martínez-Martínez, M.A. García Del Cura, J.C. Cañaveras, The influence of petrophysical properties on the salt weathering of porous building rocks, Environ. Geol. 2007, 52(2):197-206.

[22] D. Benavente, M.A. Garcia del Cura, R. Fort, S. Ordónez, Durability estimation of porous building stones from pore structure and strength, Environ. Geol. 2004, 74(1-2): 113-127.

[23] D. Benavente, J. Martínez-Martínez, N.Cueto, M.A. García-del-Cura, Salt weathering in dual-porosity building dolostones, Environ. Geol. 2007, 94(3-4): 215-226.

[24] V. Lopez-Acevedo, C. Viedma, V. Gonzalez, A. La Iglesia, Salt crystallization in porous construction materials. II. Mass transport and crystallization processes, J. Cryst. Growth 1997, 182(1-2): 103-110.

[25] Miguel Gomez-Heras, Rafael Fort, Patterns of halite (NaCl) crystallisation in building stone conditioned by laboratory heating regimes, Environ. Geol. 2007, 52(2): 239-247.

[26] C. Rodriguez-Navarro, E. Doehne, Salt weathering: influence of evaporation rate, supersaturation and crystallization pattern. Earth Surf Processes and Landforms 1999, 24(2-3): 91-209.

[27] C. Rodriguez-Navarro, L. Linares-Fernandez, E. Doehne, E. Sebastian, Effects of ferrocyanide ions on NaCl crystallization in porous stone, J. Cryst. Growth 2002, 243(3-4): 503-516.

[28] S. Charles, E. Doehne, The evaluation of crystallization modifiers for controlling salt damage to limestone, J. J Cult. Herit. 2002 3 (3) 205–216.

[29] N.R. Buenfeld, M-T. Shurafa – Daoudi, I. M. Mcloughin, Chloride transport due to wick action in concrete RILEM International Workshop on Chloride Penetration into Concrete 1995:315-324.

[30] J. Francis Yong, Sidney Mindess, Robert J. Gray, Arnon Bentur, The science and technology of civil Engineering materials, Chinese Architecture & Building Press, 2006.

[31] Y.T. Puyate, C.J. Lawrence, Steady state solutions for chloride distribution due to wick action in concrete, Chem. Eng. Sci. 2000, 55(16): 3329-3334.

[32] Y.T. Puyate, C.J. Lawrence, N.R. Buenfeld, I.M. McLoughlin, Chloride transport models for wick action in concrete at large Peclet number, Phys. Fluids. 1998, 10(3): 566-575.

[33] Y.T. Puyate, C.J. Lawrence, Wick action at moderate Peclet number, Phys. Fluids. 1998, 10(8): 2114-2116.

[34] Y.T. Puyate, C.J. Lawrence, Effect of solute parameters on wick action in concrete, Chem. Eng. Sci., 1999, 54(19):4257-4265.

[35] L. Pel, H. Huinink, K. Kopinga, Ion transport and crystallization in inorganic building materials as studied by nuclear magnetic resonance, Appl. Phys. Lett. 2002, 81(15): 2893-2895.

[36] L.Pel, H. Huinink, K. Kopinga, R.P.J.Van Hees, O.C.G. Adan, Efflorescence pathway diagram: Understanding salt weathering, Constr. Build. Mater. 2004, 18(5): 309-313.

[37] C. Cardell, D.Benavente, J. Rodríguez-Gordillo, Weathering of limestone building material by mixed sulfate solutions. Characterization of stone microstructure, reaction products and decay forms, Mater. Charact. 2008, 59 10): p 1371-1385.

[38] Nicholas Tsui, Robert J. Flatt, George W. Scherer. Crystallization damage by sodium sulfate, J. J Cult. Herit. 2003 4 (2): 109–115.

[39] Genkinger, Selma, Putnis, Andrew, Crystallisation of sodium sulfate: Supersaturation and metastable phases, Environ. Geol. 2007, 52(2): 295-303.

[40] C. Rodriguez-Navarro, E. Doehne, E. Sebastian, How does sodium sulfate crystallize? Implications for the decay and testing of building materials, Cem. Concr. Res. 2000, 30(10): 1527-1534.

[41] E.M. Winkler, P.C. Singer, Crystallization pressure of salt in stone and concrete, Geol. Soc Am. 1972, 83(11): 3509-351.

[42] E. Ruiz-Agudo, F. Mees, P. Jacobs, C. Rodriguez-Navarro, The role of saline solution properties on porous limestone salt weathering by magnesium and sodium sulfates, Environ. Geol. 2007, 52(2):305-317.

[43] F.R. Mcmillan, T.E. Stantion, I.L. Tyler, W. C. Hansen. Long-Time Study of Cement Performance in Concrete, chapter 5. Concrete exposed of sulfate solis, Portland Cement Association 1949.

[44] D. Stark. Durability of concrete in sulfate-rich soils, Research and Development Bulletin, vol. RD O97, Portland Cement Association, 1989.

[45] D. Stark Performance of Concrete in Sulfate Environments, RD129, Portland Cement Association 2002.

[46] E.F. Irassar, A. Di Maio, O.R. Batic , Sulfate attack on concrete with mineral admixtures, Cem. Concr. Res. 1996, 26(1):113-123.

[47] M. Nehdi, M. Hayek, Behavior of blended cement mortars exposed to sulfate solutions cycling in relative humidity, Cem. Concr. Res. 2005, 35(4): 731-742.

[48] Norah Crammond, The occurrence of thaumasite in modern construction – a review, Cem. Concr. Compo. 2002, 24(3-4): 393-402.

[49] P.W. Brown, April Doerr, Chemical changes in concrete due to the ingress of aggressive species, Cem. Concr. Res. 2000, 30(3): 411-418.

[50] Mingyu, Hu, Fumei, Long, Mingshu, Tang, The thaumasite form of sulfate attack in concrete of Yongan Dam, Cem. Concr. Res. 2006, 36,(10): 2006-2008.

[51] Chiara F. Ferraris, Paul E. Stutzman, Kenneth A. Snyder, Sulfate Resistance of Concrete: A New Approach, Research and Development Information PCA R&D, Serial No. 2486, 2006.

[52] Bellmann Frank, Möser Bernd, Stark Jochen, Influence of sulfate solution concentration on the formation of gypsum in sulfate resistance test specimen, Cem. Concr. Res. 2006, 36(2): 358-363.

[53] Zanqun Liu, Geert De Schutter, Dehua Deng, Zhiwu Yu, Micro-analysis of the role of interfacial transition zone in "salt weathering" on concrete, Constr. Build. Mater. 2010, 24(11): 2052–2059.

[54] Zanqun Liu, Dehua Deng, Geert De Schutter, Zhiwu Yu, Micro-analysis of "salt weathering" on cement paste, accepted by Cem. Concr. Compo. for publish.

[55] D. Benavente, García del Cura, M.A.,García-Guinea, J, Sánchez-Moral, S, Ordóñez, S, Role of pore structure in salt crystallisation in unsaturated porous stone, J. Cryst. Growth 2004, 260 (3-4): 532-544.

[56] LIU Zanqun, XIAO Jia, HUANG Hai, YUAN Qiang, DENG Dehua. Physicochemical Study on the Interface

[57] Zone of Concrete Exposed to Different Sulfate Solutions, J.Wuhan Univ.Technol. (Materials Science Edition) 2006, 21(z1):167-175.

[58] Faran J. Introduction: the transition zone – discovery and development. ITZ in concrete RILEM report 11. London: E&FN Spon; 1996.

[59] Shenyang, Zhongzi Xu, Mingshu Tang, The process of sulfate attack on cement mortars. Adv Cem Based Mater 1996;4(1):1–5.

[60] Bonen David. Micro-structural study of the effect produced by magnesium sulfate on plain and silica fume-bearing Portland cement mortars. Cem. Concr. Res. 1993;23(3):541–55.

[61] Santhanam Manu, Cohen Menashi D, Olek Jan. Mechanism of sulfate attack: afresh look – part I: summary of experimental results. Cem. Concr. Res. 2002;32(6):915–21.

[62] Zanqun Liu, Study of the basic mechanisms of sulfate attack on cementitious materials, 2010, Central and South University China and Ghent University Belgium.

[63] M. Collepardi, A state-of-the-art review on delayed ettringite attack on concrete, Cem. Concr. Compo. 2003 25(4-5): 401-407.

[64] [63]Manu Santhanam, Menashi D, Jan OLek. Sulfate attack research – whither now? Cem. Concr. Res. 2004, 31(8): 1275-1296.

[65] P. Kumar Mehta. sulfate attack on concrete: separating myths from reality, Concr. Int. 2000, 28 (8): 57– 61.

[66] Q. Zhou, J. Hill, E.A. Byars, et al, The role of pH in thaumasite sulfate attack, Cem. Concr. Res. 2006, 36(1): 160-170.

[67] Jallad, Karim N., Santhanam, Manu, Cohen, Menashi D. Stability and reactivity of thaumasite at different pH levels, Cem. Concr. Res. 2003, 33(3): 433-437.

[68] N.J. Crammond, The thaumasite form of sulfate attack in the UK, Cem. Concr. Res. 2003, 25(7): 809-818.

[69] P. Hagelia, R.G. Sibbick, N.J. Crammond, C.K. Larsen, Thaumasite and secondary calcite in some Norwegian concretes, Cem. Concr. Compo. 2003, 25(8):1131-1140.

[70] P. Hagelia, R.G.. Sibbick, Thaumasite Sulfate Attack, Popcorn Calcite Deposition and acid attack in concrete stored at the Blindtarmen test site Oslo, from 1952 to 1982, Mater. Charac. 2009, 60(7):686-699.

[71] Deng De-Hua, Xiao Jia, Yuan Qiang, On thaumasite in cementitious materials, Jianzhu Cailiao Xuebao/J. Build. Mater. 2005, 8(4): 400-409 (Chinese).

[72] AETM C 1012 – 04, Standard Test Method for Length Change of Hydraulic-Cement Mortars Exposed to a Sulfate Solution.

[73] H. Haynes, Sulfate Attack on Concrete: Laboratory versus Field Experience, Concr. Int. 2002, 24(7): 64-70.

[74] P.K. Mehta, Sulfate attack on concrete: a critical review, Materials Science of Concrete, vol.III, Amer. Ceramic Society 1993:105– 130.

[75] H.T. Cao, L. Bucea, A. Ray, S. Yozghatlian, The effect of cement composition and pH of environment on sulfate resistance of Portland cements and blended cements, Cem. Concr. Compo. 1997, 19(2): 161-171.

[76] S. Miletic, M.Ilic, S. Otovic, R.Folic, Y. Ivanov, Phase composition changes due to ammonium-sulphate: Attack on Portland and Portland fly ash cements, Constr. Build. Mater. 1999, 13(3): 117-127.

[77] Rodriguez-Camacho, Redz E., Uribe-Afif, R., Importance of using the natural pozzolans on concrete durability, Cem. Concr. Res. 2003, 32(12): 1851-1858.

[78] El Sokkary, T.M., Assal, H.H., Kandeel, A.M., Effect of silica fume or granulated slag on sulphate attack of ordinary portland and alumina cement blend, Ceram. Int. 2004, 30(2): 133-138.

[79] P. Chindaprasirt, S. Homwuttiwong, V. Sirivivatnanon, Influence of fly ash fineness on strength, drying shrinkage and sulfate resistance of blended cement mortar, Cem. Concr. Res. 2004, 34(7): 1087-1092.

[80] Hanifi Binici, Orhan Aksogan, Sulfate resistance of plain and blended cement, Cem. Concr. Compo. 2006, 28(1): 39-46.

[81] Nabil M. Al-Akhras, Durability of metakaolin concrete to sulfate attack, Cem. Concr. Res. 2005, 36(9): 1727-1734.

[82] Omar S Al-Amoudi Baghabra, Attack on plain and blended cements exposed to aggressive sulfate environments, Cem. Concr. Compo. 2002, 24(3-4): 305-316.

[83] GB/T 50476-2008, Code for durability design of concrete structures (Chinese standard).

[84] P. Kumar Mehta, Concrete: structure, properties, and materials, Second Edition, Prentice Hall College Div; 2nd Revised edition edition (November 1992);

[85] Monteiro, Paulo J.M., Kurtis, Kimberly E., Time to failure for concrete exposed to severe sulfate attack, Cem. Concr. Res. 2003, 33(7): 987-993,.

[86] R.S. Collop, H. F. W.Taylor, Microstructural and microanalytical studies of sulfate attack III: Sulfate-resisting cement: reactions with sodium and magnesium sulfate solution. Cem. Concr. Res. 1995, 25(7): 1581-1590.

[87] R.S. Collop, H. F. W.Taylor, Microstructural and microanalytical studies of sulfate attack. V. Comparison of different slag blends, Cem. Concr. Res. 1996, 26(7) 1029-1044.

[88] P. Nobst, J. Stark, Investigations on the influence of cement type on thaumasite formation, Cem. Concr. Compos.2003, 25(8): 899–906.

[89] Wu, Zichao,Naik, Tarun R., Chemically activated blended cement, ACI Mater. J. 2003, 100(5): 434-440.

[90] Shi, Caijun, and Day, Robert L. Pozzolanic reaction in the presence of chemical activators. Part I. Reaction kinetics, Cem Concr Res. 2000, 30(1):51-58.

[91] Shi, Caijun, and Day, Robert L., Pozzolanic reaction in the presence of chemical activators: Part II. Reaction products and mechanism, Cem Concr Res. 2000, 30(4): 607-613.

[92] Omar S. Baghabra Al-Amoudi, Mohammed Maslehuddin, Effect of magnesium sulfate and sodium sulfate on the durability performance of plain and blended cement, ACI Mater. J. 1995, 92(1): 15-24.

[93] Ma Kunlin, Mechanism and Evaluation Method of Salt Crystallization Attack on Concrete, Central South University PhD thesis, 2009;

[94] V.T. Ngala, C.L. Page, Effect of carbonation on pore structure and diffusional properties of hydrated cement paste, Cem Concr Res. 1997, 27(7): 995-1007.

[95] Ha-Won Song, Seung-Jun Kwon, Permeability characteristics of carbonated concrete considering capillary pore structure, Cem Concr Res. 2007, 37 (6): 909–915;

[96] GAO Rundong · ZHAO Shunbo · LI Qingbin, Deterioration Mechanisms of Sulfate Attack on Concrete under the Action of Compound Factors, Jianzhu Cailiao Xuebao/J. Build. Mater. 2009, 12(1): 41-46 (Chinese).

[97] P. Sulapha, S.F. Wong, T.H. Wee, S. Swaddiwudhipong, Carbonation of concrete containing mineral admixtures, J. Mater. Civ. Eng. 2003, 15(2): 134-143.

[98] Monkman, Sean,, Shao, Yixin, Assessing the carbonation behavior of cementitious materials, J. Mater. Civ. Eng. 2006, 18(6): 768-776.

[99] Marlova P. Kulakowski, Fernanda M. Pereira, Denise C.C. Dal Molin Carbonation-induced reinforcement corrosion in silica fume concrete, Constr. Build. Mater. 2009, 23(3): 1189-1195.

Section 2

General Issues in Crystallization

5

Synthetic Methods for Perovskite Materials – Structure and Morphology

Ana Ecija, Karmele Vidal, Aitor Larrañaga,
Luis Ortega-San-Martín and María Isabel Arriortua
*¹Universidad del País Vasco/Euskal Herriko Unibertsitatea (UPV/EHU),
Facultad de Ciencia y Tecnología, Dpto. Mineralogía y Petrología, Leioa,
²Pontificia Universidad Católica del Perú (PUCP), Dpto. Ciencias,
Sección Químicas, Lima,
¹Spain
²Perú*

1. Introduction

Solid state chemistry thrives on a rich variety of solids that can be synthesized using a wide range of techniques. It is well known that the preparative route plays a critical role on the physical and chemical properties of the reaction products, controlling the structure, morphology, grain size and surface area of the obtained materials (Cheetham & Day, 1987; Rao & Gopalakrishnan, 1997). This is particularly important in the area of ABO_3 perovskite compounds given that they have for long been at the heart of important applications. From the first uses of perovskites as a white pigments, $PbTiO_3$ in the 1930's (Robertson, 1936) to the MLC capacitors (mostly based in substituted $PbTi_{1-x}Zr_xO_3$ or $BaTiO_3$ materials) in which today's computers rely on to operate, synthetic methods have been a key factor in the optimization of their final properties (Pithan et al., 2005).

Traditionally, most of these ceramic materials have been prepared from the mixture of their constituent oxides in the so called solid state reaction, "shake and bake" or ceramic method, a preparative route for which high temperature is a must in order to accelerate the slow solid–solid diffusion (Fukuoka et al., 1997; Inaguma et al., 1993; Safari et al., 1996). Despite its extended use in practically all fields in which perovskite-structured materials are needed, not all applications are better off with this method since the low kinetics and high temperature also yield samples with low homogeneity, with the presence of secondary phases and with uncontrolled (and typically large) particle size of low surface area which are undesired for some applications such as in gas sensors or in catalysis where small particles and high surface area are needed (Bell et al., 2000). This conventional route, however, is widely employed due to its simplicity and low manufacturing cost. Nevertheless, with appropriate optimisation, as when soft-mechanochemical processing is used prior to calcinations at high temperature (Senna, 2005), the method results in high quality single phase perovskites that can be used in electroceramic applications.

Alternative routes to the solid-state reaction method are wet chemical synthetic methods such as co-precipitation (with oxalates, carbonates, cyanides or any other salt precursors), combustion (including all variants from low to high temperature), sol–gel (and all of its modifications with different chelating ligands), spray-pyrolysis, metathesis reactions, etcetera (Patil et al., 2002; Qi et al., 2003; Royer et al., 2005; Segal, 1989; Sfeir et al., 2005). In all cases the idea is to accelerate the pure phase formation, a goal that is achieved due to the liquid media which permits the mixing of the elements at the atomic level resulting in lower firing temperatures. Other advantages of these methods are the possibility of having controlled particle size, morphology and improvement in surface area.

In most cases, the final microstructure of the sample is the key issue in choosing the synthetic method, but phase purity is also a must, and is sometimes overlooked when authors praise their particular synthetic method (Kakihana & Yoshimura, 1999). This was pointed out by Polini et al. (Polini et al., 2004) in the case of the preparation of substituted LaGaO$_3$ phases for SOFC cathodes: methods that supposedly have been developed to improve the scalability and uniformity of the samples, such as the Pechini method (a particular case of the sol-gel method), do not always result in a single phase of the crystalline sample required. Similar cases are common in the literature, as in the case of La$_{1-x}$Sr$_x$MnO$_{3-\delta}$ phases (Conceição et al., 2009). This clearly indicates that there is not such a thing as the ideal synthetic method: every method has its advantages and disadvantages.

Ideally, as many as possible synthetic methods should be tried and optimised for each compound of interest in order to obtain better crystals with the proper microstructure. But this is obviously time consuming and very costly. Consequently, researchers usually choose to follow the general trends that have been observed to work in a particular area of interest. As a result, each field has its preferences. For example, the ceramic method, widely used at the beginning of the first years of the high-Tc superconductors was soon replaced because it almost always resulted in non-stoichiometric products with some undesired phases that complicated the interpretation of the superconducting properties. These materials got so much attention in the late 1980's and early 1990's that completely new synthesis methods were introduced including many modifications of sol-gel methods with the ample use of alcoxides as precursors (Petrikin and Kakyhana, 2001). In this case, the synthetic route consisted on the preparation of mixed coordination compounds with alcoxy ligands in aqueous media which ensured good distribution of all metals involved and yielded purer superconducting oxides at relatively lower temperatures than before.

On the other hand, combustion methods (glycine-nitrate, urea based, and other modifications) have been proposed as one of the most promising methods to prepare perovskite oxide powders to be used as cathode materials in Solid Oxide Fuel Cell technology. (Bansal & Zhong, 2006; Berger et al., 2007; Dutta et al., 2009; Liu & Zhang, 2008). This method consist on a highly exothermal self-combustion reaction between the fuel (usually glycine, urea or alanine) and the oxidant (metal nitrates), that produces enough heat to obtain the ceramic powders. Compared to the ceramic method this synthetic route has much faster reaction times and lower calcination temperatures leading to powders with large compositional homogeneity and nanometric particle sizes, which are desired characteristics for this type of application.

For some applications, such as in multiferroic materials based devices, it is the crystal symmetry of the multiferroic what matters. In these materials the presence or not of a centre of symmetry is crucial for the observation of ferroelectricity. With this regard, there are cases, as in some $AMnO_3$ perovskites (A=Y or Dy), in which the synthetic route determines whether an orthorhombic compound with a centre of symmetry (i.e. non ferroelectric) or a hexagonal phase without the centre (i.e. ferroelectric) is formed (Carp et al., 2003; Dho et al., 2004). Consequently, preparative conditions have to be carefully selected in order to obtain crystal phases with the adequate structure. The use of more than one synthesis method is thus worth trying in all cases.

In this work three different groups of perovskite compounds have been prepared and their crystal structure and microstructure have been studied using X-ray diffraction (XRD) and scanning electron microscopy (SEM). Each group of samples had its own structural characteristics so, prior to choosing one synthetic approach, trials were carried out using different methods. In all cases, the final method chosen was the one that maximised phase purity and resulted in better properties. Here we demonstrate that phase pure samples susceptible to be compared depending on the desired characteristics can be obtained using different synthetic methods.

2. Experimental

Up to four different synthetic methods (solid state reaction, glycine-nitrate route, sol-gel and freeze-drying) have been used to synthesize a group of 14 perovskite compounds (Figure 1), which have the potential for their use in different applications.

Compound	Synthetic method	Variable		Label
$La_{0.50}Pr_{0.30}Sr_{0.20}FeO_{3-\delta}$	Solid state reaction		0.2	LPS20
$La_{0.40}Nd_{0.30}Sr_{0.23}Ca_{0.07}FeO_{3-\delta}$	Solid state reaction		0.3	LNSC30
$La_{0.20}Pr_{0.40}Sr_{0.26}Ca_{0.14}FeO_{3-\delta}$	Solid state reaction	x in A	0.4	LPSC40
$La_{0.19}Pr_{0.31}Sr_{0.26}Ca_{0.24}FeO_{3-\delta}$	Solid state reaction		0.5	LPSC50
$La_{0.19}Pr_{0.21}Sr_{0.26}Ca_{0.34}FeO_{3-\delta}$	Solid state reaction		0.6	LPSC60
$La_{0.18}Pr_{0.12}Sr_{0.26}Ca_{0.44}FeO_{3-\delta}$	Solid state reaction		0.7	LPSC70
$La_{0.20}Sr_{0.25}Ca_{0.55}FeO_{3-\delta}$	Solid state reaction		0.8	LSC80
$La_{0.50}Ba_{0.50}FeO_{3-\delta}$	Glycine-nitrate route		1.34	LB134
$La_{0.34}Nd_{0.16}Sr_{0.12}Ba_{0.38}FeO_{3-\delta}$	Glycine-nitrate route	$<r_A>(Å)$	1.31	LNSB131
$La_{0.04}Nd_{0.46}Sr_{0.24}Ba_{0.26}FeO_{3-\delta}$	Glycine-nitrate route		1.28	LNSB128
$La_{0.05}Sm_{0.45}Sr_{0.32}Ba_{0.18}FeO_{3-\delta}$	Solid state reaction		1.25	LSSB125-ss
	Glycine-nitrate route		1.25	LSSB125-gn
$Nd_{0.8}Sr_{0.2}Mn_{0.9}Co_{0.1}O_3$	Sol-gel		0.1	NSMC10-sg
	Freeze-drying		0.1	NSMC10-fd
$Nd_{0.8}Sr_{0.2}Mn_{0.8}Co_{0.2}O_3$	Sol-gel	x in B	0.2	NSMC20-sg
	Freeze-drying		0.2	NSMC20-fd
$Nd_{0.8}Sr_{0.2}Mn_{0.7}Co_{0.3}O_3$	Sol-gel		0.3	NSMC30-sg
	Freeze-drying		0.3	NSMC30-fd

x: doping level in $Ln_{1-x}A_xFeO_{3-\delta}$ and $Nd_{0.8}Sr_{0.2}(Mn_{1-x}Co_x)O_3$ series; $<r_A>$: average ionic radius of A-cation in the $Ln_{0.5}A_{0.5}FeO_3$.

Table 1. Nominal compositions, synthetic methods, and labels of the studied perovskites.

These compounds have been divided into three groups and their compositional details are summarised in Table 1. Preparation procedures are detailed below.

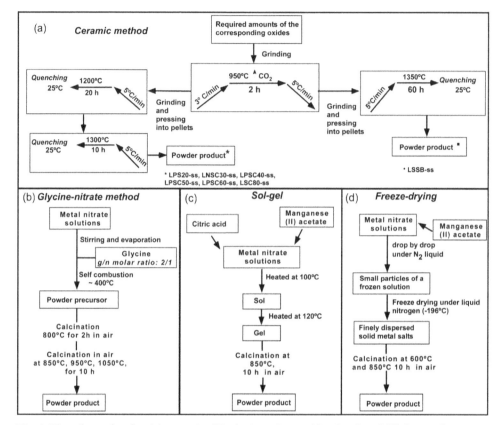

Fig. 1. Flowcharts for the: (a) ceramic, (b) glycine-nitrate, (c) sol-gel and (d) freeze-drying methods used to obtain the perovskite compounds shown in the present work.

The selection of each method and composition for each series of perovskites was based on the desired applications that are described later.

2.1 Ceramic solid state reaction

The compounds prepared via the ceramic route were obtained from mixing stoichiometric amounts of the raw oxides with 2-propanol in an agate mortar. Starting materials were always oxides of high purity such as La_2O_3 (99.99%), Sm_2O_3 (99.999%), Pr_2O_3 (99.9%), Gd_2O_3 (99.99%), BaO (99.99%), SrO (99.9%), CaO (99.9%) and Fe_2O_3 (99.98%). Afterwards, these mixtures were shaped into pellets and were fired in air at 950°C for 2h. The products obtained were ground, pelletized again and fired at higher temperatures. The flowchart shown in Figure 1a details the heat treatments required in each case until pure samples were obtained.

2.2 Glycine-nitrate route

The preparations of the perovskite compounds with the general composition $Ln_{0.5}A_{0.5}FeO_{3-\delta}$ by glycine-nitrate route involved the use of some nitrates instead of the oxides as starting materials: $Ba(NO_3)_2$ (99.99%), $Sr(NO_3)_2$ (99.9%) and $Fe(NO_3)_3$ (99.98%). The oxides used were La_2O_3 (99.99%), Sm_2O_3 (99.999%), Gd_2O_3 (99.99%). The oxides were dissolved in diluted nitric acid to obtain the corresponding nitrates and the metal nitrates dissolved in distilled water. The solutions were mixed in a 1 litre glass beaker, under constant stirring and placed on a hot plate to evaporate the water excess. After a significant reduction of the solution volume the glycine was added. The amount of glycine used was calculated in order to obtain a glycine/nitrate molar ratio of 2:1. This amino acid acts as complexing agent for metal cations and as the fuel for the combustion reaction. The resulting viscous liquid was auto-ignited by putting the glass beaker directly in a preheated plate (at ~400°C). The obtained powders were pelletized and fired at 800°C for 2 hours to remove the carbon residues. The heatings at temperatures above 800°C were repeated until pure phases were obtained. A flowchart with more details and heat treatments involved in this synthesis is shown in Figure 1b.

2.3 Sol–gel

The sol-gel method was used for the oxides of general composition $Nd_{0.8}Sr_{0.2}(Mn_{1-x}Co_x)O_3$ (x = 0.1, 0.2 and 0.3). Initially, the oxide Nd_2O_3 (99.9%) was dissolved in aqueous nitric acid followed by the addition of $Sr(NO_3)_2$ (99%), $Co(NO_3)_2.6H_2O$ (99%) and $Mn(C_2H_3O_2)_2.4H_2O$ (99%). Citric acid was then used as the quelating agent and ethylene glycol as the sol forming product. The solution was slowly evaporated in a sand bath for 24 h and the gel obtained was subjected to successive heat treatments at the temperature of 850°C (with intermediate grindings). Each firing was of 10 h and was carried out under nitrogen atmosphere. The flowchart for this synthesis is shown in Figure 1c.

2.4 Freeze-drying technique

In the freeze drying method, standardized nitrate solutions were mixed according to the stoichiometry of the final products: $Nd_{0.8}Sr_{0.2}(Mn_{1-x}Co_x)O_3$ (x = 0.1, 0.2 and 0.3). The starting materials were Nd_2O_3 (99.9%) which had to be dissolved in diluted nitric acid before the addition of the other compound; $Sr(NO_3)_2$ (99%); $Co(NO_3)_2.6H_2O$ (99%) and $Mn(C_2H_3O_2)_2.4H_2O$ (99 %).

The mixture for freeze-drying method was frozen drop-by-drop under liquid nitrogen and subjected to freeze drying at P = 5.10^{-2} mbar. Thermal decomposition was achieved by slow heating in air up to 600°C. The pure phases were obtained after repeated heatings at 850°C (with intermediate grindings), each of 10 h, under nitrogen atmosphere. A flow chart for this method is shown in Figure 1d.

2.5 Characterization

Room temperature X-ray powder diffraction data were collected in the $18 \leq 2\theta \leq 110°$ range with an integration time of 10 s/0.02° step. A Bruker D8 Advance diffractometer equipped with a Cu tube, a Ge (111) incident beam monochromator (λ = 1.5406 Å) and a Sol-X energy dispersive detector were used for the samples obtained by glycine-nitrate route and LSSB-ss. A Philips X'Pert-MPD X-ray diffractometer with secondary beam graphite monochromated Cu–Kα radiation was used for the samples obtained by ceramic solid-state, sol-gel and freeze-drying techniques.

All samples were single phase without detectable impurities. The crystal structure was refined by the Rietveld method (Rietveld, 1959) from X-ray powder diffraction data using GSAS software package (Larson & Von Dreele, 1994).

Microstructural analysis was carried out in a JEOL JSM 6400 scanning electron microscope (SEM) using a secondary electron detector at 30 kV and 1.10^{-10} A for the LPS20, LNSC30, LPSC40, LPSC50, LPSC60, LPSC70 samples and a JEOL JSM-7000F at 3 kV and 11.10^{-12} A for the rest of samples.

3. Results

3.1 Characterization of $Ln_{1-x}A_xFeO_{3-\delta}$ (Ln=La, Nd, Pr; A=Sr, Ca) perovskites with $0.2 \leq x \leq 0.8$

The $(La_{1-x}Sr_x)FeO_{3-\delta}$ (LSF) perovskite system exhibits high electronic and oxide ion conductivities at high temperatures, which make it attractive for solid oxide fuel cell (SOFC) cathodes. Several works (Ecija et al., 2011; Rodríguez-Martínez & Attfield, 1996; Vidal et al., 2007 and references therein) have shown that the physical properties of these perovskite materials are very sensitive to changes in the doping level (x) the average size of the A cations ($<r_A>$) and the effects of A cation size disorder ($\sigma^2(r_A)$).

The synthesis of these compounds allows us to study the effect of the variation of the doping level x on the properties of the perovskites with general formula $Ln_{1-x}A_xFeO_{3-\delta}$ (Ln=La, Nd, Pr; A=Sr, Ca) applied as SOFC cathodes. This has been achieved by keeping constant both the average size ($<r_A>$) and the size mismatch ($\sigma^2(r_A)$) to 1.22 Å and 0.003 Å^2, respectively.

For the preparation of this series the solid state reaction route has been chosen due to its simplicity to obtain perovskite phases in the same synthetic conditions.

3.1.1 Structural study

The room temperature X-ray powder diffraction patterns of these compounds are shown in Fig. 2a. A structural transition from orthorhombic symmetry (space group *Pnma*) for samples with $x \leq 0.4$ to rhombohedral symmetry (space group *R-3c*) for $0.5 \leq x \leq 0.7$ compounds, and finally, to a mixture of rhombohedral *R-3c* and cubic *Pm-3m* perovskite phases for the x = 0.8 composition was observed. Representative Rietveld fits to the X-ray diffraction data for the samples LPS20, LNSC50 and LNSC80 are shown in Fig. 2b, 2c and 2d, respectively.

The dependence with the doping level x of the cell parameters and cell volume per formula unit and atomic distances and bond angles for all samples are represented in Figure 3a and b, respectively. There is a systematic decrease in volume, cell parameters and <Fe-O> distances with increasing doping level across the series. Given that in the system studied the A-site mean ionic radius $<r_A>$ has been kept constant as the doping level increases, the observed effect can be solely associated to a reduction of the Fe-site mean ionic radius as it oxidises from Fe^{3+} to Fe^{4+}, with smaller radius ($<r_{Fe}>$, $r_{Fe}{}^{3+}$=0.645 Å and $r_{Fe}{}^{4+}$ =0.585 Å) (Shannon, 1976). Details of these effects are given elsewhere (Vidal et al., 2007).

The increase of the <A-O> distances and <Fe-O-Fe> bond angles with increasing doping level (x) can be explained due to the structural transition produced with x: when passing from orthorhombic (*Pnma*, LPS20) to a mixture of rhombohedral and cubic (*R-3c* + *Pm-3m*, LSC80) the octahedra that compose the perovskite structure reduce its cooperative tilting and the structure "expands".

These results are in nice agreement with other structural studies of related perovskites in which similar structural transitions with doping level were observed (Blasco et al., 2008; Dann et al., 1994; Tai et al., 1995).

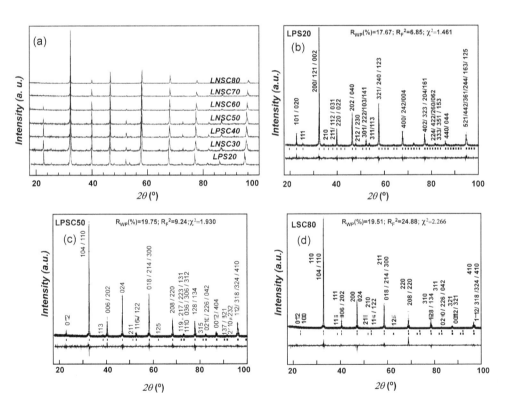

Fig. 2. (a) X-ray diffraction patterns for the series $Ln_{1-x}A_xFeO_{3-\delta}$ with x=0.2 to x=0.8, all obtained by the ceramic route. Rietveld fits to the X-ray diffraction data for samples LPS20 (b), LPSC50 (c) and LSC80 (d).

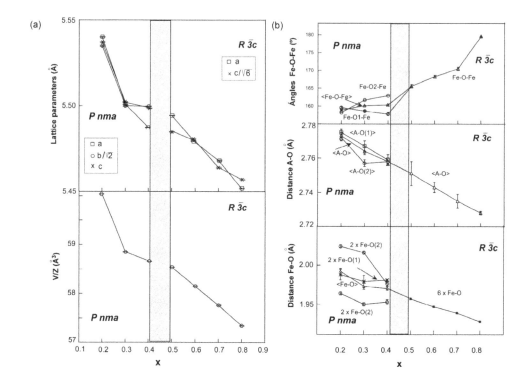

Fig. 3. (a) Variation of the unit cell parameters and volume per formula unit with doping, x. (b) Dependence of the mean atomic distances and bond angles with doping, x. Shaded area indicates the x range where the structural transition takes place.

3.1.2 Morphological study

Microstructure of bulk samples was study by scanning electron microscopy (SEM). Images of the sintered bars at 1300°C are shown in Fig. 4.

These micrographs show different particle size distributions with grain sizes ranging between 0.33 and 2.83 μm for sample LPS20, to 5 and 37 μm for sample LSC70.

The dispersion in particle sizes is larger for values of x≥0.4. As observed previously (Liou, 2004a, 2004b) this increase in particle size with the doping level seems to be a result of a change in the melting point of the samples that decreases increasing alkaline-earth cation content. According to Kharton et al. (Kharton et al., 2002) this effect results in a liquid-phase assisted by sintering and an enhanced grain growth.

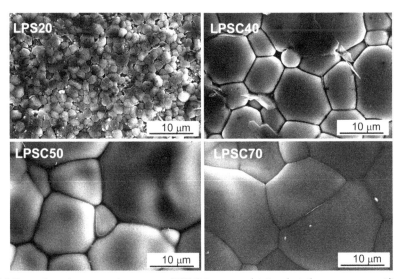

Fig. 4. (a) Scanning electron microscopy (SEM) images obtained at the same magnification for all $Ln_{1-x}A_xFeO_{3-\delta}$ compositions as a function of doping level x.

3.2 $Ln_{0.5}A_{0.5}FeO_{3-\delta}$ (Ln= La, Nd, Sm; A= Ba, Sr) 1.25≤ <r_A> ≤1.34 Å perovskites

As in the case of the previous family of iron perovskites, this new series of compounds are of interest for their use as mixed ionic electronic conducting materials, mainly from the point of view of cathodes for Solid Oxide Fuel Cells (SOFC), although they could also be used as ceramic membranes for oxygen separation. In the present case, the degree of lanthanide substitution was fixed to x=0.5 given that previous studies have shown that it is precisely at this degree of substitution when electronic and ionic conductivity are maximised (Hansen, 2010; Vidal et al., 2007).

In the present series of compounds the effect of the <r_A> variation on the properties of four different phases, $Ln_{0.5}A_{0.5}FeO_{3-\delta}$ (Ln=La, Nd, Sm; A=Ba, Sr) was evaluated. For this series (Table 1), <r_A> has been varied between 1.34 and 1.25 Å keeping x and $\sigma^2(r_A)$ constant, with values of 0.5 and 0.0161 Å^2, respectively.

Prior to choosing a synthesis method for all samples, two different methods were tried for one of the samples, $La_{0.05}Sm_{0.45}Sr_{0.18}Ba_{0.32}FeO_{3-\delta}$, for which the structural parameters and morphology were evaluated. The ceramic and glycine-nitrate routes were used in the present case.

3.2.1 Influence of the synthetic method on the structure and morphology of the $La_{0.05}Sm_{0.45}Sr_{0.18}Ba_{0.32}FeO_{3-\delta}$

Laboratory X-ray diffraction data at room temperature for $La_{0.05}Sm_{0.45}Ba_{0.5}FeO_{3-\delta}$ obtained by ceramic and glycine-nitrate routes were extremely similar, both samples being pure to the detection limits of the technique. XRD patterns were fitted by the Rietveld method (Figure 5) considering a rhombohedral symmetry (space group R-3c) in both cases.

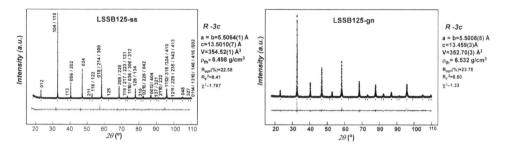

Fig. 5. Rietveld fits to room XRD patterns for LSSB-ss and LSSB-gn samples. In each case, lattice parameters (a, b, c), unit cell volume (V) and theoretical density (ρ) are included.

From the fits to the XRD data it was observed that both phases are nearly identical: they crystallise in the same space group and does not show significant difference among lattice parameters, cell volume or density values.

The morphological study, however, shows a different picture. As shown in Figure 6, where SEM micrographs taken at the same amplification for both samples are presented, their microstructure is quite different. The sample prepared following the ceramic route presents a microstructure with heterogeneous grain sizes and shapes, in which particles range from ~ 2 to approximately 8 μm in diameter. On the other hand, the average grain size of the sample obtained by glycine-nitrate route is about 200 nm, almost an order of magnitude smaller. The higher calcination temperature and longer reaction time required to obtain the samples in the ceramic process can explain the bigger grain size showed for these samples (Melo Jorge et al., 2001).

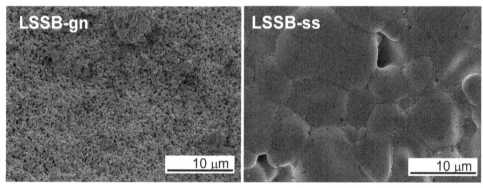

Fig. 6. SEM micrographs taken on the surface of the $La_{0.05}Sm_{0.45}Sr_{0.18}Ba_{0.32}FeO_{3-\delta}$ phases obtained by the glycine-nitrate (gn) and solid state reaction (ss) methods.

For the application as SOFC cathodes, samples with small and homogeneous particle sizes are usually preferred. As a consequence, the glycine-nitrate process was considered a more appropriate technique for preparing the perovskite samples of this series: $Ln_{0.5}A_{0.5}FeO_{3-\delta}$ (Ln= La, Nd, Sm; A= Ba, Sr) with $1.25 \leq <r_A> \leq 1.34$ Å.

3.2.2 Structural study

Room temperature X-ray diffraction patterns of the LB134, LNSB131, LNSB128, LSSB125-gn samples show that all the samples are single phase compounds (Figure 7a). A shift of the diffraction maxima to lower diffraction angles (2θ) with decreasing $<r_A>$ anticipates an increase of the lattice parameters. X-ray powder diffraction patterns were indexed using a cubic symmetry (Pm-$3m$ space group) for the LB134 and LNSB131 samples and a rhombohedral symmetry (R-$3c$ space group) in the case of LNSB128 and LSSB125-gn compounds. Figure 7b shows the Rietveld fits to the XRD patterns for two samples.

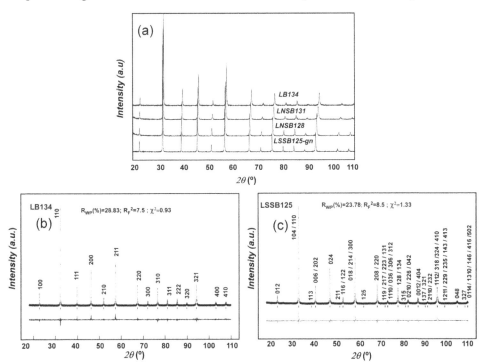

Fig. 7. (a) X-ray powder diffraction patterns at room temperature for all samples. (b) and (c) show the Rietveld fits to the X-ray powder diffraction patterns at room temperature for LB134-ss and LSS125-ss samples

The $<r_A>$ dependence of the lattice parameters, unit cell volume, main bond distances and Fe-O-Fe bond angles are shown in Figure 8. As it can be observed, lattice parameters and unit cell volume decrease with decreasing the average A-site ionic radius. Given that the doping level has been fixed (x=0.5), changes in the ratio Fe^{3+}/Fe^{4+} and, therefore, in the Fe-site ionic radius ($<r_{Fe}>$), are not expected. Consequently, the decrease in lattice parameters is ascribed to the variation of $<r_A>$. A more detailed analysis is given elsewhere (Ecija et. al, 2011). This also explains the decrease of the $<A$-$O>$ and $<A$-$Fe>$ mean distances with decreasing $<r_A>$ (Figure 8b). Although $<r_{Fe}>$ has been kept constant, there is a slight reduction of the $<Fe$-$O>$ distances as $<r_A>$ decreases, which is a consequence of the tilting in the FeO_6 octahedra due to the rhombohedral distortion. The decrease of the <Fe-O-Fe> bond angles is the result of the same effect (Woodward, 1998).

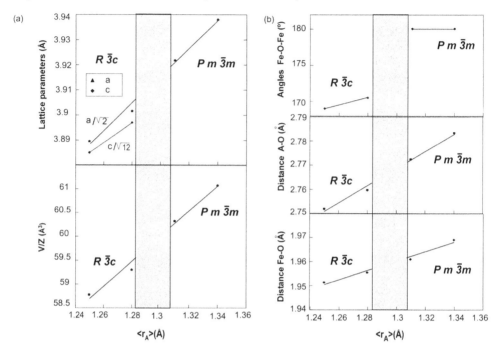

Fig. 8. $<r_A>$ dependence of (a) the unit cell parameters and volume per formula unit; and (b) the main average bond lengths (<A–Fe>, <A–O>, <Fe–O>) and Fe–O–Fe bond angle. Shaded area indicates the x range where the structural transition takes place.

3.2.3 Morphological study

Figure 9 shows the SEM micrographs of the LB134, LNSB131, LNSB128, LNSB125-gn bulk samples after the last heating at 1050°C.

In this series of samples no significant differences can be found in the morphology and average particle size. All samples present some agglomeration and fine grain size. Image analysis of the micrographs has allowed us to determine that the average grain size of the samples is in the range of 150-250 nm.

3.3 Synthesis and characterization of $Nd_{0.8}Sr_{0.2}(Mn_{1-x}Co_x)O_3$ perovskites with x = 0.1, 0.2, 0.3

The hole-doped perovskite manganese oxides with general formula $Ln_{1-x}A_xMnO_3$ (Ln= La, Pr, Nd; A= Ca, Sr, Ba, Pb; x<0.5) draw considerable attention in the late 1990′s due to their colossal magneto-resistance (CMR) effect at low temperatures (Rao, 1998). In the search for new CMR materials it was observed that doping on Mn site by other transition metal elements, such as Cr, Fe, Co and Ni, was an effective way to obtain new materials, which also helped to understand the new phenomenon (Tai, 2000; Takeuchi, 2002; Ulyanov, 2007). Phase morphology, highly dependent on preparative conditions, was also observed to play an important role in the effect: low temperature CMR was improved as the grain size was reduced (Das, 2002). In order to help in this area, it was decided to study the effects of the synthesis method in a series of perovskite compounds with the general formula $Nd_{0.8}Sr_{0.2}(Mn_{1-x}Co_x)O_3$ ($0.1 \leq x \leq 0.3$). The sol-gel and freeze-drying techniques were used in order to compare their structural, morphological and magnetic properties. Details of the later (magnetic properties) can be found elsewhere (Vidal et al., 2005).

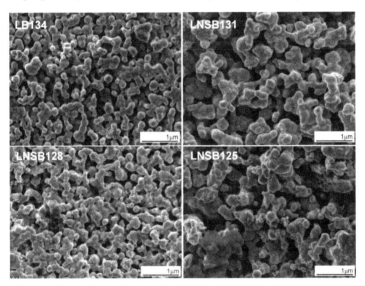

Fig. 9. SEM micrographs of the surface of the LB134, LNSB131, LNSB128, LSSB125-gn bulk samples.

3.3.1 Structural study

The X-ray powder diffraction patterns (XRPD) of all the compounds studied in this section were indexed in the orthorhombic space group *Pmna* irrespective of the synthesis method used. Fig. 10 shows the XRD and Rietveld refinement profiles for all phases with the formula $Nd_{0.8}Sr_{0.2}(Mn_{1-x}Co_x)O_3$.

When the lattice parameters and cell volume are compared a slight decrease with increasing cobalt content is observed in both cases. This effect is related to the changes in sizes of the B site atoms upon doping (Meera et al., 2001; Pollert et al., 2003). The oxidation states of Mn and Co in the $AMn_{1-x}Co_xO_3$ systems has been for long debated (Goodenough et al., 1961; 1997; Toulemonde et al., 1998; Troyanchuk et al., 2000), and different mixtures of Mn^{4+}-Mn^{3+} and Co^{3+}-Co^{2+} have been proposed. The most likely combination, based in spectroscopic and magnetic results, seems to indicate that the cobalt is introduced as Co^{2+} (and not as Co^{3+}) thus causing a mixed state (4+ and 3+) in manganese. This mixture of oxidation states has also been proved useful for Pollert et al. to explain magnetic and electrical properties of the series $Nd_{0.8}Na_{0.2}Mn_{1-x}Co_xO_3$ ($x \leq 0.1$), $Pr_{0.8}Na_{0.2}Mn_{1-x}Co_xO_3$ ($x \leq 0.2$) (Pollert et al., 2003a, 2003b) and $La_{0.8}Na_{0.2}Mn_{1-x}Co_xO_3$ (x≤0.2) (Pollert et al., 2004). On the other hand, as the Co content increases, the LSMC materials would shift from the oxygen excessive region to the oxygen deficient region, which would also contribute to the decrease in the unit volume. Wandekar et al. (Wandekar et al., 2009) carried out a detailed study on the crystal structure and conductivity of Co substituted LSM system and showed that the change in the ion radius of B-site element plays a predominant role at low Co content and the increase in oxygen vacancy becomes dominant at high Co content.

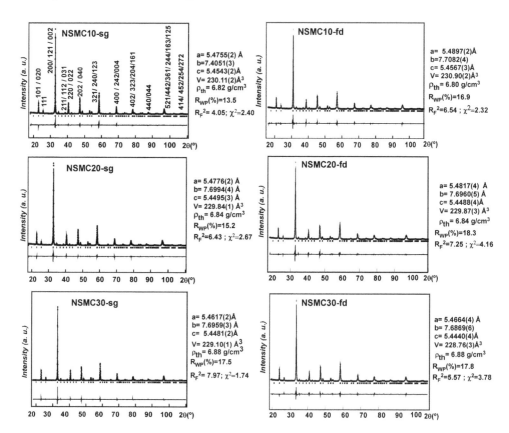

Fig. 10. Rietveld fits to the X-ray diffraction data in the orthorhombic *Pnma* space group for the $Nd_{0.8}Sr_{0.2}Mn_{1-x}Co_xO_3$ compounds. The labels "sg" and "fd" indicates that samples were prepared using the sol gel and freeze-drying techniques, respectively. In each case, lattice parameters (a, b, c), unit cell volume (V) and theoretical density (ρ) data are inset.

In consequence, considering only the high spin states of these elements and assuming nearly oxygen stoichiometric phases at room temperature, the observed reduction of the lattice volume in the present samples is consistent with the gradual appearance of Mn^{4+} (of smaller size than Mn^{3+} and Co^{2+} (Shannon, 1976) and so with the reduction of the mean B-site ionic radii. This explanation is also satisfactory for the same effect in the compounds $La_{0.7}Na_{0.3}Mn_{1-x}Co_xO_3$ where the authors had assumed the presence of only Mn^{3+}–Co^{3+} (Meera et al., 2001).

Table 2 shows details of the averge interatomic <B-O>, <A-O> and <A-B> distances together with <B-O-B> bond angles. As it can be observed <B-O> and <A-B> distances decrease as the cobalt concentration increases. These tendencies can be explained following the same reasoning indicated before, which basically concludes that cobalt enters in the structure in an oxidation state with smaller radius than manganese. No significant changes are observed in the <A-O> distances or in the <B-O-B> angles. This indicates that, despite the observed changes in the B size, the cobalt doping does not cause a noticeable distortion in the perovskite structure.

Variables		NSMC10		NSMC20		NSMC30	
		sg	fd	sg	fd	sg	fd
distance (Å)	<A–B>	3.346	3.350	3.345	3.345	3.341	3.339(1)
	<A–O1>	2.749(1)	2.756(1)	2.748(1)	2.751(2)	2.748(1)	2.747(2)
	<A–O2>	2.588(1)	2.592(1)	2.588(2)	2.593(2)	2.587(1)	2.591(2)
	<A–O>	2.668	2.674	2.668	2.672	2.667	2.671
	<B–O1>	1.954(4)	1.959(4)	1.950(4)	1.958(6)	1.958(4)	1.957(4)
	<B–O2>	1.968(1)	1.970(2)	1.967(2)	1.965(2)	1.963(2)	1.960(2)
	<B–O>	1.961	1.964	1.958	1.961	1.957	1.958
bond angle (°)	B-O1-B	160.8(4)	159.3(2)	161.5(4)	158.6(2)	158.5(3)	158.3(2)
	B-O2-B	157.9(4)	158.3(4)	158.2(4)	159.0(1)	158.5(3)	159.4(1)
	<B–O–B>	159.3	158.8	159.8	158.8	158.5	158.8

Table 2. Mean atomic distances and bond angles for the series of ABO_3 (A = $Nd_{0.8}$ $Sr_{0.2}$; B = $Mn_{1-x}Co_x$; x = 0.1, 0.2 and 0.3) samples prepared by the two methods described in the text.

In the same way, no significant changes are observed between the compounds synthesised by the sol-gel or the freeze-drying techniques.

3.3.2 Morphological study

The SEM micrographs of the present series of compounds are shown in Fig. 11. All pictures were taken on sintered bars after they were prepared by the sol-gel and freeze-drying methods.

As observed, the $Nd_{0.8}Sr_{0.2}(Mn_{1-x}Co_x)O_3$ samples prepared by the sol-gel route show a homogeneous particle size morphology distributed in small agglomerates. The grain size of the nearly spherical particles decreases with Co content from ~ 200 nm for the sample with x = 0.1 to ~ 100 nm for x = 0.3. In the case of the compounds obtained by the freeze-drying technique the morphology of the particles is also spherical and homogeneous but the grain size is quite stable in all phases and slightly higher (~250 nm) than in the previous case.

According to some studies (Kuharuangrong et al., 2004), Co doping of 20% does not usually influence the grain size of LSM, but 40 % mol Co significantly reduces the grain size (about 5-10 μm) of the $La_{0.84}Sr_{0.16}Mn_{1-x}Co_xO_3$ (x= 0, 0.2, 0.4) samples prepared by conventional oxide mixing process. However, some reduction in the particle size was observed in the case of $La_{0.67}Pb_{0.33}Mn_{1-x}Co_xO_3$ (x= 0 and 0.3) compounds, which were also prepared by the ceramic route (Dhahri et al., 2010). The present observation of a reduction in the particle size of the $Nd_{0.8}Sr_{0.2}(Mn_{1-x}Co_x)O_3$ phases in one of the methods but not in the other would point towards the synthesis method as the responsible for this effect, rather than the amount of cobalt. In this case, however, the observed decrease of the grain size is less significant than in previous studies.

Fig. 11. Secondary electron SEM micrographs obtained for all $Nd_{0.8}Sr_{0.2}(Mn_{1-x}Co_x)O_3$ compositions prepared by (a) sol-gel (sg) and (b) freeze-drying (fd) methods.

4. Conclusion

Different synthetic methods were used to prepare several perovskite type compounds. Phase-pure perovskites were obtained in all cases, irrespective of the method of synthesis and even when more than one method was used. It is to note that no differences in the overall crystal structure were observed in any of the cases when the method was changed. Moreover, changes in the crystal structure of the compounds within each series could be perfectly explained considering only compositional (and resulting structural) variables such as the mean A-site ionic radius or the A-site disorder, none of them being influenced by the synthesis route. This is true for the $Ln_{1-x}A_xFeO_{3-\delta}$ series where changes from orthorhombic to cubic are consistent with the change in the doping level x and also for the $Ln_{0.5}A_{0.5}FeO_3$ where the change was observed from rhombohedral to cubic symmetry when the mean A-site radius, $<r_A>$, was varied.

At least in the present group of compounds this seems to indicate that, properly used, and with no other factors present (such as in the case of the growth of samples on crystallographycally oriented substrates), the structural characteristics can be studied and properly addressed without having into account the synthesis method.

It is no news that, compared with the solid state reaction, the glycine-nitrate route, sol-gel and freeze-drying techniques require lower calcination temperatures to yield pure crystals of perovskite phases, with the resulting energy saving (very important nowadays). The disadvantage is, however, the fact that they are more time consuming and require more controlled synthesis conditions. In the same way, it is clearly observed that the ceramic method always yields phases with higher particle sizes but this is not a disadvantage if the final compounds are not to be used in an application that requires high specific area. It is to note that in the case of the $Nd_{0.8}Sr_{0.2}(Mn_{1-x}Co_x)O_3$, grain size was observed to change depending on the synthesis route. Although no significant differences were observed in the crystal structure, sol-gel method resulted in smaller grain sizes than the freeze drying method. In consequence, this seems to be a consequence of the preparative method which reinforces the idea that, if the microstructure of the sample is the key issue, the use of several synthesis methods is always worth trying.

5. Acknowledgments

This work has been financially supported by the Ministerio de Ciencia e Innovación PSE-12000-2009-7 (MICINN) and MAT 2010-15375; Consejería de Industria, Innovación, Comercio y Turismo (SAIOTEK 2011) and by the Consejería de Educación, Universidades e Investigación (IT-177-07) of the Basque Goverment of Gobierno Vasco/Eusko Jaurlaritza. Technical and human support provided by SGIker (UPV/EHU, MICINN, GV, EJ, ESF) and Alternative Generation Systems Group of Technological Research Centre is gratefully acknowledged.

6. References

Bansal, N.P. & Zhong, Z. (2006). Combustion Synthesis of $Sm_{0.5}Sr_{0.5}CoO_{3-x}$ and $La_{0.6}Sr_{0.4}CoO_{3-x}$ Nanopowders for Solid Oxide Fuel Cell Cathodes. *Journal of Power Sources*, Vol. 158, No. 1, (July 2006), pp. 148-153, ISSN: 0378-7753

Bell, R.J.; Millar, G.J. & Drennan. (2000). J. Influence of Synthesis Route on the Catalytic Properties of $La_{1-x}Sr_xMnO_3$. *Solid State Ionics,* (June 2000), Vol. 131, No. 3, 4, pp. 211-220, ISSN: 0167-2738

Berger, D.; Matei, C.; Papa, F.; Macovei, D.; Fruth, V. & Deloume, J.P. (2007). Pure and Doped Lanthanum Manganites Obtained by Combustion Method. *Journal of the European Ceramic Society*, Vol. 27, No. 13-15, (March 2007), pp. 4395-4398, ISSN: 0955-2219

Blasco, J.; Aznar, B.; García, J.; Subías, G.; Herrero-Martín, J. & Stankiewicz, J. (2008). Charge Disproportionation in $La_{1-x}Sr_xFeo_3$ Probed by Diffraction and Spectroscopic Experiments. *Physical Review B*, Vol. 77, N° 5, (February 2008), pp. 054107-1-054107-10, ISSN: 1098-0121

Carp, O.; Patron, L.; Ianculescu, A.; Pasuk, J. & Olar, R. (2003). New Synthesis Routes for Obtaining Dysprosium Manganese Perovskites. *Journal of Alloys and Compounds*, Vol. 351, No. 1-2, (March 2003), pp. 314-318, ISSN: 0925-8388

Cheetham, A.K. & Day, P. (1987). *Solid State Chemistry: Techniques*, Oxford University Press, ISBN: 0198551657, Oxford

da Conceicao, L.; Silva, C.R.B.; Ribeiro, N.F.P. & Souza, M.M.V.M. (2009). Influence of the Synthesis Method on the Porosity, Microstructure and Electrical Properties of $La_{0.7}Sr_{0.3}MnO_3$ Cathode Materials. *Materials Characterization*, Vol. 60, No. 12, (December 2009), pp. 1417-1423, ISSN: 1044-5803

Dann, S.E; Currie, D.B.; Weller, M.T.; Thomas, M.F. & Al-Rawwas, A.D. (1994). The Effect of Oxygen Stoichiometry on Phase Relations and Structure in the System $La_{1-x}Sr_xFeO_{3-\delta}$ ($0 \leq x \leq 1$, $0 \leq \delta \leq 0.5$). *Journal of Solid State Chemistry*, Vol. 109, No. 1, (March 1994), pp. 134-144, ISSN: 0022-4596

Das, S.; Chowdhury, P.; Gundu Rao, T. K.; Das, D. & Bahadur, D. (2002). Influence of Grain Size and Grain Boundaries on the Properties of $La_{0.7}Sr_{0.3}Co_xMn_{1-x}O_3$. *Solid State Communications*, Vol. 121, No. 12, (March 2002), pp. 691-695, ISSN: 0038-1098

Dhahri, N.; Dhahri, A.; Cherif, K.; Dhahri, J.; Taibi, K. & Dhahri, E. Structural, Magnetic and Electrical properties of $La_{0.67}Pb_{0.33}Mn_{1-x}Co_xO_3$ ($0 \Box x \Box 0.3$). *Journal of Alloys and Compounds*, Vol. 496, No. 1-2, (October 2010), pp. 69-74, ISSN: 0925-8388

Dho, J.; Leung, C.W.; MacManus-Driscoll, J.L. & Blamire, M.G. (2004). Epitaxial and Oriented $YMnO_3$ Film Growth by Pulsed Laser Deposition. *Journal of Crystal Growth*, Vol. 267, No. 3-4, (July 2004), pp. 548-553, ISSN: 0022-0248

Dutta, A.; Mukhopadhyay, J. & Basu, R.N. (2009). Combustion Synthesis and Characterization of LSCF-based Materials as Cathode of Intermediate Temperature Solid Oxide Fuel Cells. *Journal of the European Ceramic Society*, Vol. 29, No.10, (July 2009), pp. 2003-2011, ISSN: 0955-2219

Ecija, A.; Vidal, K.; Larrañaga, A.; Martínez-Amesti, A.; Ortega-San-Martín, L. & Arriortua, M.I. (2011). Characterization of $Ln_{0.5}M_{0.5}FeO_{3-\delta}$ (Ln=La, Nd, Sm; M=Ba, Sr) Perovskites as SOFC Cathodes. *Solid State Ionics*, Vol. 201, No. 1, (October 2011), pp. 35-41, ISSN: 0167-2738

Fukuoka, H.; Isami, T. & Yamanaka, S. (1997). Superconductivity of Alkali Metal Intercalated Niobate with a Layered Perovskite Structure. *Chemistry Letters*, Vol. 8, (April 1997), pp. 703-704, ISSN: 0366-7022

Goodenough, J.B.; Wold, A.; Wold, R.J.; Arnott, R.J. & Menyuk, N. (1961). Relationship Between Crystal Symmetry and Magnetic Properties of Ionic Compounds Containing Mn^{3+}. *Physical Review*, Vol. 124, (October 1961), pp. 373-384, ISSN: 0031-899X

Hansen, K.K. (2010). Electrochemical Reduction of Nitrous Oxide on $La_{1-x}Sr_xFeO_3$ Perovskites. *Materials Research Bulletin*, Vol. 45, No. 9, (May 2010), pp. 1334-1337, ISSN: 0025-5408

Inaguma, Y.; Liquan, C.; Itoh, M.; Nakamura, T.; Uchida, T.; Ikuta, H. & Wakihara, M. (1993) High Ionic Conductivity in Lithium Lanthanum Titanate. *Solid State Communications*, Vol. 86, No. 10, (June 1993), pp. 689-693, ISSN: 0038-1098

Kakihana, M. & Yoshimura, M. (1999). Synthesis and Characteristics of Complex Multicomponent Oxides Prepared by Polymer Complex Method. *Bulletin of the Chemical Society of Japan*, Vol. 72, No. 7, (1999), pp. 1427-1443, ISSN: 0009-2673

Kharton, V.V.; Shaulo, A.L.; Viskup, A.P.; Avdeev, M.; Yaremchenko, A.A.; Patrakeev, M.V.; Kurbakov, A.I.; Naumovich, E.N. & Marques, F.M.B. (2002). Perovskite-like System $(Sr,La)(Fe,Ga)O_{3-\delta}$: Structure and Ionic Transport under Oxidizing Conditions. *Solid State Ionics*, Vol. 150, No. 3-4, (October 2002), pp. 229-243, ISSN: 0167-2738

Kuharuangrong, S.; Dechakupt, T. & Aungkavattana, P. (2004). Effects of Co and Fe Addition on the Properties of Lanthanum Strontium Manganite. *Materials Letters,* Vol. 58, No. 12-13, (May 2004), pp. 1964-1970, ISSN: 0167-577X

Larson, A.C. & Von Dreele R.B. (1994). *GSAS: General Structure Analysis System, LAUR,* pp. 86–748

Liang, J.J. & Weng, H.S. (1993). Catalytic Properties of Lanthanum Strontium Transition Metal Oxides ($La_{1-x}Sr_xBO_3$; B = Manganese, Iron, Cobalt, Nickel) for Toluene Oxidation. *Industrial & Engineering Chemistry Research,* Vol. 32, No. 11, (November 1993), pp. 2563-2572, ISSN: 0888-5885

Liou, Y.C. (2004). Effect of Strontium Content on Microstructure in (La_xSr_{1-x})FeO_3 Ceramics. *Ceramics International,* Vol. 30, No. 5, (March 2004), pp. 667-669, ISSN: 0272-8842

Liou, Y.C. (2004). Microstructure Development in (La_xSr_{1-x})MnO_3 Ceramics. *Materials Science & Engineering, B: Solid-State Materials Advance Technology,* Vol. 108, No. 3, (April 2004), pp. 278-280, ISSN: 0921-5107

Liu, B. & Zhang, Y. (2008). $Ba_{0.5}Sr_{0.5}Co_{0.8}Fe_{0.2}O_3$ Nanopowders Prepared by Glycine-nitrate Process for Solid Oxide Fuel Cell Cathode. *Journal of Alloys and Compounds,* Vol. 453, No. (1-2), (April 2008), pp. 418-422, ISSN: 0925-8388

Meera, K.V.K.; Ravindranayh, V. & Rao M.S.R. (2001). Magnetotransport Studies in $La_{0.7}Ca_{0.3}Mn_{1-x}M_xO_3$ (M = Co, Ga). *Journal of Alloys and Compounds,* Vol. 326, No. 1-2, (August 2001), pp. 98-100, ISSN: 0925-8388

Melo Jorge, M.E.; Correia dos Santos, A. & Nunes, M.R. (2001). Effects of Synthesis Method on Stoichiometry, Structure and Electrical Conductivity of $CaMnO_{3-\delta}$. *International. Journal of Inorganic Materials,* Vol. 3, No. 7, (November, 2001), pp. 915-921, ISSN: 14666049

Narlikar, A. (2001). Essential Chemistry of High-Tc Cuprate Synthesis through the Solution Precursor Methods, In: *Studies of High Temperature Superconductors,* Nova Science Publishers, New York, ISBN: 1-59033-026-9

Patil, K. C.; Aruna, S. T. & Mimani, T. (2002). Combustion Synthesis: an Update. *Current Opinion in Solid State & Materials Science,* Vol. 6, No. 6, (December 2002), pp. 507-512, ISSN: 1359-0286

Pithan, C., Hennings, D. & Waser, R. (2005). Progress in the Synthesis of Nanocrystalline $BaTiO_3$ Powders for MLCC, *International Journal of Applied Ceramic Technology,* Vol. 2, No. 1, (January 2005), pp. 1–14, ISSN: 1546-542X

Polini, R.; Pamio A. & Traversa. E. (2004). Effect of Synthetic Route on Sintering Behavior, Phase Purity and Conductivity of Sr- and Mg-doped $LaGaO_3$ Perovskites. *Journal of the European Ceramic Society,* Vol. 24, No. 6, (June 2004), pp. 1365-1370. ISSN: 0955-2219

Pollert, E.; Hejtmánek, J. ; Knížek, K.; Maryško M.; Doumerc, J.P.; Grenier J.C. & Etourneau, J. (2003). Insulator-metal Transition in $Nd_{0.8}Na_{0.2}Mn_{(1-x)}Co_xO_3$ Perovskites. *Journal of Solid State Chemistry,* Vol. 170, No. 2, (February 2002), pp. 368-373, ISSN: 0022-4596

Pollert, E.; Hejtmánek, J.; Jirák, Z.; Knížek, K. & Marysko, M. (2003). Influence of Co Doping on Properties of $Pr_{0.8}Na_{0.2}Mn_{(1-y)}Co_yO_3$ Perovskites. *Journal of Solid State Chemistry,* Vol. 174. No. 2, (September 2003) pp. 466-470, ISSN: 0022-4596

Pollert, E.; Hejtmánek, J.; Jirák, Z.; Knížek, K.; Maryško, M.; Doumerc, J.P.; Grenier, J.C; & Etourneau, J. (2004). Influence of the Structure on Electric and Magnetic Properties of $La_{0.8}Na_{0.2}Mn_{1-x}Co_xO_3$ Perovskites. *Journal of Solid State Chemistry*, Vol. 177, No. 12, (December 2004), pp. 4564-4568, ISSN: 0022-4596

Qi, X; Zhou, J.; Yue, Z.X.; Gui, Z.L. & Li, L.T. (2003). A Simple Way to Prepare Nanosized $LaFeO_3$ Powders at Room Temperature. *Ceramics International*, Vol. 29, No. 3, (June 2003), pp. 347-349, ISSN: 0272-8842

Rao, C.N.R. & Gopalakrishnan, J. (1997). *New Directions in Solid State Chemistry*, 2nd ed., Cambridge University Press, ISBN: 0521495598, Cambridge

Rao, C.N.R. & Raveau, B. (1998). *Colossal Magnetoresistance, Charge Ordering and Related Properties of Manganese Oxides*, World Scientific, ISBN: 9810232764, Singapure.

Rietveld, H.M. (1969). A Profile Refinement Method for Nuclear and Magnetic Structures. *Journal of Applied Crystallography*, Vol. 2, (June 1969), pp. 65-71, ISSN: 0021-8898

Rodríguez-Martínez, L.M. & Attfield, J.P. (1996). Cation Disorder and Size Effects in Magnetoresistive Manganese Oxide Perovskites. *Physical Review B - Condensed Matter and Materials Physics*, Vol. 54, No. 22, (December 1996), pp. R15622–R15625, ISSN: 0163-1829

Royer, S.; Berube, F.S. & Kaliaguine, S. (2005). Effect Of The Synthesis Conditions on the Redox And Catalytic Properties in Oxidation Reactions of $LaCo_{1-x}Fe_xO_3$, *Applied Catalysis A: General*, Vol. 282, No. 1-2, (March 2005), pp. 273-284, ISSN: 0926-860X

Safari, A.; Panda, R. K. & Janas, V. F. (1996). Ferroelectricity. Materials, Characteristics, and Applications. *Key Engineering Materials*, Vol. 122-124, (1996), pp. 35-69, ISSN: 1013-9826

Segal, D. (1989). *Chemical Synthesis of Advanced Ceramic Materials*, Cambridge University Press, ISBN: 9780521354363, Cambridge

Senna, M. (2005). A Straight Way toward Phase Pure Complex Oxides. *Journal of the European Ceramic Society*, Vol. 25, No. 12, (March 2005), pp. 1977-1984, ISSN: 0955-2219

Sfeir, J.; Vaucher, S.; Holtappels, P.; Vogt, U.; Schindler, H.-J.; Van Herle, J.; Suvorova, E.; Buffat, P.; Perret, D.; Xanthopoulos, N. & Bucheli, O. (2005). Characterization of Perovskite Powders for Cathode and Oxygen Membranes Made by Different Synthesis Routes. *Journal of the European Ceramic Society*, Vol. 25, No.12, (March 2005), pp. 1991-1995, ISSN: 0955-2219

Shannon, R.D. (1976). Revised Effective Ionic Radii and Systematic Studies of Interatomic Distances in Halides and Chalcogenides, *Acta Crystallographica*, Vol. A32, No. 5, (September 1976), pp. 751-767, ISSN: 0567-7394

Skinner, S.J. (2001). Recent Advances in Perovskite-type Materials for Solid Oxide Fuel Cell Cathodes. *International Journal of Inorganic Materials*, Vol. 3, (March 2001), pp. 113-121, ISSN: 1466-6049

Tai, L.-W; Nasrallah, M.M.; Anderson H.U.; Sparlin, D.M & Sehlin, S.R. (1995). Structure and electrical properties of $La_{1-x}Sr_xCo_{1-y}Fe_yO_3$. Part 1. The system $La_{0.8}Sr_{0.2}Co_{1-y}Fe_yO_3$, *Solid State Ionics*, Vol. 76, No. 3-4, (March 1995), pp. 259–271, ISSN: 0167-2738

Tai, M.F.; Lee, F.Y. & Shi, J.B. (2000). Co Doping Effect on the Crystal Structure, Magnetoresistance and Magnetic Properties of an $(La_{0.7}Ba_{0.3})(Mn_{1-x}Co_x)O_3$ System with x=0-1.0., *Journal of Magnetism and Magnetic Materials*, Vol. 209, No. 1-3, (February 2000), pp. 148-150, ISSN: 0304-8853

Takeuchi, J.; Hirahara, S.; Dhakal, T.P; Miyoshi, K. & Fujiwara, K. (2002). Cr-doping Effect on the Perovskite $(Nd,Sr)MnO_3$ Single Crystals. *Physica B: Condensed Matter*, Vol. 312-313, (March 2002), pp. 754-756, ISSN: 0921-4526

Toulemonde, O.; Studer, F.; Barnabé, A.; Maignan, A. Martin, C. & Raveau, B. (1998). Charge States of Transition Metal in "Cr, Co and Ni" Doped $Ln_{0.5}Ca_{0.5}MnO_3$ CMR Manganites. *European Physical Journal B: Condensed Matter Physics*, Vol. 4, No. 2, (April 1998) pp. 159-167, ISSN: 1434-6028

Troyanchuk, I.O.; Lobanovsky, L.S.; Khalyavin, D.D.; Pastushonok, S.N. & Szymczak, H. (2000). Magnetic and Magnetotransport properties of Co-doped Manganites with Perovskite Structure. *Journal of Magnetism and Magnetic Materials*, Vol. 210, No. 1-2, (February 2000), pp. 63-72, ISSN: 0304-8853

Ulyanov, A.N.; Kim, J.S.; Shin, G.M.; Song, K.J.; Kang, Y.M. & Yoo, S.I. (2007). $La_{0.7}Ca_{0.3}Mn_{0.95}M_{0.05}O_3$ Manganites (M ¼ Al, Ga, Fe, Mn, and In): Local Structure and Electron Configuration Effect on Curie. Temperature and Magnetization. *Physica B: Condensed Matter*, Vol. 388, No. 1-2, (January 2007), pp. 16-19, ISSN: 0921-4526

Vidal, K.; Lezama, L.; Arriortua, M.I.; Rojo, T.; Gutiérrez, J. & Barandiarán. J.M. (2005). Magnetic characterization of $Nd_{0.8}Sr_{0.2}(Mn_{1-x}Co_x)O_3$ perovskites. *Journal of Magnetism and Magnetic Materials*, Vol. 290-291, No. 2, (April 2005), pp. 914-916, ISSN: 0304-8853

Vidal, K: Rodríguez-Martínez, L.M.; Ortega-San-Martín, L.; Díez-Linaza, E.; Nó, M.L.; Rojo, T.; Laresgoiti, A. & Arriortua, M.I. (2007). Isolating the Effect of Doping in the Structure and Conductivity of $(Ln_{1-x}M_x)FeO_{3-\delta}$ Perovskites. *Solid State Ionics*, Vol. 178, No. 21-22, (July 2007), pp. 1310-1316, ISSN: 0167-2738

Vidal, K.; Rodríguez-Martínez, L.M.; Ortega-San-Martin, L.; Martínez-Amesti, A.; Nó, M.L.; Rojo, T.; Laresgoiti, A. & Arriortua, M.I. (2009). The effect of doping in the electrochemical performance of $(Ln_{1-x}M_x)FeO_{3-\delta}$ SOFC cathodes. *Journal of Power Sources*, Vol. 192, No. 1, (July 2009), pp. 175-179, ISSN: 0378-7753

Vidal, K.; Rodríguez-Martínez, L.M.; Ortega-San-Martin, L.; Nó, M.L.; Rojo, T.; Laresgoiti, A. & Arriortua, M.I. (2010). $Ln_{0.5}M_{0.5}FeO_{3-\delta}$ Perovskites as Cathode for Solid Oxide Fuel Cells: Effect of Mean Radius of the A-Site Cations. *Journal of the Electrochemical Society*, Vol. 157, No. 8, (June 2010), pp. A919-A-924, ISSN: 0013-4651

Vidal, K.; Rodríguez-Martínez, L.M.; Ortega-San-Martin, L.; Nó, M.L.; Rojo, T.; Laresgoiti, A. & Arriortua, M.I. (2011). Effect of the A Cation Size Disorder on the Properties of an Iron Perovskite Series for their Use as Cathodes for SOFCs. *Fuel Cells*, Vol. 11, No. 1, (February 2011), pp. 51-58, ISSN: 1615-6846

Wandekar, R.V.; Wani, B.N. & Bharadwaj, S.R. (2009). Crystal Structure, Electrical Conductivity, Thermal Expansion and Compatibility Studies of Co-substituted Lanthanum Strontium Manganite System. *Solid State Science*, Vol. 11, No. 1, (January 2009), pp. 240-250, ISSN: 1293-2558

Woodward, P.M.; Vogt, T.; Cox, D.E.; Arulraj, A.; Rao, C.N.R.; Karen, P. & Cheetham, A. K. (1998). Influence of Cation Size on the Structural Features of $Ln_{1/2}A_{1/2}MnO_3$ Perovskites at Room Temperature. *Chemmistry of Materials*, Vol. 10, No. 11, (October 1998), pp. 3652-3665, ISSN: 0897-4756

6

Structure of Pure Aluminum After Endogenous and Exogenous Inoculation

Tomasz Wróbel

Silesian University of Technology, Foundry Department
Poland

1. Introduction

The phenomenon of crystallization following after pouring molten metal into the mould, determines the shape of the primary casting (ingot) structure, which significantly affects on its usable properties.

The crystallization of metal in the mould may result in three major structural zones (Fig.1) (Barrett, 1952; Chalmers, 1963; Fraś, 2003; Ohno, 1976):

- zone of chilled crystals (grains) formed by equiaxed grains with random crystallographic orientation, which are in the contact area between the metal and the mould,
- zone of columnar crystals (grains) formed by elongated crystals, which are parallel to heat flow and are a result of directional solidification, which proceeds when thermal gradient on solidification front has a positive value,
- zone of equiaxed crystals (grains) formed by equiaxed grains with random crystallographic orientation in the central part of the casting. The equiaxed crystals have larger size than chilled crystals and are result of volumetric solidification, which proceeds when thermal gradient has a negative value in liquid phase.

Depending on the chemical composition, the intensity of convection of solidifying metal, the cooling rate i.e. geometry of casting, mould material and pouring temperature (Fig.2), in the casting may be three, two or only one structural zone.

Due to the small width of chilled crystals zone, the usable properties of casting depend mainly on the width and length of the columnar crystals, the size of equiaxed crystals and content of theirs zone on section of ingot, as well as on interdendritic or interphase distance in grains such as eutectic or monotectic. For example, you can refer here to a well-known the Hall-Peth law describing the influence of grain size on yield strength (Fig.3) (Adamczyk, 2004):

$$\sigma_y = \sigma_0 + k \bullet d^{-\frac{1}{2}} \tag{1}$$

where:
σ_y – yield strength, MPa,

σ_o – approximate yield strength of monocrystal, for Al amount to 11,1 MPa,
k – material constant characterizing the resistance of grain boundaries for the movement of dislocations in the initial stage of plastic deformation (strength of grain boundaries), for Al amount to 0,05 $MN \cdot m^{-3/2}$,
d – grain size, mm.

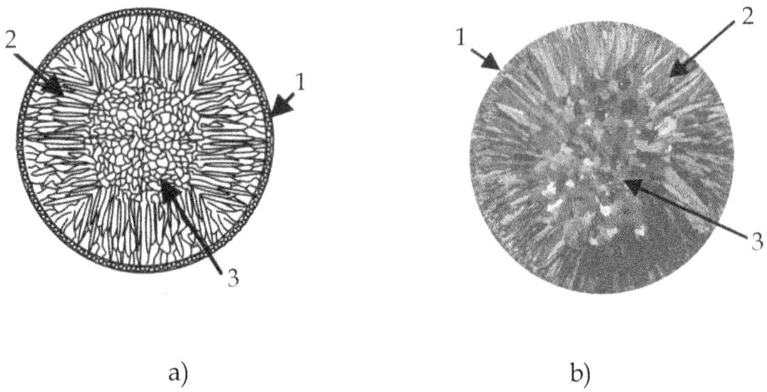

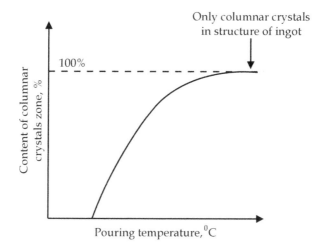

a)   b)

Fig. 1. The primary structure of ingot: a – scheme, b – real macrostructure;
1 – chilled crystals zone, 2 – columnar crystals zone, 3 – equiaxed crystals zone

Fig. 2. The influence of pouring temperature on primary structure of ingot (Fraś, 2003)

Because the Hall-Peth law concerns only metals and alloys with the structure of solid solutions, therefore the solidification of alloys with eutectic transformation for example from Al-Si group in describing the influence of refinement of structure on the value of yield strength should take into account the value of interphase distance in eutectic (Paul et al., 1982; Tensi & Hörgerl, 1994; Treitler, 2005):

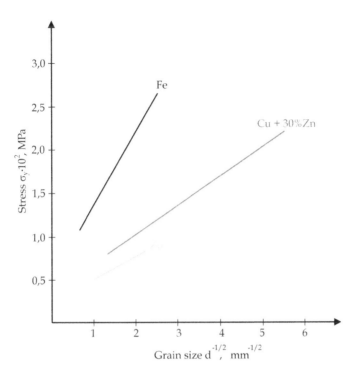

Fig. 3. The influence of grain size on yield strength of Fe, Zn and α brass (Adamczyk, 2004)

$$\sigma_y = \sigma_0 + k_1 \bullet d^{-\frac{1}{2}} + k_2 \bullet \lambda^{-\frac{1}{2}} \tag{2}$$

where:
σ_y – yield strength, MPa,
σ_o – approximately yield strength of monocrystal, MPa
k_1 and k_2 – material constants, MN·m$^{-3/2}$,
d – grain size, mm,
λ – interphase distance in eutectic, mm.

The primary structure of pure metals independently from the crystal lattice type creates practically only columnar crystals (Fig.4) (Fraś, 2003). According to presented data, this type of structure gives low mechanical properties of castings and mainly is unfavourable for the plastic forming of continuous and semi-continuous ingots, because causing forces extrusion rate reduction and during the ingot rolling delamination of external layers can occur (Szajnar & Wróbel, 2008a, 2008b).

This structure can be eliminated by controlling the heat removal rate from the casting, realizing inoculation, which consists in the introduction of additives to liquid metal and/or influence of external factors for example infra- and ultrasonic vibrations or electromagnetic field.

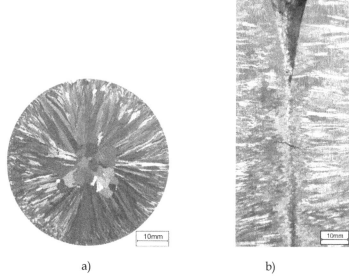

a)   b)

Fig. 4. Macrostructure of ingot of Al with a purity of 99,7%: a – transverse section,
b – longitudinal section

2. Endogenous inoculation of pure aluminum structure

In aim to obtain an equiaxed and fine-grained structure, which gives high mechanical properties of castings, can use inoculation, which occurs in introducing into metal bath of specified substances, called inoculants (Fraś, 2003). Inoculants increase grains density as result of creation of new particles in consequence of braking of grains growth velocity, decrease of surface tension on interphase boundary of liquid – nucleus, decrease of angle of contact between the nucleus and the base and increase of density of bases to heterogeneous nucleation (Fraś, 2003; Jura, 1968). The effectiveness of this type of inoculation depends significantly on crystallographic match between the base and the nucleus of inoculated metal. This crystallographic match is described by type of crystal lattice or additionally by index (Fraś, 2003):

$$\xi = \left((1 - \frac{x_b - x_n}{x_n}) \right) \bullet 100\% \tag{3}$$

where:

ξ - match index,

x_b, x_n – parameter of crystal lattice in specified direction, suitable for base and nucleus.

When the value of index (ξ) is closer to 100%, it the more effective is the base to heterogeneous nucleation of inoculated metal.

Therefore active bases to heterogeneous nucleation for aluminum are particles which have high melting point i.e. TiC, TiN, TiB, TiB_2, AlB_2 i Al_3Ti (Tab.1) (Easton & StJohn, 1999a,

1999b; Fjellstedt et al., 2001; Fraś, 2003; Guzowski et al., 1987; Hu & H. Li, 1998; Jura, 1968; Kashyap & Chandrashekar, 2001; H. Li et al., 1997; P. Li et al., 2005; McCartney, 1988; Murty et al., 2002; Naglič et al., 2008; Pietrowski, 2001; Sritharan & H. Li, 1996; Szajnar & Wróbel, 2008a, 2008b; Whitehead, 2000; Wróbel, 2010; Zamkotowicz et al., 2003).

Phase	Melting point (circa) [°C]	Type of crystal lattice	Parameters of crystal lattice [nm]
Al	660	Cubical A1	a = 0,404
TiC	3200	Cubical B1	a = 0,431
TiN	3255	Cubical B1	a = 0,424
TiB	3000	Cubical B1	a = 0,421
TiB$_2$	2900	Hexagonal C32	a = 0,302 c = 0,321
AlB$_2$	2700	Hexagonal C32	a = 0,300 c = 0,325
Al$_3$Ti	1400	Tetragonal D0$_{22}$	a = 0,383 c = 0,857

Table 1. Characteristic of bases to heterogeneous nucleation of aluminum (Donnay & Ondik, 1973)

Moreover the effectiveness of inoculants influence can be assessed on the basis of the hypothesis presented in the paper (Jura, 1968). This hypothesis was developed at the assumption that the fundamental physical factors affecting on the crystallization process are the amount of give up heat in the crystallization process on the interphase boundary of liquid - solid and the rate of give up heat of crystallization. After analyzing the results of own researches, the author proposed to determine the index (α), which characterizes the type of inoculant.

$$\alpha = \frac{\left(\Delta E_k / v\right)_s}{\left(\Delta E_k / v\right)_p} \bullet W \qquad (4)$$

where:

ΔE_k – heat of crystallization of 1 mol of inoculant or inoculated metal, J/mol,
v – characteristic frequency of atomic vibration calculated by the Lindemman formula, 1/s,
s – symbol of inoculant, p – symbol of inoculated metal.
W – parameter dependent on the atomic mass of inoculant and inoculated metal.
On the basis of equation (4) additives can be divided into three groups:
At α > 1 – additives which inhibit crystals growth by the deformation of the crystallization front, thus are effective inoculants.
At α = 1 – additives do not affect on structure refinement.
At α < 1 – additives which accelerate crystals growth, favoring consolidation of the primary structure of the metal, thus are deinoculants.

In case of inoculation of Al the index α = 2.35 for inoculant in form of Ti and 1.76 for inoculant in form of B.

In case of aluminum casting inoculants are introduced in form of master alloy AlTi5B1. This inoculant has Ti:B ratio equals 5:1. This Ti:B atomic ratio, which corresponds to the mass content of about 0.125% Ti to about 0.005% B, assures the greatest degree of structure refinement (Fig.5). For this titanium and boron ratio bases of type TiB_2 and Al_3Ti are created (Fig.6) (Easton & StJohn, 1999a, 1999b; Fjellstedt et al., 2001; Guzowski et al., 1987; Hu & H. Li, 1998; Kashyap & Chandrashekar, 2001; H. Li et al., 1997; P. Li et al., 2005; Murty et al., 2002; Naglič et al., 2008; Pietrowski, 2001; Sritharan & H. Li, 1996; Szajnar & Wróbel, 2008a, 2008b; Whitehead, 2000; Wróbel, 2010). Type and amount of bases to heterogeneous nucleation of aluminum depend on Ti:B ratio. For example given in paper (Zamkotowicz et al., 2003) the possibility of application of master alloy AlTi1.7B1.4, which has Ti:B ratio equals 1.2:1 is presented. This ratio allows to increase in amount of fine phases TiB_2 and AlB_2 along with the Al_3Ti phase decrease.

Moreover minimum quantities of carbon and nitrogen, which come from metallurgical process of aluminum, create with inoculant the bases in form of titanium carbide TiC and titanium nitride TiN (Fig.7) (Pietrowski, 2001; Szajnar & Wróbel, 2008a, 2008b).

Additionally, because there is a possibility of creation the bases to heterogeneous nucleation of aluminum in form of TiC phase without presence of bases in form of borides, in the practice of casting the inoculation with master alloy AlTi3C0.15 is used (Naglič et al., 2008; Whitehead, 2000). However, on the basis of results of own researches was affirmed that assuming of introducing to Al with a purity of 99,5% the same quantity of Ti i.e. 25ppm, the result of structure refinement caused by master alloy AlTi3C0.15 is weaker than caused by master alloy AlTi5B1 (Fig.8).

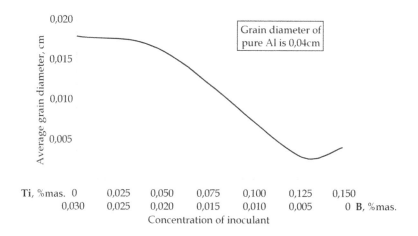

Fig. 5. Influence of Ti and B contents on the average size of Al ingots (H. Li et al., 1997)

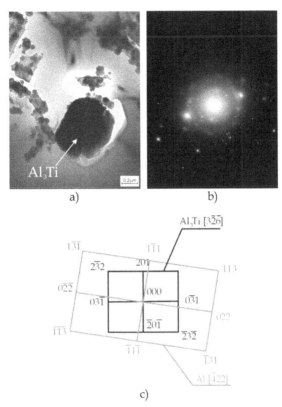

a)   b)

AlₜTi [3̄2̄6̄]

13̄1̄   1̄1̄1
    2̄3̄2   201̄   113
02̄2̄   000
    031̄   031̄   022
1̄1̄3̄   2̄01̄   2̄3̄2̄
        1̄1̄1   1̄31

Al [4̄22]

c)

Fig. 6. Structure of thin foil from pure Al after inoculation with (Ti+B), a) TEM bright field mag. 30000x , b) diffraction pattern from the area as in Fig. a, c) analysis of the diffraction pattern from Fig. b

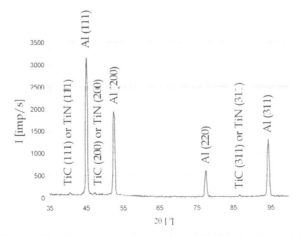

Fig. 7. Result of X-ray diffraction of Al with a purity of 99,5% after inoculation with Ti

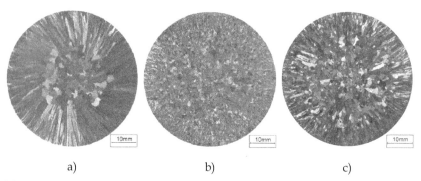

a)   b)   c)

Fig. 8. Macrostructure of ingot of Al with a purity of 99,5%: a – in as-cast condition,
b – after inoculation with (Ti+B), c – after inoculation with (Ti+C)

However, this undoubtedly effective method of inoculation of primary structure of ingot is
limited for pure metals, because inoculants decrease the degree of purity specified in
European Standards, and Ti with B introduced as modifying additives are then classify as
impurities. Moreover, inoculants, mainly Ti which segregates on grain boundary of Al
(Fig.9) influence negatively on physical properties i.e. electrical conductivity of pure
aluminum (Fig.10) (Wróbel, 2010).

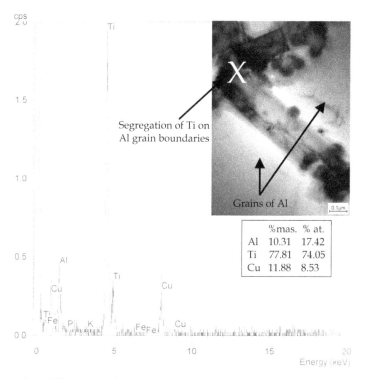

Fig. 9. Segregation of Ti on grain boundaries of Al

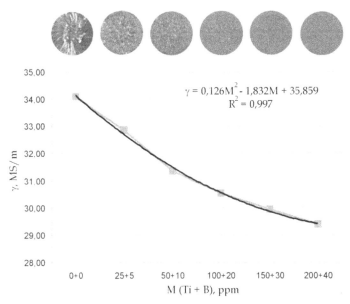

Fig. 10. The influence of quantity of inoculants in form of Ti and B on electrical conductivity γ of Al with a purity of 99,5%

Moreover the presence of the bases to heterogeneous nucleation in form of hard deformable phases for example titanium borides in structure in aluminum, generate possibility of point cracks formation (Fig. 11) and in result of this delamination of sheet (foil) during rolling (Keles & Dundar, 2007).

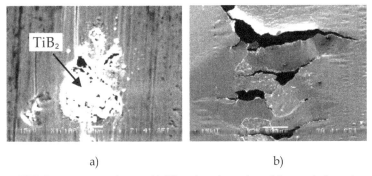

a)   b)

Fig. 11. Phase TiB$_2$ in structure of pure Al (Fig. a) and produced in result from its present crack in sheet (foil) during rolling (Fig. b) (Keles & Dundar, 2007)

Therefore important is the other method of inoculation, which consists of influence of electromagnetic field (Asai, 2000; Campanella et al., 2004; Doherty et al., 1984; Gillon, 2000; Griffiths & McCartney, 1997; Harada, 1998; Szajnar, 2004, 2009; Szajnar & Wróbel, 2008a, 2008b, 2009; Vives & Ricou, 1985; Wróbel, 2010) or mechanical vibrations (Abu-Dheir et al., 2005; Szajnar, 2009) on liquid metal in time of its solidification in mould.

3. Exogenous inoculation of pure aluminum structure

First research works on the application of stirring of liquid metal at the time of its solidification in order to improve the castings quality were carried out by Russ Electroofen in 1939 and concerned the casting of non-ferrous metals and their alloys (Wróbel, 2010). In order to obtain the movement of the liquid metal in the crystallizer in the researches carried out at this period of time and also in the future, a physical factor in the form of a electromagnetic field defined as a system of two fields i.e. an electric and magnetic field was introduced. The mutual relationship between these fields are described by the Maxwell equations (Sikora, 1998).

Generated by the induction coil powered by electric current intensity (I_0) electromagnetic field affects the solidifying metal induces a local electromotive force (E_m), whose value depends on the local velocity of the liquid metal (V) and magnetic induction (B) (Gillon, 2000).:

$$E_m = \overline{V} \times \overline{B} \tag{5}$$

This is a consequence of the intersection of the magnetic field lines with the current guide in form of liquid metal. It also leads to inducing an eddy current of intensity (I) in liquid metal (Gillon, 2000; Vives & Ricou, 1985):

$$\overline{I} = \sigma\left(\overline{V} \times \overline{B}\right) \tag{6}$$

where:

σ - electrical conductivity proper to the liquid metal.

The influence of the induced current on the magnetic field results in establishing of the Lorenz (magnetohydrodynamic) force (F) (Gillon, 2000; Vives & Ricou, 1985):

$$\overline{F} = \overline{I} \times \overline{B} \tag{7}$$

that puts liquid metal in motion e.g. rotary motion in the direction consistent with the direction of rotation of the magnetic field. Strength (F) has a maximum value when the vector (V) and (B) are perpendicular (Fig.12).

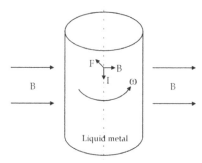

Fig. 12. Scheme of electromagnetic field influence on the liquid metal

In addition, as presented in the paper (Szajnar, 2009) the rotating velocity of the liquid metal (V) is inversely proportional to the density of the metal (ρ), because with some approximation we can say that (Fig.13):

$$\overline{V} \approx \frac{\overline{F}}{\rho} \, or \, \frac{\overline{B}}{\rho} \qquad (8)$$

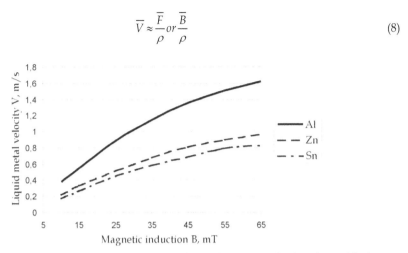

Fig. 13. Dependence of peripheral velocity of liquid metal (V) in a cylindrical mould of inside diameter 45mm on magnetic induction (B) for example metals (Szajnar, 2009)

Forced liquid metal movement influences by diversified way on changes in structure of casting i.e. by changes of thermal and concentration conditions on crystallization front, which decrease or completely stops the velocity of columnar crystals growth (Szajnar, 2004, 2009) and by (Campanella et al., 2004; Doherty et al., 1984; Fraś, 2003; Ohno, 1976; Szajnar, 2004, 2009; Szajnar & Wróbel, 2008a, 2008b, 2009; Wróbel, 2010):

- tear off of crystals from mould wall, which are transferred into metal bath, where they can convert in equiaxed crystals,
- fragmentation of dendrites by coagulation and melting as result of influences of temperature fluctuation and breaking as a result of energy of liquid metal movement,
- crystals transport from the free surface to inside the liquid metal,
- crystals from over-cooled outside layer of the bath are transported into liquid metal.

One of the hypotheses regarding the mechanism of dendrites fragmentation caused by the energy of the movement of liquid metal is presented in work (Doherty et al., 1984). It is based on the assumption of high plasticity of growing dendrites in the liquid metal, which in an initial state are a single crystal with specified crystallographic orientation (Fig.14a). The result of liquid metal movement is deformation (bending) of plastic dendrite (Fig.14b), which caused creation of crystallographic misorientation angle Θ (Fig.14c). Created high-angle grain boundary ($\Theta > 20°$) has the energy γ_{GZ} much greater than double interfacial energy of solid phase - liquid phase γ_{S-L}. In result of unbalancing and satisfying the dependence $\gamma_{GZ} > 2 \gamma_{S-L}$ the grain boundary is replaced by a thin layer of liquid metal. This leads to dendrite shear by liquid metal along the former grain boundary (Fig.14d). Dendrite fragments of suitable size after moving into the metal bath can transform into equiaxed crystals.

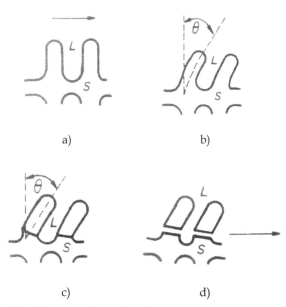

a)   b)

c)   d)

Fig. 14. Schematic model of the grain boundary fragmentation mechanism: a – an undeformed dendrite, b – after bending, c – the reorganization of the lattice bending to give grain boundaries, d – for $\gamma_{GZ} > 2\ \gamma_{S-L}$ the grain boundaries have been "wetted" by the liquid phase (Doherty et al., 1984).

The influence of electromagnetic field on liquid metal in aim of structure refinement (Fig.15), axial and zonal porosity elimination and obtaining larger homogeneity of structure, was applied in permanent mould casting (Griffiths & McCartney, 1997; Szajnar & Wróbel, 2008a, 2008b, 2009; Wróbel, 2010) and mainly in technologies of continuous (Adamczyk, 2004; Gillon, 2000; Harada, 1998; Miyazawa, 2001; Szajnar et al., 2010; Vives & Ricou, 1985) and semi-continuous casting (Guo et al., 2005).

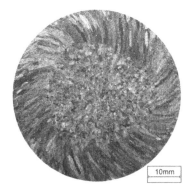

Fig. 15. Macrostructure of ingot of Al with a purity of 99,5% after cast with influence of rotating electromagnetic field

In case of continuous ingots of square and circular transverse section, rotating electromagnetic field induction coils are used. Rotating electromagnetic field forces rotational movement of liquid metal in perpendicular planes to ingot axis (Fig.16a). Whereas, mainly for flat ingots, longitudinal electromagnetic field induction coils are used, which forced oscillatory movement of liquid metal in parallel planes to ingot axis (Fig.16b) (Adamczyk, 2004; Miyazawa, 2001).

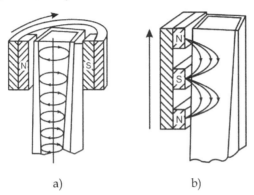

a)   b)

Fig. 16. Scheme of an electromagnetic stirrer (induction coil) forced rotational (a) and oscillatory movement of liquid metal (Adamczyk, 2004)

Whereas the authors of paper (Szajnar & Wróbel, 2008a, 2008b) suggests the use of reversion in the direction of electromagnetic field rotation during permanent mould casting. The advantage of casting in rotating electromagnetic field with reversion compared to casting in rotating field, mainly based on the fact that the liquid metal located in the permanent mould and put in rotary-reversible motion practically does not create a concave meniscus, and thus is not poured out off the mould under the influence of centrifugal forces. Moreover, the influence of this type of field combines impact of high amplitude and low frequency vibration with action of rotating electromagnetic field. Also important is double-sided bending of growing crystals, causing the creation in the columnar crystals zone of ingot characteristic crystals so-called corrugated (Fig.17).

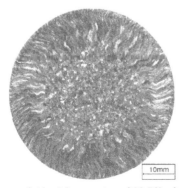

Fig. 17. Macrostructure of ingot of Al with a purity of 99,5% after cast with influence of rotating electromagnetic field with reversion

However in papers (Szajnar, 2004, 2009; Szajnar & Wróbel, 2008a, 2008b, 2009) was shown that influence of forced movement of liquid metal by use of electromagnetic field to changes in structure of pure metals, which solidify with flat crystallization front is insufficient. The effective influence of this forced convection requires a suitable, minimal concentration of additives i.e. alloy additions, inoculants or impurities in casting. Suitable increase of additives concentration causes at specified thermal conditions of solidification, occurs in change of morphology of crystallization front according to the scheme shown in Figure 18.

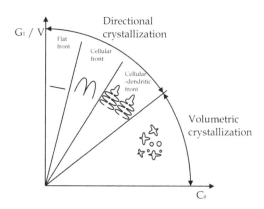

Fig. 18. Scheme of relationship between thermal and concentration conditions and type of crystallization; C_0 – concentration of additives, G_T – thermal gradient on crystallization front, V – velocity of crystallization (Fraś, 2003)

However it should be noted that, based on the latest results of author researches was affirmed that in some cases it is possible to obtain a sufficient refinement degree of pure aluminum structure in result of inoculation carried out only with the use of an electromagnetic field. Because it shows a possibility of increasing the force, which creates movement of liquid metal and in result of this the velocity of its rotation in mould, not only by increasing the value of magnetic induction according to the dependences (7) and (8), but also by increasing the frequency of the current supplied to the induction coil (Fig.19).

The effect of refinement of structure of Al with a purity of 99,5% caused by the rotating electromagnetic field produced by the induction coil supplied by current with frequency different from the network i.e. 50Hz is presented in Table 2. On the basis of macroscopic metallographic researches, which lead to the calculation of the equiaxed crystals zone content on transverse section of ingot (SKR) and average area of macro-grain in this zone (PKR) was affirmed, that application of frequency of supply current f ≤ 50Hz does not guarantee favourable transformation of pure aluminum structure (Fig.20 and 21). Whereas induction coil supplied with frequency of current larger than power network, mainly 100Hz generates rotating electromagnetic field, which guarantees favourable refinement of structure, also in comparison to obtained after inoculation with small, acceptable by European Standards amount of Ti and B i.e. 25 and 5ppm (Tab.2).

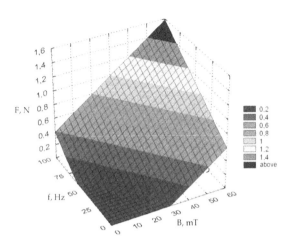

Fig. 19. The influence of magnetic induction (B) and frequency (f) of the current supplied to the induction coil on force value (F), which creates movement of liquid metal

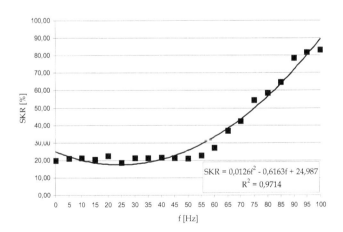

$$SKR = 0,0126f^2 - 0,6163f + 24,987$$
$$R^2 = 0,9714$$

Fig. 20. The influence of current frequency (f) supplied to the induction coil on equiaxed crystals zone content (SKR) on transverse section of pure Al ingot

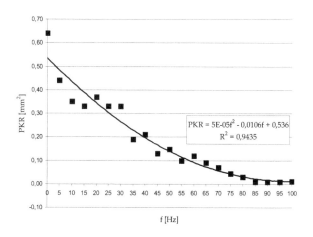

Fig. 21. The influence of current frequency (f) supplied to the induction coil on average area of equiaxed crystal (PKR) of pure Al ingot

No.	Cast parameters			Refinement parameters		Macrostructure of ingot
	B [mT]	f [Hz]	(Ti+B) [ppm]	SKR [%]	PKR [mm²]	
-1-	-2-	-3-	-4-	-5-	-6-	-7-
1	-	-	-	19,94	0,64	
2	-	-	25+5	80,30	0,42	

No.	Cast parameters			Refinement parameters		Macrostructure of ingot
	B [mT]	f [Hz]	(Ti+B) [ppm]	SKR [%]	PKR [mm²]	
-1-	-2-	-3-	-4-	-5-	-6-	-7-
3		5		21,01	0,44	10mm
4	60	10	–	21,36	0,35	10mm
5		15		20,66	0,33	10mm
6	60	20	–	22,63	0,37	10mm

No.	Cast parameters			Refinement parameters		Macrostructure of ingot
	B [mT]	f [Hz]	(Ti+B) [ppm]	SKR [%]	PKR [mm²]	
-1-	-2-	-3-	-4-	-5-	-6-	-7-
7		25		18,90	0,33	10mm
8		30		21,42	0,33	10mm
9		35		21,44	0,19	10mm
10		40		21,68	0,21	10mm

No.	Cast parameters			Refinement parameters		Macrostructure of ingot
	B [mT]	f [Hz]	(Ti+B) [ppm]	SKR [%]	PKR [mm²]	
-1-	-2-	-3-	-4-	-5-	-6-	-7-
11	60	45	-	21,46	0,13	10mm
12		50		21,21	0,15	10mm
13		55		22,87	0,10	10mm
14		60		27,22	0,12	10mm

No.	Cast parameters			Refinement parameters		Macrostructure of ingot
	B [mT]	f [Hz]	(Ti+B) [ppm]	SKR [%]	PKR [mm²]	
-1-	-2-	-3-	-4-	-5-	-6-	-7-
15		65		37,05	0,09	
16		70		42,53	0,07	
17	60	75	-	54,63	0,04	
18		80		58,56	0,03	

No.	Cast parameters			Refinement parameters		Macrostructure of ingot
	B [mT]	f [Hz]	(Ti+B) [ppm]	SKR [%]	PKR [mm²]	
-1-	-2-	-3-	-4-	-5-	-6-	-7-
19		85		64,70	0,01	
20		90		78,67	0,01	
21		95		81,78	0,01	
22	60	100	-	83,36	0,01	

Table 2. The influence of rotating electromagnetic field on structure of Al with a purity of 99,5%

4. The influence of exogenous inoculation on the result of endogenous inoculation in pure aluminum

In the practice of casting is also a problem of connection of endogenous inoculation i.e. realized by use of additives, for example Ti and B with exogenous inoculation i.e. realized by use of electromagnetic field. However as presented in papers (Szajnar & Wróbel, 2008a, 2008b, Wróbel, 2010) occurs that the phenomenon of convection transport (rejection) of impurities for example Cu and inoculants for example Ti from crystallization front into metal bath volume in result of intensive, forced by electromagnetic field the movement of liquid metal. This leads to an increase in density of bases to heterogeneous nucleation of aluminum and in consequence to increase in density of grains in the central area of ingot. Results of determination of Cu and Ti concentration in near-surface and central areas of investigated ingots with use of emission optical spectrometry is a proof of such reasoning. On their basis was affirmed, that in ingot of Al with a purity of 99,5% which was cast under the influence of electromagnetic field and with (Ti + B) inoculation, the Cu and Ti concentration in central area increase was observed (Fig.22a). Whereas in Al, which was cast only with (Ti + B) inoculation, the Cu and Ti concentrations in the near-surface and central areas of ingot are similar (Fig.22b).

The second proof of convection transport (rejection) of Cu and Ti from crystallization front into liquid metal volume is the analysis of macrostructure of investigated ingots and counting of all macro-grains in equiaxed crystals zone. Macrostructure of ingot of Al with a purity of 99,5%, which was cast with the combined effect of the electromagnetic field and with (Ti+B) inoculation has smaller equiaxed crystals zone than the ingot which was cast only with the influence of endogenous inoculation, but the first ingot has a smaller size of macro-grain in its equiaxed crystals zone than the ingot which was cast only with (Ti+B) inoculation (Fig.23).

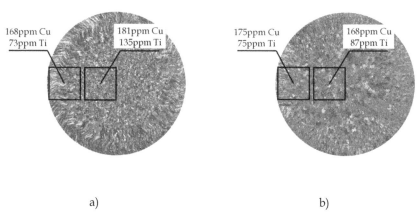

a)   b)

Fig. 22. Concentration of Cu and Ti in near-surface and central areas of Al with a purity of 99,5%ingots: a – after common exogenous (electromagnetic field) and endogenous (Ti + B) inoculation, b – only after endogenous (Ti + B) inoculation

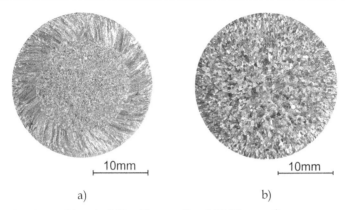

a)   b)

Fig. 23. Macrostructure of ingot of Al with a purity of 99,5%: a – after common exogenous (electromagnetic field) and endogenous (Ti + B) inoculation, b – only after endogenous (Ti + B) inoculation

Based on conducted calculations of number of macro-grains in equiaxed crystals zone following formula was formulated:

$$n_{ex+en} > n_{ex} + n_{en} \tag{9}$$

where:

n_{ex+en} – number of macro-grains in equiaxed crystals zone of ingot which was cast with common influence of exogenous (electromagnetic field) and endogenous (Ti + B) inoculation,

n_{ex} – number of macro-grains in equiaxed crystals zone of ingot which was cast only with influence of exogenous (electromagnetic field) inoculation,

n_{en} – number of macro-grains in equiaxed crystals zone of ingot which was cast only with influence of endogenous (Ti + B) inoculation.

Summarize, was affirmed that application of common exogenous (electromagnetic field) and endogenous (Ti + B) inoculation strengthens effect of structure refinement in comparison with application of one type of inoculation, only if is used skinning of ingot surface i.e. machining in aim of columnar crystals zone elimination.

5. Conclusions

In conclusion can say, that even endogenous inoculation with small amount of (Ti + B) strongly increase on refinement in pure aluminum structure. It results from reactions, which proceed between inoculating elements and inoculated metal or charge impurities. These reactions lead to formation of active bases to heterogeneous nucleation of aluminum as high melting small particles of type TiB, TiB$_2$, AlB$_2$, Al$_3$Ti and TiC or TiN, which have analogy in crystal lattice with Al.

However on the basis of conducted analysis of the literature and results of authors researches it was affirmed, that the rotating electromagnetic field generated by induction coil supplied by current with frequency larger than power network, influences liquid metal in time of its solidification in mould, guarantees refinement of structure of pure Al without necessity of application of inoculants sort Ti and B.

This method of exogenous inoculation is important, because Ti and B decrease the degree of purity and electrical conductivity of pure aluminum. Moreover Ti and B are reason of point cracks formation during rolling of ingots.

Presented method of inoculation by use of electromagnetic field is possible to apply in conditions of continuous casting because it allows producing of ingots from aluminum of approx. 99,5% purity with structure without columnar crystals, which are unfavourable from point of view of usable properties.

6. Acknowledgment

Project financed from means of National Science Centre.

7. References

Abu-Dheir, N.; Khraisheh, M.; Saito, K. & Male, A. (2005). Silicon morphology modyfication in the eutectic Al-Si alloy using mechanical mold vibration. *Materials Science and Engineering:A*, Vol.393, No.1-2, (September 2004), pp. 109-117, ISSN 0921-5093

Adamczyk, J. (2004). *Engineering of metallic materials*, Publishers of Silesian University of Technology, ISBN 83-7335-223-6, Gliwice, Poland

Asai, S. (2000). Recent development and prospect of electromagnetic processing of materials. *Science and Technology of Advanced Materials*, Vol.1, No.4, (September 2000), pp. 191-200, ISSN 1468-6996

Barrett, C. (1952). *Structure of metals, Metallurgy and Matallurgical Engineering Series*, McGraw-Hill Book Co. Inc., New York, USA.

Campanella, T.; Charbon, C. & Rappaz, M. (2004). Grain refinement induced by electromagnetic stirring: a dendrite fragmentation criterion. *Metallurgical and Materials Transactions A*, Vol.35, No.10, (December 2003), pp. 3201-3210, ISSN 1073-5623

Chalmers, B. (1963). The structure of ingot. *Journal of the Australian Institute of Metals*, Vol.8, No.6, pp. 255-263, ISSN 0004-9352

Doherty, R.; Lee, H. & Feest, E. (1984). Microstructure of stir-cast metals. *Materials Science and Engineering*, Vol.65, (January 1984), pp. 181-189, ISSN 0025-5416

Donnay, J. & Ondik, H. (1973). *Crystal date – Determinative Tables*, NSRDS – Library of Congress, Washington, USA

Easton, M. & StJohn, D. (1999). Grain refinement of aluminum alloys: Part I. The nucleant and solute paradigms – a review of the literature. *Metallurgical and Materials Transactions A*, Vol.30, No.6, (February 1998), pp. 1613-1623, ISSN 1073-5623

Easton, M. & StJohn, D. (1999). Grain refinement of aluminum alloys: Part II. Confirmation of, and a mechanism for, the solute paradigm. *Metallurgical and Materials Transactions A*, Vol.30, No.6, (February 1998), pp. 1625-1633, ISSN 1073-5623

Fjellstedt, J.; Jarfors, A. & El-Benawy, T. (2001). Experimental investigation and thermodynamic assessment of the Al-rich side of the Al-B system. *Materials & Design*, Vol.22, No.6, (February 2000), pp. 443-449, ISSN 0261-3069

Fraś, E. (2003). *Crystallization of metals*, WNT, ISBN 83-204-2787-8, Warsaw, Poland

Gillon, P. (2000). Uses of intense d.c. magnetic fields in materials processing. *Materials Science and Engineering:A*, Vol.287, No.2, (August 2000), pp.146-152, ISSN 0921-5093

Griffiths, W. & McCartney, D. (1997). The effect of electromagnetic stirring on macrostructure and macrosegregation in the aluminium alloy 7150. *Materials Science and Engineering:A*, Vol.222, No.2, (May 1996), pp.140-148, ISSN 0921-5093

Guo, S.; Cui, J.; Le, Q. & Zhao, Z. (2005) The effect of alternating magnetic field on the process of semi-continuous casting for AZ91 billets. *Materials Letters*, Vol.59, No.14-15, (November 2004), pp. 1841-1844, ISSN 0167-577X

Guzowski, M.; Sigworth, G. & Sentner, D. (1987). The role of boron in the grain refinement of aluminum with titanium. *Metallurgical and Materials Transactions A*, Vol.18, No.5, (May 1985), pp. 603-619, ISSN 1073-5623

Harada, H.; Takeuchi, E.; Zeze, M. & Tanaka, H. (1998). MHD analysis in hydromagnetic casting process of clad steel slabs. *Applied Mathematical Modeling*, Vol.22, No.11, (July 1997), pp. 873-882, ISSN 0307-904X

Hu, B. & Li, H. (1998) Grain refinement of DIN226S alloy at lower titanium and boron addition levels, *Journal of Materials Processing Technology*, No.74, (October 1996), pp. 56-60, ISSN 0924-0136

Jura, S. (1968). *Modeling research of inoculation process in metals*, Publishers of Silesian University of Technology, Gliwice, Poland

Kashyap, K. & Chandrashekar, T. (2001). Effects and mechanism of grain refinement in aluminium alloys. *Bulletin of Materials Science*, Vol.24, No.4, (April 2001), pp. 345-353, ISSN 0250-4707

Keles, O. & Dundar, M. (2007). Aluminum foil: its typical quality problems and their causes. *Journal of Materials Processing Technology*, Vol. 186, No.1-3, (December 2006), pp. 125-137, ISSN 0924-0136

Li, H.; Sritharan, T.; Lam, Y. & Leng, N. (1997). Effects of processing parameters on the performance of Al grain refinement master alloy Al-Ti and Al-B in small ingots. *Journal of Materials Processing Technology*, Vol.66, No.1-3, (October 1995), pp. 253-257, ISSN 0924-0136

Li, P.; Kandalova, E. & Nikitin, V. (2005). Grain refining performance of Al-Ti master alloy with different microstructures. *Materials Letters*, Vol.59, No.6, (December 2004), pp. 723-727, ISSN 0167-577X

McCartney, D. (1988). Discussion of "The role of boron in the grain refinement of aluminum with titanium". *Metallurgical and Materials Transactions A*, Vol.19, No.2, (July 1987), pp. 385-387, ISSN 1073-5623

Miyazawa, K. (2001). Continuous casting of steels in Japan. *Science and Technology of Advanced Materials*, Vol.2, No.1, (June 1999), pp.59-65, ISSN 1468-6996

Murty, B.; Kori, S. & Chakraborty, M. (2002). Grain refinement of aluminium and its alloys by heterogeneous nucleation and alloying. *International Materials Reviews*, Vol.47, No.1, pp. 3-29, ISSN 1743-2804

Naglič, I.; Smolej, A. & Doberšek, M. (2008). Remelting of aluminium with the addition of AlTi5B1 and AlTi3C0.15 grain refiners. *Metalurgija*, Vol.47, No.2, (January 2007), pp. 115-118, ISSN 0543-5846

Ohno, A. (1976). *The solidification of metals*, Chijin Shokan Co. Ltd, Tokyo, Japan

Paul, J.; Exner, H. & Müller-Schwelling, D. (1982). Microstructure and mechanical properties of cast and heat-treated eutectic Al-Si alloys, *Metallkunde*, Vol.1, No.43, pp. 50-55, ISSN 0044-3093

Pietrowski, S. (2001). *High-silicon aluminum alloys*, Publishers of Technical University of Lodz, ISBN 83-7283-029-0, Łódź, Poland

Sikora, R. (1998). *Theory of electromagnetic field*, WNT, ISBN 83-204-2226-4, Warsaw, Poland

Sritharan, T. & Li, H. (1996). Optimizing the composition of master alloys for grain refining aluminium, *Scripta Materialia*, Vol.35, No.9, (February 1996), pp. 1053-1056, ISSN 1359-6462

Szajnar, J. (2004). The columnar crystals shape and castings structure cast in magnetic field. *Journal of Materials Processing Technology*, Vol.157-158, (December 2004), pp. 761-764, ISSN 0924-0136

Szajnar, J. (2009). *The influence of selected physical factors on the crystallization process and casting structure*, Archives of Foundry Engineering, ISBN 1897-3310, Katowice-Gliwice, Poland

Szajnar, J.; Stawarz, M.; Wróbel, T.; Sebzda, T.; Grzesik, B. & Stępień, M. (2010). Influence of continuous casting conditions on grey cast iron structure, *Archives of Materials Science and Engineering*, Vol.42, No.1, (January 2010), pp. 45-52, ISSN 1897-2764

Szajnar, J. & Wróbel, T. (2008). Inoculation of pure aluminium aided by electromagnetic field. *Archives of Foundry Engineering*, Vol.8, No.1, (July 2007), pp. 123-132, ISSN 1897-3310

Szajnar, J. & Wróbel, T. (2008). Inoculation of pure aluminum with an electromagnetic field, *Journal of Manufacturing Processes*, Vol.10, No.2, (September 2008), pp. 74-81, ISSN 1526-6125

Szajnar, J. & Wróbel, T. (2009). The use of electromagnetic field in the process of crystallization of castings, In: *Progress in the theory and practice of foundry*, Szajnar, J., pp. 399-418, Archives of Foundry Engineering, ISBN 978-83-929266-0-3, Katowice-Gliwice, Poland

Tensi, H. & Hörgerl, J. (1994) Metallographic studies to quality assessment of alloys Al-Si. *Metallkunde*, Vol.10, No.73, pp. 776-781, ISSN 0044-3093

Treitler, R. (2005). *Calculating the strenght of casting and extruded alloys aluminum - magnesium*, Universitätsverlag Karlsruhe, ISBN 3-937300-94-5, Karlsruhe, Germany

Vives, C. & Ricou, R. (1985). Experimental study of continuous electromagnetic casting of aluminum alloy. *Metallurgical and Materials Transactions B*, Vol.16, No.2, (July 1983), pp. 377-384, ISSN 1073-5615

Whitehead, A. (2000). Grain refiners (modifiers) of the Al-Ti-C type – their advantages and application. *Foundry Review*, Vol.50, No.5, pp. 179-182, ISSN 0033-2275

Wróbel, T. (2010). Inoculation of pure aluminum structure with use of electromagnetic field, In: *The tendency of optimization of production system in foundries*, Pietrowski, S., pp. 253-262, Archives of Foundry Engineering, ISBN 978-83-929266-1-0, Katowice-Gliwice, Poland

Zamkotowicz, Z.; Stuczński, T.; Augustyn, B.; Lech-Grega, M. & Wężyk, W. (2003). Sedimentation of intermetallic compounds in liquid aluminum alloys of type AlSiCu(Ti), In: *Nonferrous Metals – Science and Technology*, Bonderek Z., CCNS, pp. 77-82

7

Phosphoramidates: Molecular Packing and Hydrogen Bond Strength in Compounds Having a P(O)(N)n(O)3-n (n = 1, 2, 3) Skeleton

Mehrdad Pourayoubi[1], Fahimeh Sabbaghi[2],
Vladimir Divjakovic[3] and Atekeh Tarahhomi[1]

[1]*Department of Chemistry, Ferdowsi University of Mashhad, Mashhad,*
[2]*Department of Chemistry, Zanjan Branch, Islamic Azad University, Zanjan,*
[3]*Department of Physics, Faculty of Sciences, University of Novi Sad, Novi Sad,*
[1,2]*Iran*
[3]*Serbia*

1. Introduction

Compounds containing the $P(O)(N)_n(O)_{3-n}$ (n = 1, 2, 3), $P(O)(N)_m(O)_{2-m}X$ (m = 1, 2, X = C, Cl, F, S etc.) and $P(O)(O)_3$ moieties are among the well-studied inorganic compounds [an interested reader may find many examples of compounds with the mentioned skeletons through a CSD search, [1]]. *N,N,N',N',N'',N''*-hexamethyl phosphoric triamide (HMPA, Scheme 1) is an important polar aprotic solvent with a high-dielectric constant [2] and an excellent ligand for interaction with hard metal-cations [3].

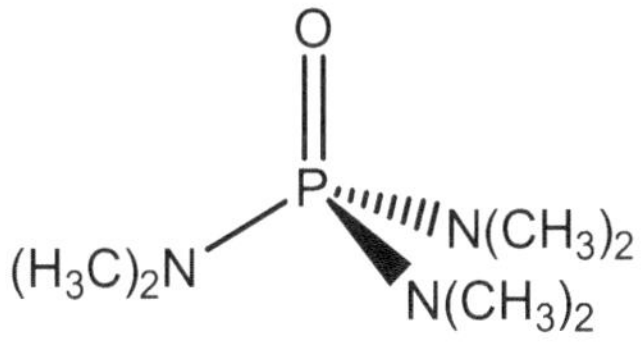

Scheme 1. *N,N,N',N',N'',N''*-hexamethyl phosphoric triamide

Tabun, $NCP(O)[N(CH_3)_2][OCH_2CH_3]$ (Scheme 2), Sarin, $FP(O)(CH_3)[OCH(CH_3)_2]$ and Soman, $CH_3P(O)(F)[OCH(CH_3)(C(CH_3)_3)]$ are among the well-known "nerve agents" that act as acetylcholinesterase enzyme (AChE) inhibitors in human body and mammals [4].

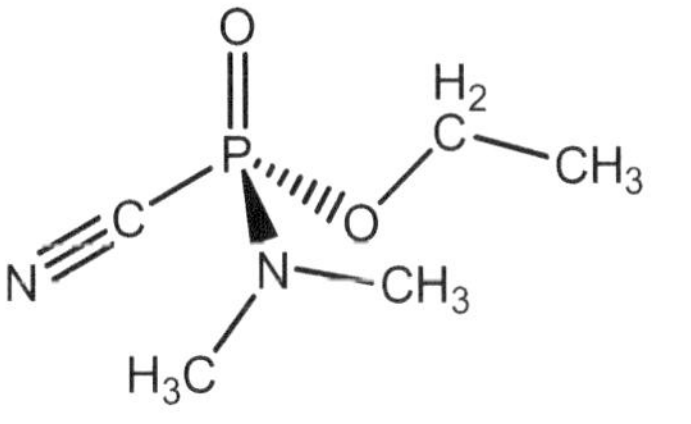

Scheme 2. Tabun, a nerve agent

Some researchers focus on decontamination of such compounds under UV-irradiation or in the presence of nano-oxides or nano-photocatalysts under sun-light [5]. The flame retardancy of some phosphoric esters was studied [6] and some phosphoramidates have therapeutic applications in the treatments of HIV and cancer [7]. Some pure chemists have interested to the NMR consideration [8], chemical calculation [9] and crystallography [10] of such compounds. A few bio-inorganic chemists have worked on the prediction of the biological properties of compounds based on their structures, with the related software programs such as PASS [11], and the evaluation of some relationships between structural features and biological activities [12]. In our laboratory, we centralize on the synthesis of new phosphorus-nitrogen and phosphorus-oxygen compounds and on obtaining their suitable single crystals for the X-ray crystallography experiments [13-64].

A schematic classification for the compounds having a $P(O)(N)_n(O)_{3-n}$ (n = 1, 2, 3) skeleton is shown in Scheme 3.

The numbers of the reported crystal structures in each family are presented in Scheme 3. The central box (blue) indicates the overall number of phosphoramidates having a $P(O)(N)_n(O)_{3-n}$ (n = 1, 2, 3) skeleton; the more well-studied categories of phosphoramidates are shown as green boxes in the top and bottom of the central box namely: a) phosphoric triamides (having a $P(O)(N)(N)(N)$ or $P(O)(NHC(O))(N)(N)$ fragment), and b) amidophosphoric acid esters (containing a $P(O)(O)(N)(N)$ or $P(O)(O)(O)(N)$ skeleton).

As a nitrogen bonded H atom is very important in the H-bond pattern consideration, in the sub-categories, the presence or the absence of this H atom is clarified. In the applied notation, for example, the $P(O)(NH)_3$ and $P(O)(N)_3$ denote to the presence of secondary and tertiary nitrogen atoms, respectively. The phosphoramidates containing a $P(O)NH_2$ moiety are distinguished in the left side box directly related to the central box.

The less-studied (so far) related compounds i.e. c) the proton-transfer and phosphate salts and the acids, and d) the anhydride compounds with a $P(O)(O)P(O)$ skeleton are shown in the right and the left of the central box.

In this flowchart, the skeletons of 643 compounds -which their crystal structures were deposited- have been collected. In this classification, the phosphoramidates containing the phosphorus-carbon and the phosphorus-halogen bonds have not been considered.

2. Synthesis and purification of phosphoramidates and phosphoric acid esters

The reaction of phosphorus(V)-halogen compounds of the type $P(O)X_{3-n}Y_n$ (X = halide, Y = another group such as amide, alkoxide and so on, and n = 0, 1 and 2) with primary or secondary amines leads to the formation of phosphorus(V)-nitrogen compounds. The promotion of this reaction needs to the presence of an excess amount of amine as an HX scavenger or the presence of another acid scavenger such as tertiary amines [61] or pyridine [8] (Scheme 4). In this strategy, removing of the hydrohalide salt of the organic base is a challenging task in the purification process.

The purification may be performed by stirring the crude product in water to remove the amine hydrohalide or pyridinium halide and/or may be done by selecting the solvent

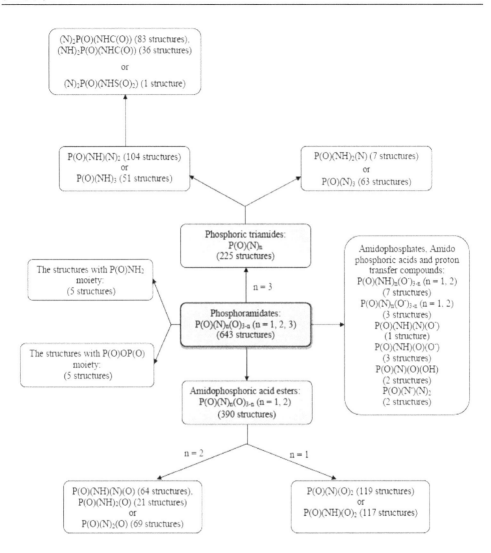

Scheme 3. The classification of compounds having a P(O)(N)$_n$(O)$_{3-n}$ (n = 1, 2, 3) skeleton

which the salt is as precipitate (and the product is soluble) and then the filtering off the salt. Moreover, if more than twice mole ratio of amine relative to each P-X bond is used, removing the un-reacted amine should be done in the purification process, too, which may be performed by stirring the crude product in a diluted hydrochloric acid [65].

Setzer and co-workers reported the synthesis of 1,3,2-oxazaphospholane from the reaction between (1R,2R)-(-)-pseudoephedrine, phenyl dichlorophosphate and triethylamine in ethyl acetate. Triethylamine hydrochloride was filtered off and the solvent removed from the filtrate under reduced pressure [66].

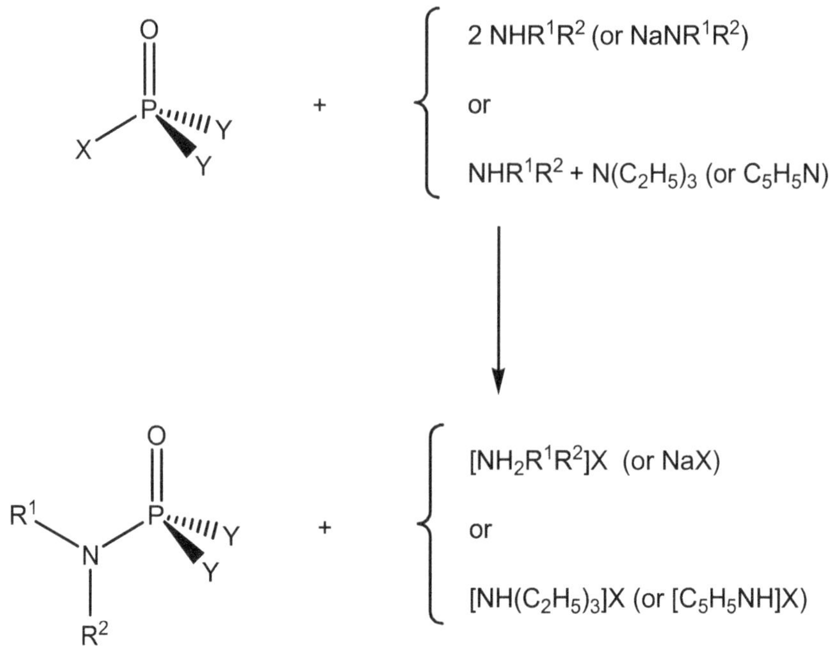

Scheme 4. A common route for the synthesis of phosphoramidates

This method may be developed to the reaction between phenols and phosphorus-chlorine compounds. Selecting of a suitable solvent, which triethylamine hydrochloride or the other salt is low-soluble, develops the synthesis of some initial phosphorus-chlorine compounds such as $[(CH_3)(C_6H_{11})N]P(O)Cl_2$ [67], $[4\text{-}CH_3C_6H_4NH]P(O)Cl_2$ [68], $[C_6H_5O][4\text{-}CH_3C_6H_4NH]P(O)Cl$ [69] and so on. For example, *para*-toluidine hydrochloride is relatively insoluble in CH_3CN; so, the reaction of $P(O)Cl_3$ or $[C_6H_5O]P(O)Cl_2$ with $4\text{-}CH_3\text{-}C_6H_4NH_2$ (1:2 mole ratio) respectively leads to the formation of $[4\text{-}CH_3C_6H_4NH]P(O)Cl_2$ [68] and $[C_6H_5O][4\text{-}CH_3C_6H_4NH]P(O)Cl$ [69] which are soluble in acetonitrile, whereas *para*-toluidine hydrochloride is simply filtered off.

Selection of a suitable solvent for such reactions leads to avoid from the time tedious purification methods such as column chromatography. Recently, we are developing this simple strategy for the synthesis of new phosphorus-chlorine compounds such as $[C_6H_5O]P(O)[NHC_6H_{11}]Cl$, $CF_3C(O)NHP(O)[NHC_6H_4(4\text{-}CH_3)]Cl$, $[C_6H_{11}NH]P(O)Cl_2$ and $[(C_6H_5CH_2)_2N]P(O)Cl_2$ [70].

With starting from $P(O)Cl_3$ or PCl_5 as initial phosphorus-chlorine compounds to reaction with an amine, surely a dry solvent is needed. A fully de-watered solvent is obtained by refluxing of a relatively dry solvent in the presence of a very efficient drying agent such as P_2O_5 (for CCl_4 and $CHCl_3$) or sodium (for CH_3OH, C_2H_5OH, C_6H_6 and $C_6H_5CH_3$) and distilling the totally dried solvent. However, it seems that the sensitivity of a $YP(O)Cl_2$ starting material (Y = amide, alkoxy, phenoxy and so on) is very reduced to the moisture and the solvent which was dried with a moderate desiccant (such as $CaCl_2$) is good for the synthesis. For a bulky amine such as *iso*-propylbenzyl amine or di-cyclohexyl amine as a nucleophile, it seems that a totally-dried solvent is better; of course, it needs to approve with further experiments.

In the case of *iso*-propylbenzyl amine as nucleophile, the reactions with [C$_6$H$_5$O]P(O)Cl$_2$, [C$_6$H$_5$O]$_2$P(O)Cl or 4-F-C$_6$H$_4$C(O)NHP(O)Cl$_2$ were not successful to prepare the pure [C$_6$H$_5$O]P(O)[N(CH(CH$_3$)$_2$)(CH$_2$C$_6$H$_5$)]$_2$, [C$_6$H$_5$O]$_2$P(O)[N(CH(CH$_3$)$_2$)(CH$_2$C$_6$H$_5$)] and 4-F-C$_6$H$_4$C(O)NHP(O)[N(CH(CH$_3$)$_2$)(CH$_2$C$_6$H$_5$)]$_2$; however, the crystal structures of two polymorphs of [NH$_2$(CH(CH$_3$)$_2$)(CH$_2$C$_6$H$_5$)]Cl were obtained [71,72]. With using this amine, the compounds [4-NO$_2$-C$_6$H$_4$C(O)NH]P(O)[N(CH(CH$_3$)$_2$)(CH$_2$C$_6$H$_5$)]$_2$ [51], [NH$_2$(CH(CH$_3$)$_2$)(CH$_2$C$_6$H$_5$)][CCl$_3$C(O)NHP(O)(O)[OCH$_3$]] [40] and [NH$_2$(CH(CH$_3$)$_2$)(CH$_2$C$_6$H$_5$)][CF$_3$C(O)NHP(O)(O)(N(CH(CH$_3$)$_2$)(CH$_2$C$_6$H$_5$))] [73] were prepared which structurally studied, too.

In the case of [NH$_2$(CH(CH$_3$)$_2$)(CH$_2$C$_6$H$_5$)][CCl$_3$C(O)NHP(O)(O)[OCH$_3$]] salt, for example, it seems that the presence of a few amount of H$_2$O in solvent (or environment) leads to the formation of CCl$_3$C(O)NHP(O)(OH)Cl which the proton-transfer reaction with the amine produces [NH$_2$(CH(CH$_3$)$_2$)(CH$_2$C$_6$H$_5$)][CCl$_3$C(O)NHP(O)(O)Cl] and then crystallization in methanol replaces the Cl with OCH$_3$. Moreover, from the reaction of P(O)(OC$_6$H$_5$)Cl$_2$ and NH(C$_6$H$_{11}$)$_2$, the related pure amido phosphoric acid ester was not achieved; however, the crystal structure of [(C$_6$H$_{11}$)$_2$NH$_2$]$^+$Cl$^-$ was obtained [74].

We are going to try to synthesize neutral phosphoramidate compounds with this and the other bulky amines. A similar feature was observed for the reaction of POCl$_3$ with *tert*-butyl cyclohexyl amine in CHCl$_3$ under reflux condition, where the salt [NH$_2$(*tert*-C$_4$H$_9$)(C$_6$H$_{11}$)][PO$_2$Cl$_2$] was obtained [75].

The moisture led to the formation of some undesirable but interesting products such as X$_2$P(O)OP(O)X$_2$ (X = (CH$_3$)$_3$CNH [76], C$_6$H$_4$(2-CH$_3$)NH [48] and C$_6$H$_4$(4-CH$_3$)NH [77]) from the reaction of P(O)Cl$_3$ and corresponding amine (1 to 6 or more mole ratio) , and also formation of [3-F-C$_6$H$_4$C(O)NH][(CH$_3$)$_3$CNH]P(O)(O)P(O)[NHC(CH$_3$)$_3$][NHC(O)C$_6$H$_4$(3-F)] [78] from the reaction of 3-F-C$_6$H$_4$C(O)NHP(O)Cl$_2$ and *tert*-butyl amine. Another salt, [*tert*-C$_4$H$_9$NH$_2$][CF$_3$C(O)NHP(O)(O)NH(*tert*-C$_4$H$_9$)].0.333CH$_3$CN.0.333H$_2$O, was also obtained [79].

N-methyl cyclohexyl amine showed an interesting feature in some examples which may be accidental needing to further considerations. In the reaction of 4-CH$_3$C$_6$H$_4$S(O)$_2$NHP(O)Cl$_2$ with an excess amount of NH(CH$_3$)(C$_6$H$_{11}$) (1:5 mole ratio), the product is a proton-transfer compound, [NH$_2$(CH$_3$)(C$_6$H$_{11}$)][4-CH$_3$-C$_6$H$_4$S(O)$_2$NP(O)[N(CH$_3$)(C$_6$H$_{11}$)]$_2$] [23]; furthermore, the crystal structures of [NH$_2$(CH$_3$)(C$_6$H$_{11}$)][CF$_3$C(O)NP(O)[N(CH$_3$)(C$_6$H$_{11}$)]$_2$] [80] and [NH$_2$(CH$_3$)(C$_6$H$_{11}$)][CCl$_3$C(O)NP(O)[N(CH$_3$)(C$_6$H$_{11}$)]$_2$] [81] were obtained, but in an effort to preparation of their alkaline complexes.

Synthesis of such proton-transfer compounds through stirring a mixture of a few examined amines (NHR^{1}R^2) and a synthesized phosphoric triamide (CF$_3$C(O)NHP(O)[NR^{1}R^2]$_2$) were not successful; however, this also needs to some further experiments.

Another strategy for preparation of phosphoramidate compounds is the application of sodium amide salts which produces sodium halide as a by-product [82] (Scheme 4).

The P-N bond formation between an amide, of the type RC(O)NH$_2$, and a phosphorus(V) site may be performed *via* a two-stages reaction, which is shown for PCl$_5$ in Scheme 5, showing the reaction of PCl$_5$ with an amide and then the treatment of HCOOH. Moreover, a few efforts have been devoted to the synthesis of RS(O)$_2$NHP(O)Cl$_2$ by a similar procedure [23].

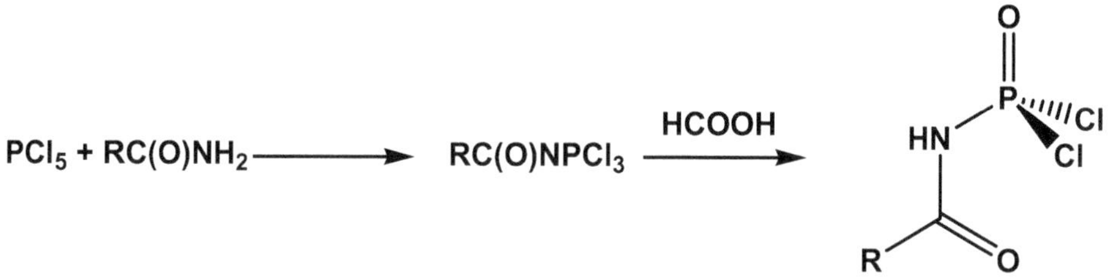

Scheme 5. Synthesis of RC(O)NHP(O)Cl$_2$

The simple mentioned methods for the preparation of phosphoramidates from the reaction of phosphorus-chlorine compounds and amines may be extended to the diamines or amino alcohols to produce cyclic [83] or bridged compounds [61].

The preparation of some compounds containing the P-Cl bonds, such as 4-CH$_3$C$_6$H$_4$OP(O)Cl$_2$ and [(CH$_3$)$_2$N]P(O)Cl$_2$, were performed through the reaction between corresponding phenol derivatives or amine hydrochloride salts [for example *para*-cresol or dimethylamine hydrochloride salt for the mentioned phosphorus-chlorine compounds] with an excess amount of POCl$_3$ and then the removal of the remaining POCl$_3$ in a reduced pressure [32,47].

Some compounds were synthesized by the reaction of P-H compounds (such as dimethylphosphine oxide, (CH$_3$)$_2$P(O)H and *N,N*-disubstituted derivatives of 5,6-benzo-2H-2-oxo-1,3,2λ^4-diazaphosphorinan-4-one, Scheme 6) with ketones [84].

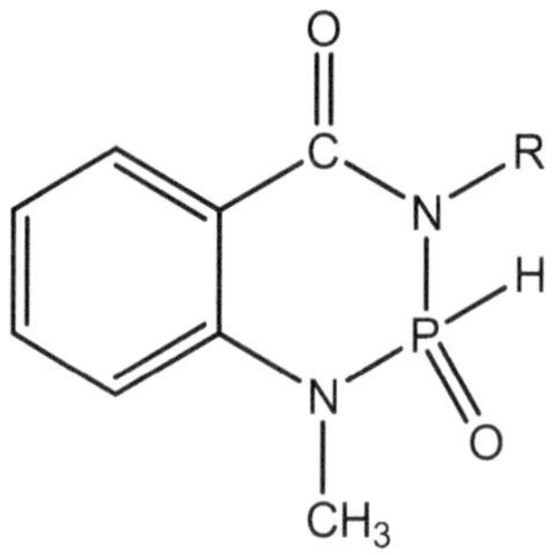

Scheme 6. 5,6-benzo-2H-2-oxo-1,3,2λ^4-diazaphosphorinan-4-one

Wan and Modro developed the synthesis of a bicyclic phosphoric triamide (Scheme 7) *via* the base-promoted cyclization of the corresponding 3-(2-chloroethyl)-2-oxo-1-aryl-2-arylamino-1,3,2-diazaphospholidine [85].

Mbianda and co-workers reported the solvolysis of 1-oxo-2,8-diphenyl-2,5,8-triaza-1 λ^5-phosphabicyclo[3.3.0]octane under base-promoted alcoholysis and acid-catalyzed alcoholysis. Scheme 8 shows the two different products of such solvolysis reactions [86].

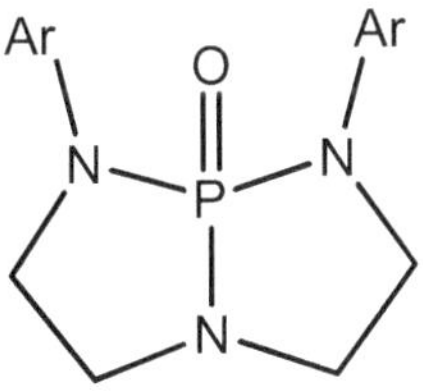

Scheme 7. A bicyclic phosphoric triamide

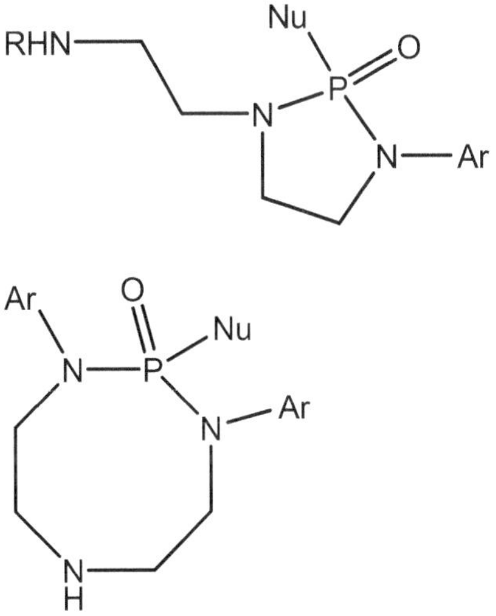

Base-promoted alcoholysis

Acid-catalyzed alcoholysis

Scheme 8. The obtained products from the solvolysis of 1-oxo-2,8-diphenyl-2,5,8-triaza-1 λ^5-phosphabicyclo[3.3.0]octane under base-promoted alcoholysis and acid-catalyzed alcoholysis

3. Crystallization of phosphoramidates

The convenient solvents for obtaining suitable single crystals for the studied compounds may be $CH_3C(O)CH_3$, $CHCl_3$, $CHCl_3/n-C_7H_{16}$, CH_2Cl_2, CH_3CN, CH_3CN/CH_3OH, $CH_3CN/CHCl_3$, CH_3OH, CH_3OH/H_2O, $C_2H_5OH/n-C_6H_{14}$, $(CH_3)_2CHOH/n-C_6H_{14}$, $(CH_3)_2NC(O)H/CHCl_3$, $(CH_3)_2NC(O)H/CH_3OH$ and $n-C_6H_{14}$. The crystal may be obtained at room temperature after slow evaporation of the solvent.

4. General features of phosphoramidate compounds

Compounds with formula RC(O)NHP(O)[NR¹R²]₂ and RC(O)NHP(O)[NHR¹]₂

The four different groups linked to the P atom result in a distorted tetrahedral configuration; as one instance, the bond angles around the P atom of $P(O)[NHC(O)CF_3][NHCH_2C_6H_4(2-Cl)]_2$ range from 102.67(12)° to 117.60(12)° [87]. In the

C(O)NHP(O) moiety, however, the carbonyl and phosphoryl groups are separated from each other with one N-H unit, but the terms *syn*, *gauche* and *anti* were used for the description of the C(O) orientation versus P(O) in the literatures [10,26]. Up to now, both *gauche* and *anti* orientations were found for phosphoric triamide compounds having a C(O)NHP(O) skeleton, respectively in 14 and 98 structures (from the 119 structurally reported compounds, 7 cifs are not available). Among them, for the acyclic compounds of the type $RC(O)NHP(O)[NHR^1]_2$, merely, the *anti* orientation was reported, so far.

Fig. 1. indicates a general view of compounds with formula $CF_3C(O)NHP(O)[NHR]_2$.

In the C(O)NHP(O) moiety, the P−N bond is longer and the O−P−N angle is contracted compared with the respective values in the $[P(O)NHR]_2$ section. For the phosphoramidate compounds, each N atom bonded to phosphorus has a sp^2 character which is reflected in the C−N−P angles of the C(O)NHP(O) or C−NH−P moiety or sum of the surrounding angles around the tertiary nitrogen atom (C−N−C + C−N−P + P−N−C). The deviation of this summation from 360° (to a lower value) has been used to show the deviation of nitrogen atom environment from planarity. This may be also illustrated with the distance between the position of N atom from the plane crossing from the directly attached atoms to nitrogen, i.e. C, C and P. In $[C_6H_5O]_2P(O)[NC_4H_8N]P(O)[OC_6H_5]_2$ (Fig. 2) which belongs to the amidophosphoric acid ester family, the N atom shows some deviation from planarity and it is 0.25(1) Å above (or below) the CCP plane [61]. For the phosphoramidate compounds, the P-N bonds are shorter than the P-N single bond and the P=O bond are longer than the normal P=O bond [88].

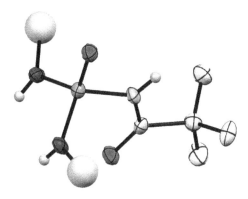

Fig. 1. A typical view for a compound with formula $CF_3C(O)NHP(O)[NHR]_2$ [Color key: O atoms are red, the N atom of C(O)NHP(O) is light blue, the other amido N atoms are dark blue, F atoms are yellowish green., C and H atoms are light grey and P atom is orange]; the R substituents are shown as big balls.

In 1,3-diazaphosphorinane compounds, the P=O bond is placed in an equatorial position and the aliphatic six-membered rings adopt conformation between chair and envelope (Fig. 3).

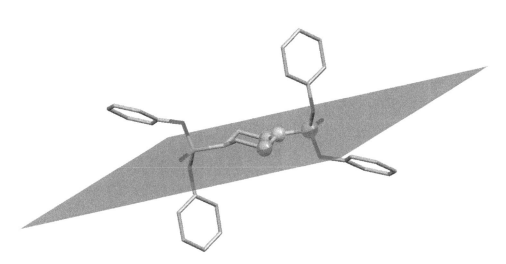

Fig. 2. A view of the PCC mean plane which is crossed from the phosphorus and carbon atoms shown as balls in the right-side [C$_6$H$_5$O]$_2$P(O)(NC$_2$H$_4$) moiety of [C$_6$H$_5$O]$_2$P(O)[NC$_4$H$_8$N]P(O)[OC$_6$H$_5$]$_2$ (the molecule is organized around an inversion center located at the centre of the piperazine ring), the N atom environment shows some deviation from planarity. The balls representation denote to the P (orange), N (blue) and C (grey) atoms.

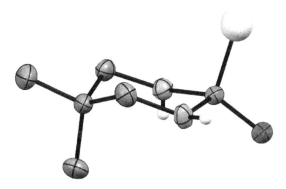

Fig. 3. A general view of a 1,3-diazaphosphorinane, a six-membered ring heterocyclic phosphorus compound (the carbon-bonded H atoms were omitted for clarity). The grey big ball in the figure may be RC(O)NH, RNH or the other moieties.

The hydrogen bond pattern of compounds having the C(O)NHP(O)(N)$_2$ and C(O)NHP(O)(NH)$_2$ skeletons may be predictable with considering the following "empirical rules":

1. In the reported compounds, the nitrogen atoms bonded to P don't involve in hydrogen bonding interaction as an acceptor (due to their low Lewis base characteristic). Scheme 9 illustrates the possible H-donor sites and H-acceptor centers in the structure of compounds having the C(O)NHP(O)(N)$_2$ and C(O)NHP(O)(NH)$_2$ skeletons.

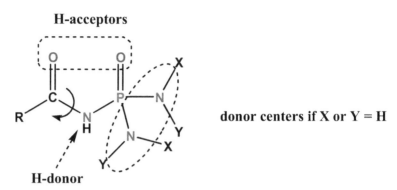

H-acceptors

donor centers if X or Y = H

H-donor

Usually the N atoms don't involve in hydrogen bonding as an acceptor

Scheme 9. The possible H-donor sites and H-acceptor centers in the C(O)NHP(O)(N)$_2$ and C(O)NHP(O)(NH)$_2$ skeletons (the R, X or Y groups may also be contained the additional H-donor or H-acceptor sites in their structures which may be involved in the H-bond pattern, the curved arrow shows that the orientation of C=O versus P=O may change)

2. The P=O is a better H-acceptor than the C=O counterpart.
3. In compounds having a C(O)NHP(O)(N)$_2$ skeleton, i.e. with formula RC(O)NHP(O)[NR'R"]$_2$, both *gauche* and *anti* orientations of P=O *versus* C=O have been found, so, two kinds of packing are expectable which are seen: a) a 1-D chain for a *gauche* orientation, and b) a dimeric aggregate (as an R$_2^2$(8) loop; for H-bond motifs of phosphoric triamide, see: ref. [10]) with C$_i$ or C$_1$ symmetry for *anti*. The unique NH proton interacts with the oxygen atom of PO, whereas the CO does not cooperate in HB. Such H-bond patterns may also be expectable for the other phosphoramidate compounds having a P(O)NH group, Scheme 10; however, the other H-bond patterns have also been observed in the other sub-categories with a P(O)NH moiety which will be noted, later.
4. In compounds having a C(O)NHP(O)(NH)$_2$ skeleton, only an *anti* situation has been found in acyclic molecules; however, in diazaphosphorinane molecules both conformations were found.
5. In the crystal packing of acyclic compounds having a C(O)NHP(O)(NH)$_2$ skeleton, adjacent molecules are often linked *via* N$_{C(O)NHP(O)}$−H...O=P and N−H...O=C (or (N−H)$_2$...O=C) hydrogen bonds, building R$_2^2$(8) and R$_2^2$(12) rings (or R$_2^2$(8) and R$_2^2$(12)/R$_2^1$(6)) in a linear arrangement (Scheme 11). However, the existence of a PO...HNR interaction has been observed for a few compounds as tri-centered PO[...H$_{C(O)NHP(O)}$N][...HNR] and PO[...HNR][...HNR] hydrogen bonding, where the oxygen atom of the phosphoryl group acts as a double H-acceptor (for a definition of a double-H bond acceptor, see: ref. [89]).

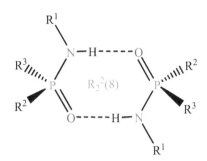

R^1 = alkyl, aryl or C(O)R'
R^2 and R^3 = OR, NR'R"

R^1 = alkyl, aryl or C(O)R, R^2 = OR or NR'R", R^3 = OR, NHR' or NR"R'''

Scheme 10. The observed hydrogen bond patterns in compounds having a P(O)NH skeleton (such as C(O)NHP(O)[N]₂)

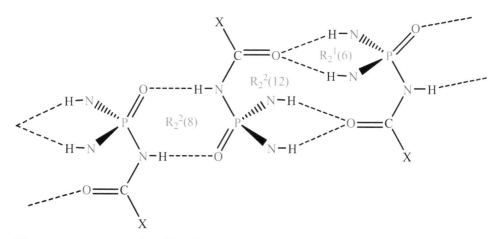

Scheme 11. A sequence of $R_2^2(8)$ and $R_2^2(12)$ (top), a sequence of $R_2^2(8)$ and $R_2^2(12)/R_2^1(6)$ (bottom) rings in compounds having a C(O)NHP(O)(NH)$_2$ skeleton: in these H-bond patterns, the P=O...H-N$_{C(O)NHP(O)}$ and C=O...H-N or C=O...(H-N)$_2$ exist

In most cases of compounds having a C(O)NHP(O)(NH)$_2$ skeleton (containing two H-acceptors–three H-donors), the HBs lead to a 1-D chain. Different 1-D ladder arrangements with tetramer motifs and a linear arrangement with two different kinds of motifs (dimer and tetramer) were also observed. Therefore, two H-donor sites (HN$_{C(O)NHP(O)}$ and one of the HNR) participate with two O atoms in the intermolecular HBs, the other HNR may act as the three following manners: (a) in an intramolecular HB with C(O), (b) in a weaker HB with P(O) as the above mentioned tri-centered HB and (c) without cooperation in any HB.

6. A sequence of $R_2^2(10)$ (or $R_2^2(10)/R_2^1(6)$ or $R_2^2(10)/S_6$) may be expected when the C=O is hydrogen-bonded to the N$_{C(O)NHP(O)}$-H unit and the P=O interacts with the N$_{amide}$-H unit. If the remaining N$_{amide}$-H unit is involved in an intramolecular hydrogen bond with the oxygen of carbonyl, the $R_2^2(10)/S_6$- graph-set is formed, in this case the oxygen atom of carbonyl acts as a double-H acceptor. In the case of involving this N-H unit in the H-bonding interaction with the oxygen of phosphoryl, the $R_2^2(10)/R_2^1(6)$ is formed (Scheme 12).

7. In a solvated molecule [90], the hydrogen-bond pattern may not be predictable, but some previously mentioned rules may be beneficial, Scheme 13.

8. The investigation for the phosphoric triamide containing a C(O)NHP(O)(NH)$_2$ skeleton shows that in eleven structures the carbonyl oxygen atom acts as a double-H acceptor *via* C(O)...(H−N)(H−N) grouping (in the $R_2^2(12)/R_2^1(6)$ motifs) and in ten structures the phosphoryl oxygen atom acts as a double-H acceptor *via* P(O)...(H−N)(H−N) or P(O)...(H−N)(H−N$_{C(O)NHP(O)}$) groups in the $R_2^2(12)/R_2^1(6)$ rings or $R_2^2(10)/R_2^1(6)$ ring or in the 1-D ladder arrangement). In the other such phosphoramidate compounds the remaining N−H unit doesn't cooperate in H-bonding interaction. The unique found structure of phosphoramidate with a C(O)NHP(O)Cl$_2$ skeleton is CF$_3$C(O)NHP(O)Cl$_2$ [91].

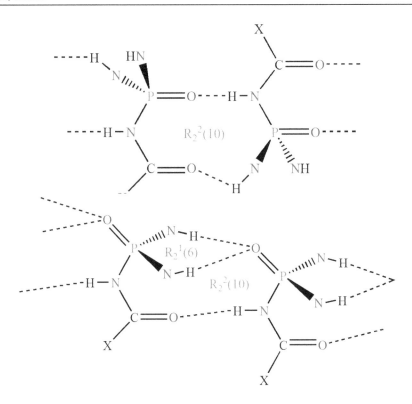

Scheme 12. A sequence of $R_2^2(10)$ (top) and $R_2^2(10)/R_2^1(6)$ rings (bottom) in compounds having a $C(O)NHP(O)(NH)_2$ skeleton; in the existing examples of $R_2^2(10)/S_6$, the N-H...O angle is less than 110°

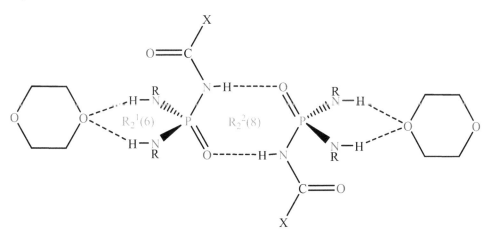

Scheme 13. A four-component cluster in the solvated molecule
$CCl_3C(O)NHP(O)[NHC_6H_{11}]_2 \cdot C_4H_8O_2 (X = CCl_3, R = C_6H_{11})$

Proton transfer compounds

In the proton transfer compounds containing a $\{[C(O)NP(O)][N]_2\}^-$ skeleton (Scheme 14), the P-N bond in the [C(O)NP(O)] fragment is shorter than the two other P-N bonds, one example for such compounds is $[C_6H_{11}NH_2CH_3][CF_3C(O)NP(O)[N(CH_3)(C_6H_{11})]_2]$ in which the phosphoryl and carbonyl groups are staggered [O-P-N-C = 64.8(3)°] [80].

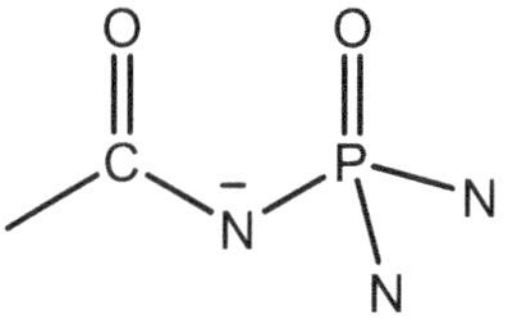

Scheme 14. The $\{[C(O)NP(O)][N]_2\}^-$ skeleton

The hydrogen bonds (Scheme 15) in such compounds, of the type charge-assisted and also polarization-assisted HBs [92], are strong, reflecting in the distances between the donor and acceptor atoms. Scheme 16 shows a polarization-assisted hydrogen bond in a neutral phosphoramidate, for comparison.

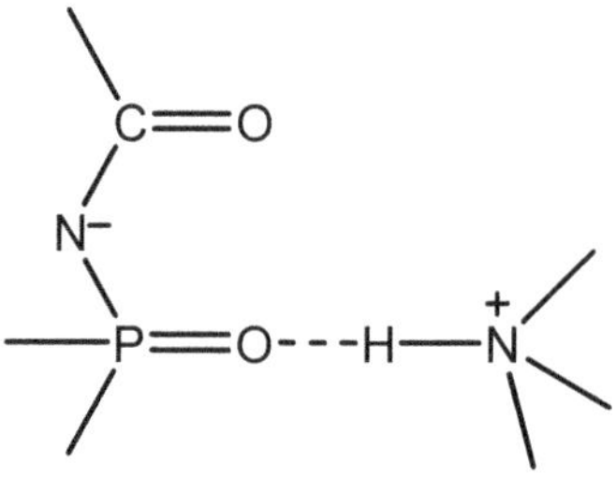

Scheme 15. Contribution of two factors in strengthening of hydrogen bonds (charge and polarization) in the proton-transfer compounds: charge and polarization-assisted hydrogen bonds

Two reported crystal structures of this category, $[C_6H_{11}NH_2CH_3][CF_3C(O)NP(O)[N(CH_3)(C_6H_{11})]_2]$ [80] and $[C_6H_{11}NH_2CH_3][4-CH_3-C_6H_4S(O)_2NP(O)[N(CH_3)(C_6H_{11})]_2]$ [20], have a similar HB pattern as a centrosymmetric four-component cluster involving two anions and two cations which interact through N−H…O hydrogen bonds (Fig. 4).

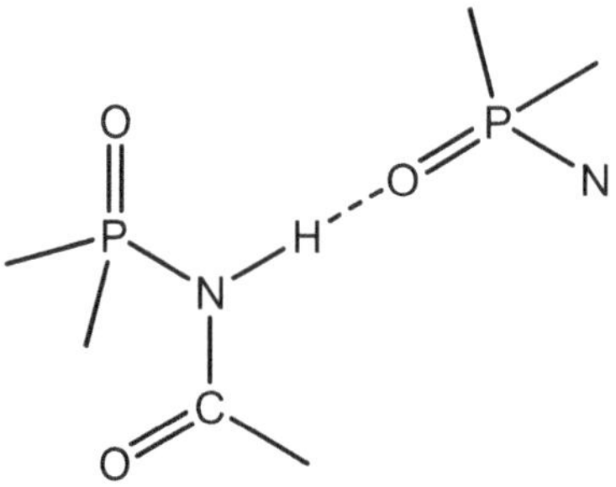

Scheme 16. A polarization-assisted hydrogen bond in a neutral phosphorus compound

Phosphoramidates: Molecular Packing and Hydrogen Bond Strength in Compounds Having a P(O)(N)n(O)3-n (n = 1, 2, 3) Skeleton

175

Compounds with formula [R^{1}R^{2}N][R^{3}R^{4}N][R^{5}R^{6}N]P(O), [RNH][R^{1}R^{2}N][R^{3}R^{4}N]P(O), [R^{1}NH][R^{2}NH][R^{3}R^{4}N]P(O), [R^{1}NH][R^{2}NH][R^{3}NH]P(O) or more complicated phosphoric triamide compounds

Tris-alkyl (aryl) amido phosphates of the formula [R^{1}R^{2}N]$_3$P(O), with three equal amido substituents linked to the P atom, are easily prepared from a one-pot reaction between phosphoryl chloride and corresponding amine. The single crystal X-ray determinations were performed for [R^{1}R^{2}N] = NHCH$_3$ (KABVAL) [93], N(CH$_3$)$_2$ (POTJAJ) [94], NHC(CH$_3$)$_3$ (KABVEP) [93], NHC$_6$H$_5$ (KEQLUO) [95], NHCH$_2$C$_6$H$_5$ (TOKXIB) [96] and NHC$_6$H$_4$(4-OCH$_3$) (WAWNIS) [97] and also for the substituents shown in Scheme 17 [98-105]. Moreover, a few other phosphoramide compounds which each contains a triamido moiety (like for example compounds with refcodes EDEVAK [106] and NUVSEC [107] (Schemes 18 and 19) and some co-crystal compounds (for example BARHMP [108] and VAFRIE [109], see Schemes 20 and 21) were reported.

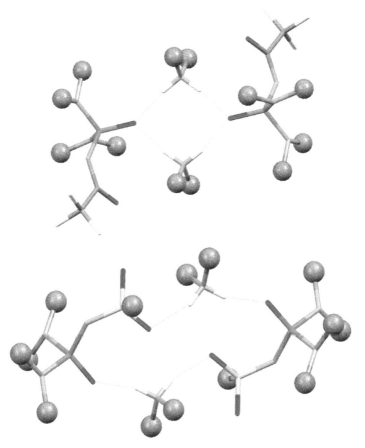

Fig. 4. A view of the H-bonded centrosymmetric four-component cluster in the crystal packing of {C$_6$H$_{11}$NH$_2$CH$_3$}$^+${CF$_3$C(O)NP(O)[N(CH$_3$)(C$_6$H$_{11}$)]$_2$}$^-$ (top) and {C$_6$H$_{11}$NH$_2$CH$_3$}$^+${4-CH$_3$C$_6$H$_4$S(O)$_2$NP(O)[N(CH$_3$)(C$_6$H$_{11}$)]$_2$}$^-$ (bottom); the N−H···O hydrogen bonds are shown

as dotted lines. The H atoms not involved in hydrogen bonding have been omitted for the sake of clarity and the 4-CH_3-C_6H_4 (bottom) and C_6H_{11} and CH_3 substituents are shown as balls (the N...O distances are 2.771(3) & 2.804(3) Å and 2.648(4) & 2.864(4) Å, respectively).

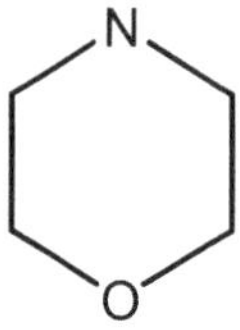

	Ph Ph	NH	
BIVYAG [98]	JEKLER [99]	LAFNAI [100]	QONBEB [101]
	NHC_6H_4(4-CH_3)	NHC_6H_4(3-CH_3)	NH
[1]QUFJIL [102]	[2]TAZPOX [103]	[3]TOLZPO [104]	[4]OLOCEY [105]

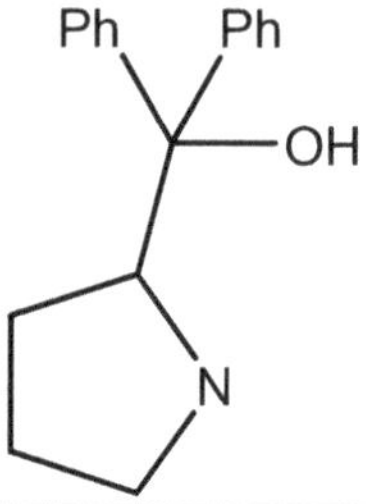

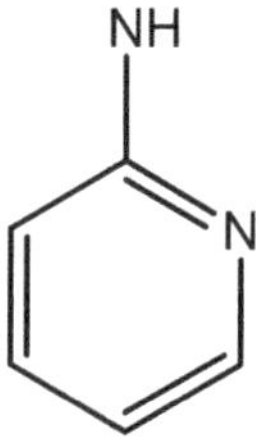

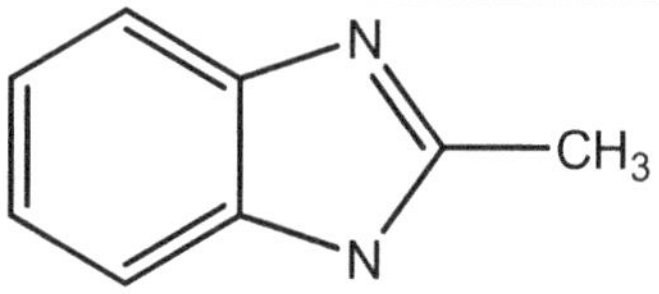

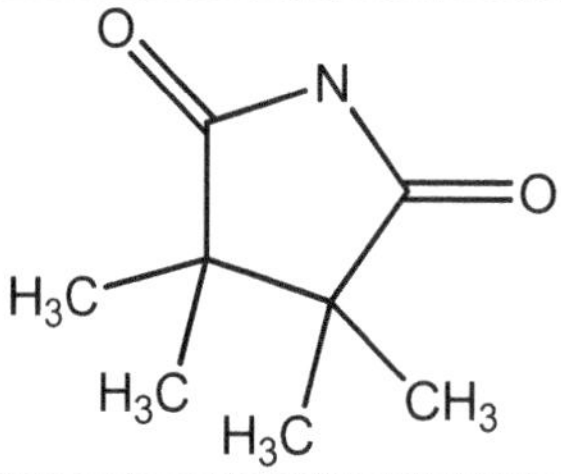

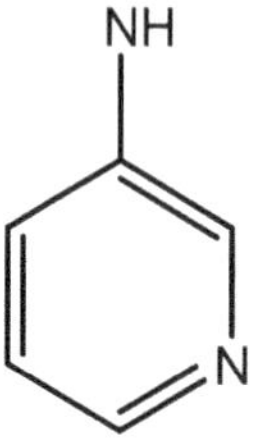

[1]There is the hydrogen-bonded amide molecule in the structure, i.e. the formula is $C_{24}H_{36}N_3O_7P_1,C_8H_{13}N_1O_2$. [2]There is the solvent C_2H_5OH molecule in the structure, i.e. the formula is $C_{21}H_{24}N_3O_1P_1,C_2H_6O_1$. [3]The formula is $C_{21}H_{24}N_3O_1P_1,2(C_7H_{10}N_1^{1+}),2(Cl_1^{1-})$. [4]The formula is $C_{15}H_{15}N_6O_1P_1,H_2O_1$.

Scheme 17. The structurally investigated compounds of the formula $[R^1R^2N]_3P(O)$ or $[R^1R^2N]_3P(O).B$, where B is a hydrogen-bonded species to phosphoric triamide (the related amido moieties and the CSD refcode are presented)

In the crystal packing of molecules having a P(O)(NH)$_3$ skeleton, hydrogen bonded 1-D chain, 1-D ladder, 2-D layer and 3-D arrangements were found. Three different types of 1-D arrangement are formed respectively through a P=O...(H-N)$_3$ or P=O...(H-N)$_2$ groups or via the P=O...H-N hydrogen bond. In the two latter cases, respectively one and two N-H units don't cooperate in the hydrogen bond interaction. One example of a linear arrangement, involving the N−H··O and N−H··N HBs, is also found in the structure of P(O)(NH−$C_5H_4N)_3$ (LAFNAI) [100] in which the pyridine nitrogen atom is involving in the HB pattern as an acceptor, too. As, the phosphoryl oxygen atom may cooperate in H-bonding interaction as a double- or a triple- acceptor, some examples of 2-D and 3-D arrangements have also been found in this class of compounds.

Some phosphoric triamides $[R^1R^2N][R^3R^4N][R^5R^6N]P(O)$ (where merely tertiary nitrogen atoms exist in the structure of molecule) have been reported (for example see: Scheme 22 [110]). Structures with a P(O)(N)$_3$ skeleton, where "N" is a tertiary nitrogen atom, do not show any classical (normal) hydrogen bonding in their crystal packing if the substituent involving the N atom doesn't contain the hydrogen linked to an electronegative atom.

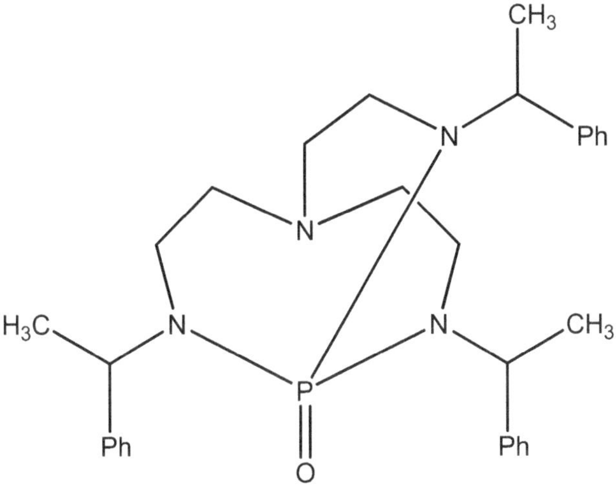

Phosphoramidates: Molecular Packing and Hydrogen Bond Strength in Compounds Having a P(O)(N)$_n$(O)$_{3-n}$
(n = 1, 2, 3) Skeleton

177

Scheme 18. 4,6,9-Tris(1-phenylethyl)-1,4,6,9-tetraaza-5-phosphabicyclo(3.3.3) undecane P-oxide (refcode: EDEVAK [106])

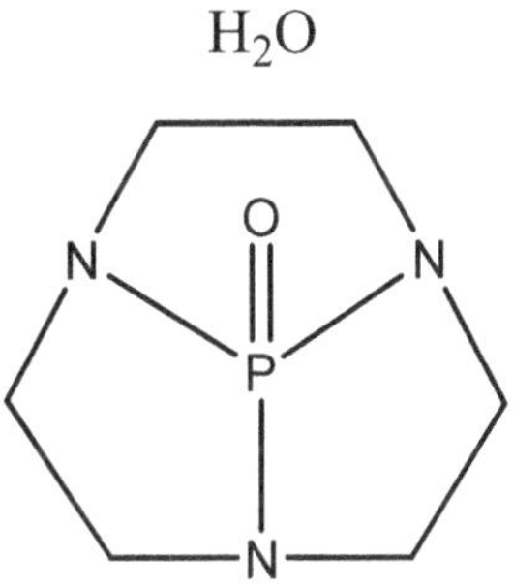

Scheme 19. 10-Oxo-10-phospha-1,4,7-triazatricyclo(5.2.1.0^{4,10})decane monohydrate (NUVSEC [107])

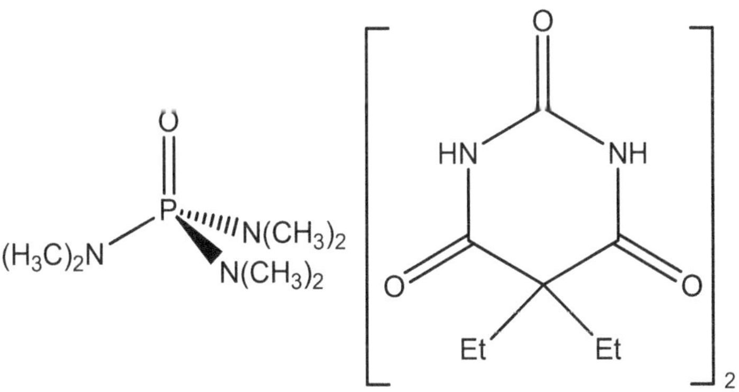

Scheme 20. Bis(barbital)-hexamethylphosphoramide complex (BARHMP [108])

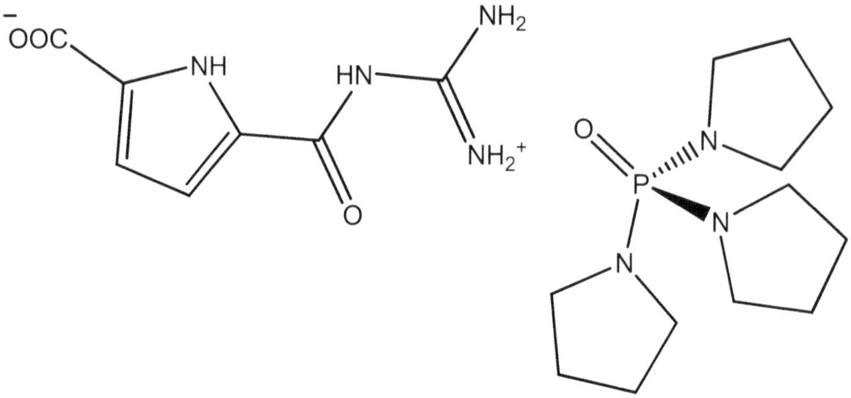

Scheme 21. 5-(Guanidiniocarbonyl)pyrrole-2-carboxylate tris(pyrrolidino)phosphine oxide solvate (VAFRIE [109])

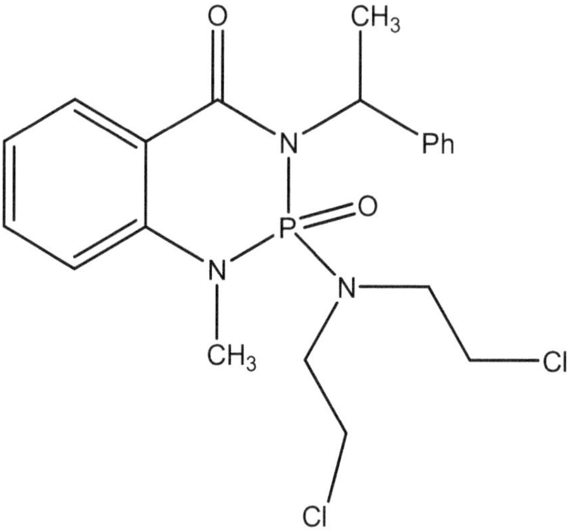

Scheme 22. Refcode BEJNEJ [110]

A search on the CSD shows that the nitrogen atoms bound to phosphorus in phosphoramidate compounds aren't involved in normal H-bonding interaction as an acceptor due to the deviation of each N atom environment from pyramidality after binding to P and decreasing its Lewis base character with respect to the initial amine; so that, merely one example (refcode: HESCEO [111], Scheme 23), belonging to the diazaphosphorinane family, is observed so far with the donor...acceptor (N...N) distance of 3.258(8) Å in which it may be considered as a weak $N-H...N-P$ hydrogen bond.

In compounds having a $P(O)(NH)(N)_2$ skeleton, only the P(O)NH unit cooperates in a HB interaction; so, two expectable HB patterns are the H-bonded dimer (with C_i symmetry; a dimer with C_1 symmetry has not reported, so far) and the 1-D chain (Scheme 10). Usually, an H-bonded dimer forms when the P(O) group and the N−H unit have a *syn* orientation with respect to one another. However, in one structure (refcode: DIYMED [112], Scheme 24) with the *syn* orientation of P(O) *versus* N−H, the molecules are aggregated as a one dimensional H-bonded chain.

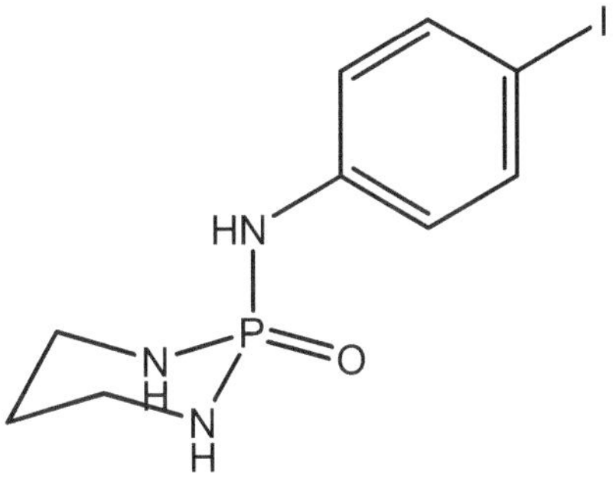

Scheme 23. One N atom of the diazaphosphorinane ring cooperates in hydrogen bonding interaction as an H-acceptor [111]

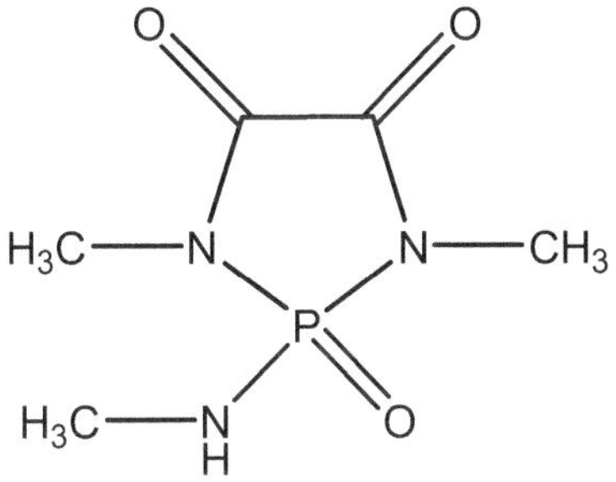

Scheme 24. Refcode DIYMED [112]

If the P(O) adopts an *anti* orientation *versus* NH, an extended 1-D chain arrangement is expectable through the intermolecular PO⋯HN hydrogen bond which is found for the most of reported compounds. One example without any N−H⋯O HB (BIFDUP [113], Scheme 25) and one example as H-bonded tetramer (XAVXEY, Scheme 26) were also found.

In compounds having an (N)P(O)(NH)$_2$ skeleton, three different linear arrangements were observed: a) through P(O)⋯H−N hydrogen bonds in which one N−H unit doesn't cooperate in H-bonding (NUVROL [107], Scheme 27 and HIVLII [67], Scheme 28), b) through $R_2^1(6)$ (in [(CH$_3$)$_2$N]P(O)[NHC$_6$H$_5$]$_2$ [52] and the compound with refcode MIFYIJ [114], Scheme 29 (top)) and c) through $R_2^2(8)$ rings ([(CH$_3$)$_2$N]P(O)[NHC$_5$H$_9$]$_2$ [62] and the compound with refcode IKASAP [47], Scheme 29 (bottom)), two latter cases *via* P(O)⋯(H−N)(H−N) grouping in which the phosphoryl oxygen atom acts as a double-H acceptor.

In this series, some other H-bond motifs were observed in compounds having an NH$_2$ moiety instead of NHR moiety (BIXFOE [115], GOMDOB [116]), Scheme 30.

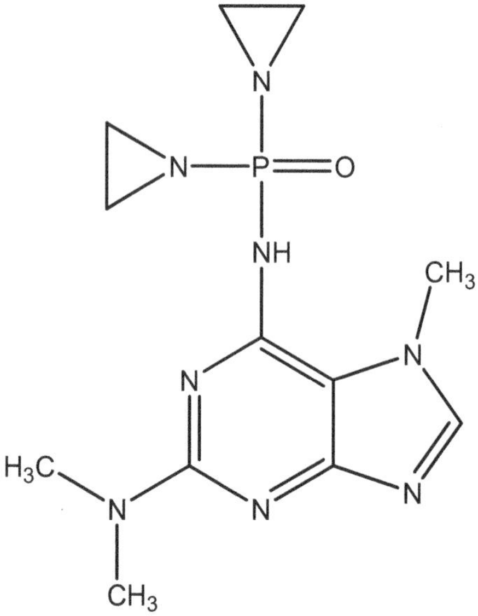

Scheme 25. Refcode BIFDUP [113]

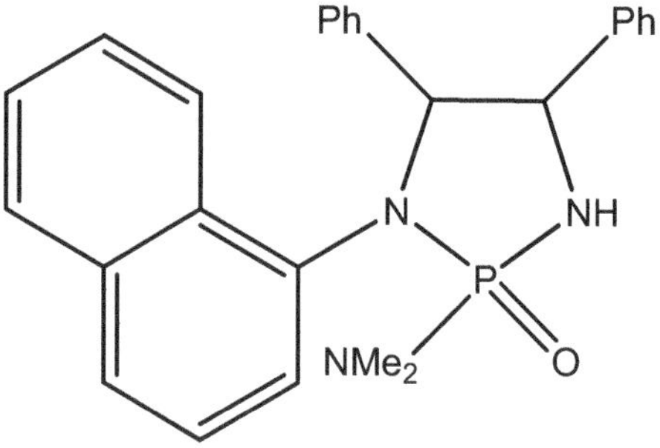

Scheme 26. Refcode XAVXEY (any reference to a journal, book and so on was not found for this structure)

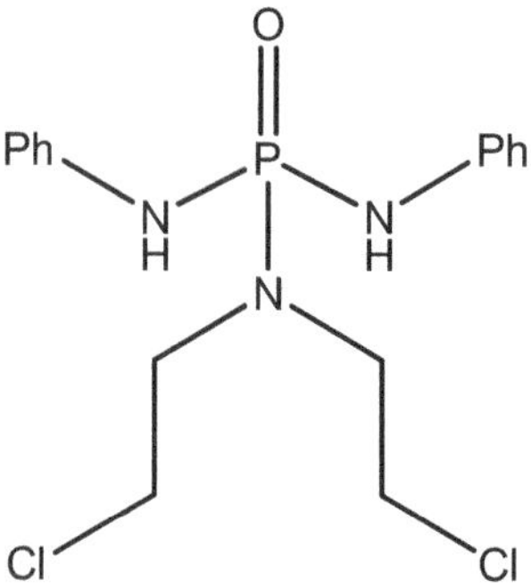

Scheme 27. Refcode NUVROL [107]

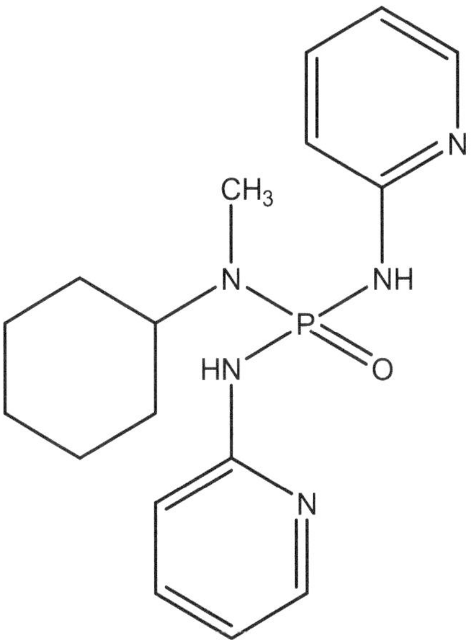

Phosphoramidates: Molecular Packing and Hydrogen Bond Strength in Compounds Having a P(O)(N)$_n$(O)$_{3-n}$ (n = 1, 2, 3) Skeleton

181

Scheme 28. Refcode HIVLII [67]

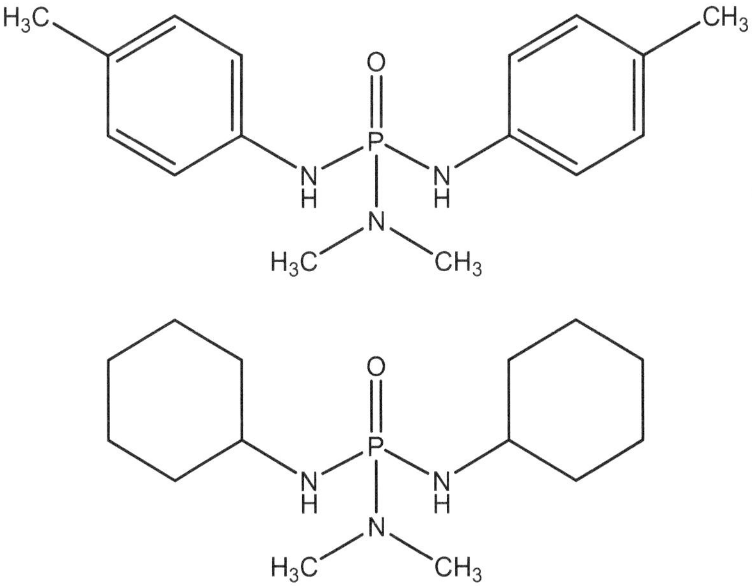

Scheme 29. Refcodes MIFYIJ (top) [114] and IKASAP (bottom) [47]

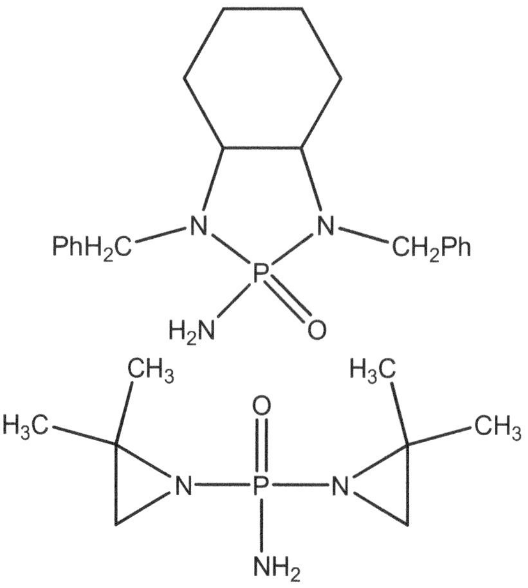

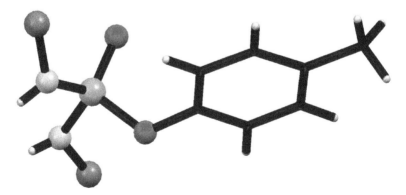

Scheme 30. Refcodes BIXFOE (top) [115] and GOMDOB (bottom) [116]

Compounds with formula (R¹O)(R²O)(R³R⁴N)P(O), (R¹O)(R²O)(R³NH)P(O), (R¹O)(R²R³N) (R⁴R⁵N)P(O), (R¹O)(R²NH)(R³R⁴N)P(O) and (R¹O)(R²NH)(R³NH)P(O)

The tetrahedral configuration of phosphorus atom is significantly distorted as it has been noted for the other phosphoramides and their chalco-derivatives [117]. For example, the bond angles around the P atom of $(4\text{-}CH_3\text{-}C_6H_4O)(C_6H_{11}NH)_2P(O)$ [116] vary in the range from $101.48(10)°$ [for $O_{phenoxy}-P-N1$ angle] to $118.58(9)°$ [for $O_{phosphoryl}-P-N2$ angle]. The $C-O-P$ angle is $123.52(15)°$. A general view of a compound with formula $(4\text{-}CH_3\text{-}C_6H_4O)(RNH)_2P(O)$ is shown in Fig. 5.

Fig. 5. A general view of a compound with formula $(4\text{-}CH_3\text{-}C_6H_4O)(RNH)_2P(O)$, the R moieties are shown as grey balls.

Phosphoramidates: Molecular Packing and Hydrogen Bond Strength in Compounds Having a P(O)(N)$_n$(O)$_{3-n}$ (n = 1, 2, 3) Skeleton

183

Similar to the other phosphoramidates, each N atom bonded to phosphorus doesn't involve in any HB as an acceptor, showing its low Lewis-base character. In most cases, it has a nearly planar environment [26]. Of course, the nitrogen atom' environment of some substituents, such as aziridinyl, like for example in the compound with refcode GOMDOB [118] (Scheme 30 (bottom)) shows some deviation from planarity, but such N atom doesn't cooperate in hydrogen bonding interaction, too. Moreover, the oxygen atom of the phenoxy or alkoxy groups in the 47 structures with an (O)P(=O)(NH)(N) skeleton (like for example in 4-CH$_3$-C$_6$H$_4$OP(O)[N(CH$_3$)$_2$][NHC(CH$_3$)$_3$], GUDGIW: [119]) doesn't cooperate in the HB interaction, as it can not compete with the phosphoryl oxygen atom for H-accepting from the unique H-donor site in the molecule. Furthermore, among the 106 deposited structures with an (O)(O)P(=O)(NH) skeleton, only the structure of [CH$_3$O]$_2$P(O)[NHCH(CH(CH$_3$) (OC(O)CH$_3$))(C(O)(C(NN)(COOC$_2$H$_5$)))] (IJUMAB: [120]) shows N-H...O(CH$_3$) not N-H...O(P) hydrogen bond (in this consideration, some structures with unavailable cifs were not enumerated). So, such compounds [if the substituents linked to the N or/and O atoms don't contain any H-acceptor or H-donor centers] may be almost always considered as compounds with "one H-acceptor (the oxygen of phosphoryl) and one H-donor sites", both in the P(O)NH group.

The oxygen atom of OR moiety in some examples of compounds with a higher H-donor sites, such as compounds containing an (O)P(=O)(NH)(NH) skeleton, however, it has a lower H-acceptability than the phosphoryl oxygen atom, is enforced to involve in the HB interaction ([4-CH$_3$-C$_6$H$_4$O]P(O)[NHC$_6$H$_4$(4-CH$_3$)]$_2$: MUBPIJ, [36]). The better H-acceptability of the phosphoryl O atom than that of the RO moiety, in some cases for example in [C$_6$H$_5$O]P(O)[NHC$_6$H$_{11}$][NHC$_6$H$_4$(4-CH$_3$)] (ERUFIH: [69]), leads to act it as a double-H acceptor. In the molecular packing of [C$_6$H$_5$O]P(O)[NHC$_6$H$_{11}$]$_2$.CH$_3$OH (HIVLOO, [67]), a linear arrangement is formed through a P(O)[...H−O][...H−N] grouping, where, the P(O) group acts as a double H-acceptor, the OH unit belongs to the solvent methanol.

In the 2-D H-bonded arrangement for diazaphosphorinane 4-CH$_3$C$_6$H$_4$OP(O)X [X = NHCH$_2$CH$_2$CH$_2$NH (KIVXIX)] and C$_6$H$_5$OP(O)Y [Y = NHCH$_2$C(CH$_3$)$_2$CH$_2$NH (KIVXOD)], the P(O) functions as a double-H acceptor [121]. In the other cases, both O atoms are involved in the HB interactions with two N−H units (or the other H-donor site(s) in the molecule or in the crystal), in which the P(O) forms a stronger HB. Typically, in the crystal packing of [4-CH$_3$-C$_6$H$_4$O]P(O)[NHC$_6$H$_4$(4-CH$_3$)]$_2$ (MUBPIJ, [36]), the N...O(P) = 2.805(2) Å & N...O(C$_6$H$_4$-4-CH$_3$) = 3.068(2) Å and of C$_6$H$_5$OP(O)[NHC$_6$H$_4$(4-CH$_3$)][NHCH$_2$C$_6$H$_5$], in a recently published paper by Pourayoubi *et al.*, 2011 [43], these distances are 2.761(3) Å & 3.127(3) Å, respectively.

Scheme 31 illustrates the contribution of P(O) as a double-H atom acceptor (top) and cooperation of both oxygen atoms (bottom) in hydrogen bond pattern in compounds having a P(O)(O)(NH)(NH) moiety.

In the crystal packing of compounds with the general formula (R^{1}O)P(O)[NHR^2]$_2$, both linear and 2-D hydrogen-bonded arrangements were observed; for example, C$_6$H$_5$OP(O)[NHC$_6$H$_{11}$]$_2$.CH$_3$OH (HIVLOO [67]), 4-CH$_3$-C$_6$H$_4$OP(O)[NHC$_6$H$_4$-4-CH$_3$]$_2$ (MUBPIJ [36]), 4-CH$_3$-C$_6$H$_4$OP(O)[NHC$_6$H$_4$-2-CH$_3$]$_2$ (YUPVEL [32]) and 4-CH$_3$-C$_6$H$_4$OP(O)X (X = NHCH$_2$C(CH$_3$)$_2$CH$_2$NH, NIBNOC [83]-a) exist as a linear H-bonded arrangement, whereas a 2-D array is found for instance in each of 4-CH$_3$−C$_6$H$_4$OP(O)X (X =

NHCH$_2$CH$_2$CH$_2$NH, KIVXIX [121]), C$_6$H$_5$OP(O)X (X = NHCH$_2$C(CH$_3$)$_2$CH$_2$NH, KIVXOD [83]-a), C$_6$H$_5$OP(O)(NH$_2$)$_2$ (PPOSAM [122]) and C$_6$H$_5$OP(O)X (X = NHNHP(O)(OC$_6$H$_5$) NHNH (FIMVUS [123])).

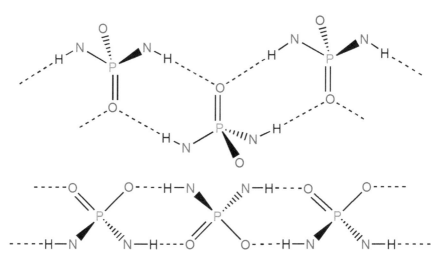

Scheme 31. A view of contribution of phosphoryl oxygen atom as a double-H acceptor (top) and a view of contribution of both oxygen atoms (bottom) in hydrogen bond pattern of compounds having a P(O)(O)(NH)(NH) skeleton

In summary, the cif files of all published compounds with the (O)P(=O)(NH)$_2$, (O)$_2$P(=O)(NH) and (O)P(=O)(NH)(N) skeletons were investigated and the following "empirical rules" were obtained:

1. In none of the reported structures, the nitrogen atom doesn't cooperate in HB interaction as an acceptor.
2. Almost in all of the compounds having the (O)$_2$P(=O)(NH) and (O)P(=O)(NH)(N) skeletons, the oxygen atom of the phenoxy (or alkoxy) group doesn't cooperate in the HB interaction, as it can't compete with the phosphoryl oxygen atom to H-accepting from the unique H-donor site in the molecule. There is only one example of hydrogen bond of the type N−H...O(R) in this family of compounds in one compound containing some H-acceptor centers in addition to one phosphoryl group.
3. The oxygen atom of OR moiety in some examples of compounds with a higher number of H-donor sites relative to the H-acceptor centers, such as compounds containing an (O)P(=O)(NH)$_2$ skeleton, however, it has a lower H-acceptability than the phosphoryl oxygen atom, is enforced to involve in the HB interaction.
4. In compounds having an (O)P(=O)(NH)$_2$ skeleton, the better H-acceptability of the phosphoryl O atom than that of the RO moiety, in some cases, leads to act it as a double-H acceptor.

Amidophosphoric acid and amido phosphate compounds

Compounds with refcodes PHOXBP (Scheme 32, left) [124] and TMPMET (Scheme 32, right) [125] are respectively the examples of an acid and a solvated acidic-salt belonging to the phosphorus-nitrogen compounds' family. The crystal packing of the latter compound contains some various HBs such as two very strong homo-conjugated [O−H...O]- hydrogen bonds (O...O = 2.43 & 2.50 Å) and relatively strong hetero-conjugated [N−H...O]$^{\delta+}$ hydrogen bond (N...O = 2.93 Å). In the hydrated zwitterionic compound shown in Scheme 33 (refcode: GAHFUS) [126], the N+ which is lack of the lone electron pair doesn't involve in H-bonding interaction; whereas, the [P(O)(O)]- unit cooperates in some N−H...O and O−H...O hydrogen bonds (Fig. 6). Compounds with refcodes IGASUF (Scheme 34) [127] and WIYFAL (Scheme 35) [128] are respectively a simple phosphate salt and an HCl-water absorbed phosphorus-nitrogen compound in which the Cl- ion is hydrogen-bonded to the N−H units of two-neighboring phosphoramidates.

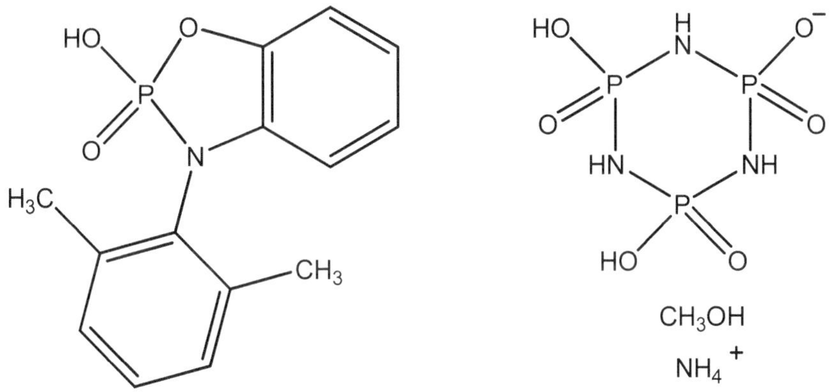

Scheme 32. Refcodes PHOXBP (left) [124] and TMPMET (right) [125]

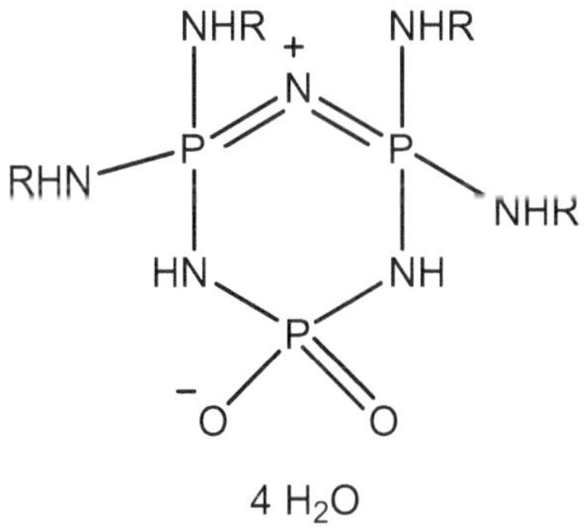

Scheme 33. Refcode GAHFUS, R = cyclo-hexyl [126]

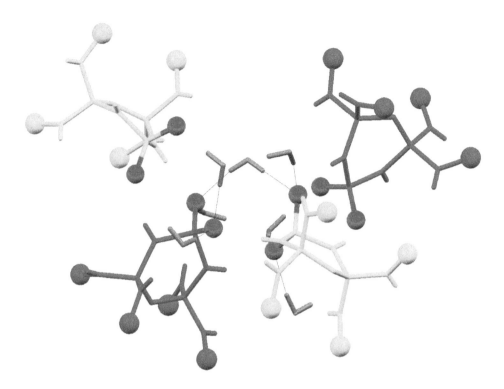

Fig. 6. Fragment of the crystal packing of the hydrated zwitterionic compound with refcode GAHFUS [126] showing the involvement of [P(O)(O)]⁻ units in N−H...O (blue dotted lines) and O−H...O (black dotted lines) hydrogen bonds, two symmetrically independent zwitterionic compounds in the structure are shown as blue and green apart the oxygen atoms of [P(O)(O)]⁻ units which are shown as red balls, the water molecules are represented with grey color. The cyclohexyl groups are shown as balls (green and blue).

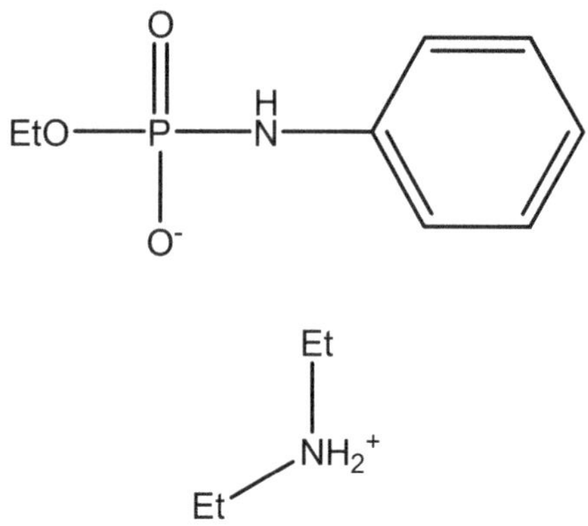

Scheme 34. Refcode IGASUF [127]

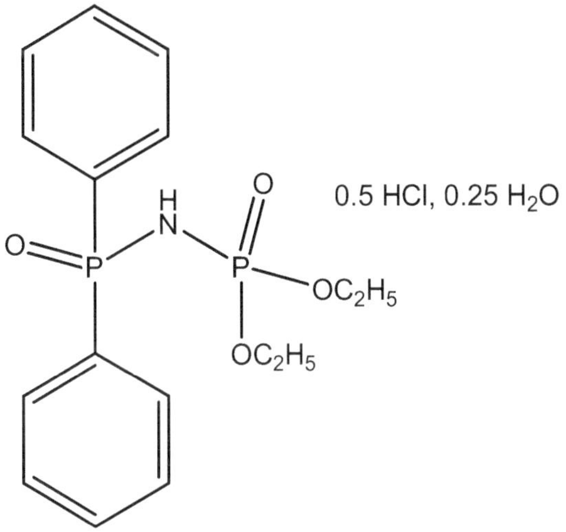

Scheme 35. Refcode WIYFAL [128]

5. Crystallographically independent molecules and ions

Different orientations resulting from non-rigid units in some molecules and ions and the presence of different H-bonds or the other short contacts may result in two or more conformers (or symmetrically independent molecules (or ions)) in solid state. Compound $C_6H_5C(O)NHP(O)[NH(tert-C_4H_9)]_2$, exists as two conformers in crystalline lattice (which are detectable in solution, too by NMR experiment) [27]. They are due to different spatial orientations of *tert*-butyl amido groups. One of the two conformers has two NH units (of *tert*-butyl amido moieties) which are *syn*, but not in the other. Another example is the presence of disorder in the cyclic amido moiety. For example, $C_6H_5C(O)NHP(O)[NC_4H_8]_2$ appears as two crystallographically independent molecules [55]. This is based on the conformational forms of the pyrrolidinyl groups and the orientation of the phenyl ring. The dimmeric aggregate in this case, between two independent molecules, is not centrosymmetric. The structure of $[NH_2(C_6H_{11})(tert-C_4H_9)][PO_2Cl_2]$ consists of two symmetrically independent dichlorophosphate anions as well as cyclohexyl-*tert*-butylammonium cations [75]. In the crystal structure of [*tert*-$C_4H_9NH_3][CF_3C(O)NHP(O)(O)(tert-C_4H_9NH)].0.333CH_3CN.0.333H_2O$ [79], there are three symmetrically independent trifluoroacetyl-N-(*tert*-butylamino) phosphate anions and three independent cations of *tert*-butyl- ammonium; one of the anion indicates disorder in the *tert*-C_4H_9 moiety. There are some other examples of disordered components for the groups such as *tert*-C_4H_9, cyclopentyl and cyclohexyl etc. in the deposited cifs.

6. Hydrogen bond strengths in phosphoramidates

Histogram of the N...O distances in the N−H...O hydrogen bonds in compounds having a $P(O)(N)_n(O)_{3-n}$ (n = 1, 2, 3) skeleton is given in Fig. 7. In this figure, the distribution of H-bond strength in different families of phosphoramidates are shown with different colored columns: compounds having a $P(O)(NH)_n(N)_m(O)_{3-(n+m)}$ skeleton (n = 1, 2; n+m < 3) as black columns, C(O)NHP(O)(NH)$_2$ and C(O)NHP(O)(N)$_2$ as blue and P(O)(NH)$_x$(N)$_{3-x}$ (x = 1, 2, 3) as red columns.

In phosphoramidates having a P(O)(NH)$_n$(N)$_m$(O)$_{3-(n+m)}$ skeleton (balck), the strongest and weakest N−H...O hydrogen bonds are found for hydrogen bonds in the range of 2.65 to 2.75 Å and 3.20 to 3.30 Å.

In compounds containing a P(O)(NH)$_x$(N)$_{3-x}$ skeleton, the strongest N−H...O hydrogen bonds are seen for the HBs in the range of 2.70 to 2.80 Å. The phosphoryl group' involvement in a multi-centered P(O)$\cdots$[H−N]$_n$ (n = 2 & 3) grouping may lead to some weak H-bonds; for example in P(O)[NHC(CH$_3$)$_3$]$_3$ (KABVEP [93]), N...O distances & N−H...O angles are 3.255(4) Å & 111.1(2)°, 3.294(4) Å & 93.4(2)° and 3.159(4) Å & 123.0(2)°, and in P(O)[NH(C$_6$H$_5$)]$_3$ (KEQLUO [95]) these parameters are 3.06 Å & 110° and 3.06 Å & 108°; this weakening of H-bond strength is attributed to the *anti*-cooperativity effect [89].

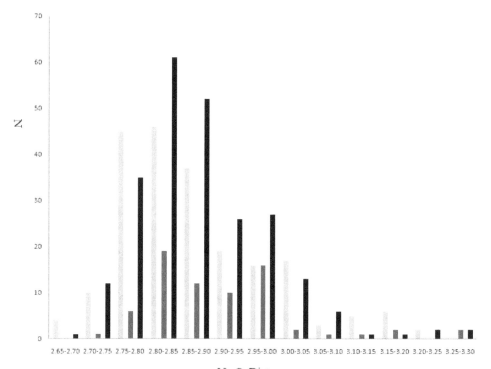

Fig. 7. Histogram of the N...O distances in the N−H...O hydrogen bonds in compounds having P(O)(NH)$_n$(N)$_m$(O)$_{3-(n+m)}$ (n = 1, 2; n+m < 3) (black), C(O)NHP(O)(NH)$_2$ and C(O)NHP(O)(N)$_2$ (blue), and P(O)(NH)$_x$(N)$_{3-x}$ (x = 1, 2, 3) (red) skeletons (the co-crystals and solvated compounds and the compounds having a disorder in the sites involving HB interaction were not enumerated).

In this family, a long donor$\cdots$acceptor distance (3.477(2) Å) is observed with a relatively linear N$-$H...O angle (171 (2)°) for the N$-$H$\cdots$O(CH$_3$) hydrogen bond in the packing of P(O)[NHC$_6$H$_4$(4-OCH$_3$)]$_3$ (WAWNIS [97]).

In compounds having a C(O)NHP(O) fragment, the strongest N$-$H...O hydrogen bonds are found for the P=O...H$-$N$_{C(O)NHP(O)}$ hydrogen bonds, especially in the R$_2^2$(8) rings of some molecules [in the case of a *syn* orientation of P=O versus N$-$H which allows the building of the cyclic motif through a pair of P=O...H$-$N$_{C(O)NHP(O)}$ hydrogen bonds]. In this sub-category of compounds, the strongest and weakest hydrogen bonds are observed for the N...O distances in the ranges 2.70–2.80 Å [for N$_{C(O)NHP(O)}-$H...O hydrogen bonds] and 3.00–3.25 Å [for N$_{amide}-$H...O hydrogen bonds], while in the range 2.80–3.00 Å for donor–acceptor distances both types of hydrogen bonds are found. In a recently published structure [129] with a NHC(O)NHP(O)[NH]$_2$ skeleton, the intramolecular N$-$H...O hydrogen bonds are found in the range 2.65-2.70 Å (Fig. 7). The asymmetric unit contains four independent molecules and the hydrogen bond pattern is different from all phosphoric triamides having a C(O)NHP(O)[NH]$_2$ fragment.

7. Some future aims and proposals

Phosphorus has a very deep and widespread chemistry and understanding its nature in the compounds is noticeable interest. This may be achieved through the study on collective behaviors of phosphorus compounds in point of view of their different aspects.

The structural investigations and the study on the hydrogen-bond patterns may help to predict the molecular packing from the molecular structure. Moreover, as the biological activity of phosphorus compounds is very important, finding a relationship between the structure and a biological property is beneficial. For example, as well-known, a biological property may be related to the three important factors: lipophilicity, electronic and steric parameters. Probably, part of these factors could be well-understand by considering the crystal structure' study of such compounds; the electronic parameters may be related to the nature of chemical bonds or to the electron density or valence bond in different parts of the molecule. It is believed that, in the first step of interaction with acetylcholinesterase, the phosphoryl group is involving with the enzyme active site through a non-covalent bond; so, considering the non-covalent interaction of phosphoryl group with different atoms such as hydrogen helps to understand that the molecule how much could close to the enzyme active site. The steric parameters may be elucidated from the volume of molecule or considering the V(volume of unit cell)/Z(number of molecules in the unit cell). This may be in fact the practical volume which the molecule has, as the molecule usually cannot be closer to the neighboring molecules from this frontier boundary.

The solubility of molecules in different solvents must be checked in the crystal growth process; so, elucidation of lipophilicity is easy; however, the best method for measuring of this parameter is using of the spectrophotometer. And finally, the steric parameters can simply accessible through a structural study.

Preparation of phosphorus acids of the formula RP(O)(OH)$_2$ may develop the synthesis of the functionalized nano-phosphate materials and also polyoxometalate-based organic/inorganic hybrid compounds in which the phosphorus atom is trapping between the R group from one side and a cluster containing metal-oxygen framework in the other side. These acids may develop the extraction process of cations, too.

Application of phosphoryl donor ligands in preparation of oxo-centered clusters, in which their terminal ligands are replaced by phosphorus compounds, may be interesting for consideration. Preparation of single- enantiomer phosphoramidates by using a chiral primary or secondary amine is easy; it may extend the strategies for the synthesis of chiral phosphoramidates, phosphoric acids, nano-phosphates and so on.

Synthesis of phosphoramidate-based hybrid compounds by using polyoxoanions may be valuable for spending the time on its consideration and experiment. Some of the well-known hybrids contain the molecule-cation components of the type $[B-H...B]^+$, where the B may be a base such as amide. Designing of such molecule-cation pairs with phosphoramidates, $[PO-H...OP]^+$, may extend the experimental data about the ^{31}P-^{31}P coupling constant through the hydrogen-bond.

The NMR experiments on phosphoramidate-based compounds may develop the study on coupling constants of phosphorus and the other atoms, such as $^2J(^{31}P$-$^{127}Tl)$ or $^2J(^{31}P$-$^{39}K)$. These values may apply to evaluate the strengths of P=O$-$Tl or P=O$-$K bonds in their complexes.

We wish to develop the spectroscopic features and chemical calculations on phosphorus compounds, preparation of N-deuterated compounds in order to a good assignment of IR and Raman absorption bands, collecting the NMR data such as chemical shifts and short and long-range coupling constants, and finally chemical calculations on hydrogen-bonded molecules in the crystals.

8. Conclusion

In this chapter, the common methods for the synthesis and crystallization of phosphoramidates, their molecular structural features and the hydrogen-bond patterns and strengths were reviewed; the important structural aspects may be classified as follows:

1. The four different groups linked to the P atom result in a distorted tetrahedral configuration.
2. In the C(O)NHP(O) unit, the P=O is a better H-acceptor than the C=O counterpart, moreover, the *anti* orientation of P=O versus C=O is more common than the *gauche* orientation; in acyclic compounds with formula RC(O)NHP(O)[NHR']$_2$, a *gauche* situation has not been reported, so far. In the C(O)NHP(O)[N]$_2$ fragment, the P$-$N bond of the C(O)NHP(O) moiety is longer than the two other P$-$N bonds; whereas, in the [C(O)NP(O)[N]$_2$]$^-$ fragment, similar P$-$N bond are shorter than the two others.
3. The nitrogen atom of the P(O)N unit has a sp^2 character and virtually doesn't involve in hydrogen-bond pattern as an H-acceptor.
4. In the C$-$O$-$P(=O) fragment, the oxygen of phosphoryl is a better H-acceptor than the other oxygen atom; the C$-$O$-$P angle is about 120°.
5. In the diazaphosphorinane ring, the P=O bond is placed in an equatorial position.
6. The hydrogen bond in a neutral phosphoramidate is of the type polarization-assisted hydrogen bond; whereas, in the proton-transfer and phosphate compounds two factors help to strength of hydrogen bond: polarization-assisted and charge-assisted.

Phosphoramidates: Molecular Packing and Hydrogen Bond Strength in Compounds Having a P(O)(N)$_n$(O)$_{3-n}$
(n = 1, 2, 3) Skeleton

191

7. In the multi-centered hydrogen-bond of the type P=O[...H-N]$_n$ (n = 2 & 3) the hydrogen-bond is weak due to the anticooperativity effect.

We wish to collect more structural data about this class of compounds and study the collective behavior of this family in the other domains such as spectroscopy and chemical calculations.

9. Acknowledgements

M. P. wishes to thank Ferdowsi University of Mashhad for the Research University Grant (No. 15144/2). The authors also thank Dr. Karla Fejfarová for the CSD searches and Monireh Negari (MSc student) for some computer works.

10. References

[1] Allen, F. H. (2002). *Acta Cryst.* B58, 380.
[2] Müller, P. & Siegfried, B. (1972). *Helvetica Chimica Acta*, 55, 2965.
[3] (a) Le Carpentier, J.-M., Schlupp, R. & Weiss, R. (1972). *Acta Cryst.* B28, 1278. (b) Crawford, M.-J., Mayer, P., Nöth, H. & Suter M. (2004). *Inorg. Chem.* 43, 6860.
[4] (a) Aas, P. (2004). *Prehospital and Disaster Medicine*, 18, 208. (b) Bhattacharjee, A. K., Kuča, K., Musilek, K. & Gordon, R. K. (2010). *Chem. Res. Toxicol.* 23, 26.
[5] (a) Rodriguez, J. A. & Fernandez-Garcia, M. (2007). *Synthesis, properties and applications of oxide nanomaterials*, Wiley-interscience, A John Wiley & Sons, INC., Publication. (b) Hirakawa, T., Sato, K., Komano, A., Kishi, Sh., Nishimoto, Ch. K., Mera, N., Kugishima, M., Sano, T., Ichinose, H., Negishi, N., Seto, Y. & Takeuchi, K. (2010). *J. Phys. Chem.* C114, 2305.
[6] (a) Nguyen, C. & Kim, J. (2008). *Polym. Degrad. Stabil.* 93, 1037. (b) Levchik, S. V., Levchik, G. F. & Murashko, E. A. (2001). *Phosphorus-Containing Fire Retardants in Aliphatic Nylons, Fire and Polymers,* Chapter 17, pp 214.
[7] Roush, R. F., Nolan, E. M., Löhr, F. & Walsh, C. T. (2008). *J. Am. Chem. Soc.* 130, 3603.
[8] Gholivand, K., Ghadimi, S., Naderimanesh, H. & Forouzanfar, A. (2001). *Magn. Reson. Chem.* 39, 684.
[9] Gholivand, K., Mostaanzadeh, H., Koval, T., Dusek, M., Erben, M. F., Stoeckli-Evans, H. & Della Védova, C. O. (2010). *Acta Cryst.* B66, 441.
[10] Pourayoubi, M., Tarahhomi, A., Saneei, A., Rheingold, A. L. & Golen, J. A. (2011). *Acta Cryst.* C67, o265.
[11] Ghadimi, S., Mousavi, S. L. & Javani, Z. (2008). *J. Enz. Inhibit. Med. Chem.* 23, 213.
[12] Ekstrom, F., Akfur, C., Tunemalm, A. & Lundberg, S. (2006). *Biochemistry*, 45, 74.
[13] Pourayoubi, M., Rostami Chaijan, M., Torre-Fernández, L. & García-Granda, S. (2011). *Acta Cryst.* E67, o1360.
[14] Pourayoubi, M., Rostami Chaijan, M., Torre-Fernández, L. & García-Granda, S. (2011). *Acta Cryst.* E67, o1031.
[15] Pourayoubi, M. & Saneei, A. (2011). *Acta Cryst.* E67, o665.
[16] Pourayoubi, M., Tarahhomi, A., Rheingold, A. L. & Golen, J. A. (2010). *Acta Cryst.* E66, o2524.

[17] Pourayoubi, M., Tarahhomi, A., Rheingold, A. L. & Golen, J. A. (2010). *Acta Cryst.* E66, o3159.

[18] Pourayoubi, M., Tarahhomi, A., Rheingold, A. L. & Golen, J. A. (2011). *Acta Cryst.* E67, o934.

[19] Pourayoubi, M., Tarahhomi, A., Rheingold, A. L. & Golen, J. A. (2011). *Acta Cryst.* E67, o3027.

[20] Pourayoubi, M., Fadaei, H., Tarahhomi, A. & Parvez, M. (2011). *Acta Cryst.* E66, o2795.

[21] Pourayoubi, M., Tarahhomi, A., Rheingold, A. L. & Golen, J. A. (2011). *Acta Cryst.* E67, o2643.

[22] Pourayoubi, M., Tarahhomi, A., Rheingold, A. L. & Golen, J. A. (2011). *Acta Cryst.* E67, o2444.

[23] Pourayoubi, M., Sadeghi Seraji, S., Bruno, G. & Amiri Rudbari, H. (2011). *Acta Cryst.* E67, o1285.

[24] Pourayoubi, M., Toghraee, M. & Divjakovic, V. (2011). *Acta Cryst.* E67, o333.

[25] Tarahhomi, A., Pourayoubi, M., Rheingold, A. L. & Golen, J. A. (2011). *Struct. Chem.* 22, 201.

[26] Toghraee, M., Pourayoubi, M. & Divjakovic, V. (2011). *Polyhedron*, 30, 1680.

[27] Gholivand, K. & Pourayoubi, M. (2004). *Z. Anorg. Allg. Chem.* 630, 1330.

[28] Pourayoubi, M., Golen, J. A., Rostami Chaijan, M., Divjakovic, V., Negari, M. & Rheingold, A. L. (2011). *Acta Cryst.* C67, m160.

[29] Tarahhomi, A., Pourayoubi, M., Rheingold, A. L. & Golen, J. A. (2011). *Acta Cryst.* E67, o2643.

[30] Pourayoubi, M., Eshtiagh-Hosseini, H., Negari, M. & Nečas, M. (2011). *Acta Cryst.* E67, o2202.

[31] Sabbaghi, F., Pourayoubi, M., Toghraee, M. & Divjakovic, V. (2010). *Acta Cryst.* E66, o344.

[32] Sabbaghi, F., Mancilla Percino, T., Pourayoubi, M. & Leyva, M. A. (2010). *Acta Cryst.* E66, o1755.

[33] Pourayoubi, M., Ghadimi, S. & Ebrahimi Valmoozi, A. A. (2010). *Acta Cryst.* E66, o450.

[34] Pourayoubi, M., Ghadimi, S. & Ebrahimi Valmoozi, A. A. (2007). *Acta Cryst.* E63, o4631.

[35] Ghadimi, S., Ebrahimi Valmoozi, A. A. & Pourayoubi, M. (2007). *Acta Cryst.* E63, o3260.

[36] Pourayoubi, M., Ghadimi, S., Ebrahimi Valmoozi, A. A. & Banan, A. R. (2009). *Acta Cryst.* E65, o1973.

[37] Sabbaghi, F., Rostami Chaijan, M. & Pourayoubi, M. (2010). *Acta Cryst.* E66, o1754.

[38] Pourayoubi, M., Eshtiagh-Hosseini, H., Zargaran, P. & Divjakovic, V. (2010). *Acta Cryst.* E66, o204.

[39] Pourayoubi, M., Ghadimi, S. & Ebrahimi Valmoozi, A. A. (2007). *Acta Cryst.* E63, o4093.

[40] Pourayoubi, M. & Sabbaghi, F. (2007). *Acta Cryst.* E63, o4366.

[41] Sabbaghi, F., Pourayoubi, M., Negari, M. & Nečas, M. (2011). *Acta Cryst.* E67, o2512.

[42] Pourayoubi, M., Eshtiagh-Hosseini, H., Negari, M. & Nečas, M. (2011). *Acta Cryst.* E67, o1870.

[43] Pourayoubi, M., Karimi Ahmadabad, F. & Nečas, M. (2011). *Acta Cryst.* E67, o2523.

[44] Pourayoubi, M., Zargaran, P., Rheingold, A. L. & Golen, J. A. (2011). *Acta Cryst.* E67, o5.

[45] Raissi Shabari, A., Pourayoubi, M. & Saneei, A. (2011). *Acta Cryst.* E67, o663.

[46] Pourayoubi, M., Keikha, M. & Nečas, M. (2011). *Acta Cryst.* E67, o2439.

[47] Sabbaghi, F., Pourayoubi, M., Karimi Ahmadabad, F., Azarkamanzad, Z. & Ebrahimi Valmoozi, A. A. (2011). *Acta Cryst.* E67, o502.

[48] Pourayoubi, M., Padělková, Z., Rostami Chaijan, M. & Růžička, A. (2011). *Acta Cryst.* E67, o450.

[49] Pourayoubi, M., Fadaei, H. & Parvez, M. (2011). *Acta Cryst.* E67, o2046.

[50] Pourayoubi, M., Keikha, M. & Parvez, M. (2011). *Acta Cryst.* E67, o2792.

[51] Pourayoubi, M. & Sabbaghi, F. (2009). *J. Chem. Crystallogr.* 39, 874.

[52] Pourayoubi, M., Yousefi, M., Eslami, F., Rheingold, A. L. & Chen, C. (2011). *Acta Cryst.* E67, o3220.

[53] Gholivand, K., Pourayoubi, M., Shariatinia, Z. & Mostaanzadeh, H. (2005) *Polyhedron*, 24, 655.

[54] Gholivand, K., Shariatinia, Z. & Pourayoubi, M. (2005). *Z. Anorg. Allg. Chem.* 631, 961.

[55] Gholivand, K., Pourayoubi, M. & Mostaanzadeh, H. (2004). *Anal. Sci.* 20, x51.

[56] Gholivand, K., Hosseini, Z., Pourayoubi, M. & Shariatinia, Z. (2005). *Z. Anorg. Allg. Chem.* 631, 3074.

[57] Gholivand, K., Shariatinia, Z., Pourayoubi, M. & Farshadian, S. (2005). *Z. Naturforsch. Teil B*, 60, 1021.

[58] Gholivand, K., Shariatinia, Z. & Pourayoubi, M. (2006). *Z. Anorg. Allg. Chem.* 632, 160.

[59] Gholivand, K., Shariatinia, Z. & Pourayoubi, M. (2006). *Polyhedron*, 25, 711.

[60] Gholivand, K., Mahzouni, H. R., Pourayoubi, M. & Amiri, S. (2010). *Inorg. Chim. Acta*, 363, 2318.

[61] Pourayoubi, M. & Zargaran, P. (2010). *Acta Cryst.* E66, o3273.

[62] Raissi Shabari, A., Pourayoubi, M., Ghoreishi, F. & Vahdani, B. (2011). *Acta Cryst.* E67, o3401.

[63] Pourayoubi, M., Shoghpour, S., Torre-Fernández, L. & García-Granda, S. (2011). *Acta Cryst.* E67, o3425.

[64] Pourayoubi, M., Elahi, B. & Parvez, M. (2011). *Acta Cryst.* E67, o2848.

[65] Cameron, T. S., Cordes, R. E. & Jackman F. A. (1978). *Z. Naturforsch. Teil B*, 33, 728.

[66] Setzer, W. N., Black, B. G., Hovanes, B. A. & Hubbard, J. L. (1989). *J. Org. Chem.* 54, 1709.

[67] Gholivand, K., Della Védova, C. O., Erben, M. F., Mahzouni, H. R., Shariatinia, Z. & Amiri, S. (2008). *J. Molec. Struct.* 874, 178.

[68] Gholivand, K., Pourayoubi, M., Farshadian, S., Molani, S. & Shariatinia, Z. (2005). *Anal. Sci.* 21, x55.

[69] Sabbaghi, F., Pourayoubi, M., Karimi Ahmadabad, F. & Parvez, M. (2011). *Acta Cryst.* E67, o1502.

[70] Pourayoubi, M., Tarahhomi, A., Keikha, M. & Karimi Ahmadabad, F. unpublished results.

[71] Pourayoubi, M. & Negari, M. (2010). *Acta Cryst.* E66, o708.

[72] Pourayoubi, M., Eshtiagh-Hosseini, H. & Negari, M. (2010). *Acta Cryst.* E66, o1180.

[73] Yazdanbakhsh, M. & Sabbaghi, F. (2007). *Acta Cryst.* E63, o4318.

[74] Pourayoubi, M., Negari, M. & Nečas, M. (2011). *Acta Cryst.* E67, o332.

[75] Gholivand, K. & Pourayoubi, M. (2004). *Z. Kristallogr. New Cryst. Struct.* 219, 314.

[76] Pourayoubi, M., Tarahhomi, A., Karimi Ahmadabad, F., Fejfarová, K., van der Lee, A. & Dušek, M. (2012). Acta Cryst. C68, In press.

[77] Pourayoubi, M., Fadaei, H., Tarahhomi, A. & Parvez, M. (2012). *Z. Kristallogr. New Cryst. Struct.* 227, In press.

[78] Pourayoubi, M. & Shoghpour, S. unpublished results.

[79] Gholivand, K., Pourayoubi, M., Shariatinia, Z. & Molani, S. (2005). *Z. Kristallogr. New Cryst. Struct.* 220, 387.

[80] Yazdanbakhsh, M., Eshtiagh-Hosseini, H. & Sabbaghi, F. (2009). *Acta Cryst.* E65, o78.

[81] Pourayoubi, M. & Fadaei, H. unpublished results.

[82] Gholivand, K., Shariatinia, Z. & Tadjarodi, A. (2005). *Main Group Chem.* 4, 111.

[83] (a) Gholivand, K., Pourayoubi, M. & Shariatinia, Z. (2007). *Polyhedron*, 26, 837. (b) Gholivand, K., Mojahed, F., Mohamadi, L. & Bijanzadeh, H. R. (2007). *Phosphorus, Sulfur, Silicon, Relat. Elem.* 182, 631.

[84] Kadyrov, A. A., Neda, I., Kaukorat, T., Fischer, A., Jones, P. G. & Schmutzler, R. (1995). *J. Fluorine Chemistry*, 72, 29.

[85] Wan, H. & Modro, T. A. (1996). *Synthesis*, 1227.

[86] Mbianda, X. Y., Modro, T. A. & Van Rooyen, P. H. (1998). *Chem. Commun.* 741.

[87] Pourayoubi, M., Nečas, M. & Negari, M. (2012). *Acta Cryst.* C68, o51.

[88] Corbridge, D. E. C. (1995). *Phosphorus, an Outline of its Chemistry, Biochemistry and Technology*, 5th ed., p. 1179. New York: Elsevier Science.

[89] Steiner, T. (2002). *Angew. Chem. Int. Ed.* 41, 48.

[90] Amirkhanov, V. M., Ovchynnikov, V. A., Glowiak, T. & Kozlowski, H. (1997). *Z. Naturforsch. Teil B*, 52, 1331.

[91] Narula, P. M., Day, C. S., Powers, B. A., Odian, M. A., Lachgar, A., Pennington, W. T. & Noftle, R. E. (1999). *Polyhedron*, 18, 1751.

[92] Gilli, P., Bertolasi, V., Ferretti, V. & Gilli, G. (1994). *J. Am. Chem. Soc.* 116, 909.

[93] Chivers, T., Krahn, M., Schatte, G. & Parvez, M. (2003). *Inorg. Chem.* 42, 3994.

[94] Hartmann, F., Dahlems, T. & Mootz, D. (1998). *Z. Kristallogr. New Cryst. Struct.* 213, 639.

[95] Anjum, S., Atta-ur-Rahman & Fun, H.-K. (2006). *Acta Cryst.* E62, o4569.

[96] Gholivand, K., Mostaanzadeh, H., Shariatinia, Z. & Oroujzadeh, N. (2006). *Main Group Chem.* 5, 95.

[97] Li, C., Dyer, D. J., Rath, N. P. & Robinson, P. D. (2005). *Acta Cryst.* C61, o654.

[98] Romming, C. & Songstad, J. (1982). *Acta Chem. Scand.* A36, 665.

[99] Du, D.-M., Fang, T., Xu, J. & Zhang, S.-W. (2006). *Org. Lett.* 8, 1327.

[100] Gholivand, K., Alavi, M. D. & Pourayoubi, M. (2004). *Z. Kristallogr. New Cryst. Struct.* 219, 124.

[101] Benincori, T., Marchesi, A., Pilati, T., Ponti, A., Rizzo, S. & Sannicolô, F. (2009). *Chem. Eur. J.* 15, 94.

[102] Vyšvařil, M., Dastych, D., Taraba, J. & Nečas, M. (2009). *Inorg. Chim. Acta*, 362, 4899.

[103] Cameron, T. S., Magee, M. G. & McLean, S. (1976). *Z. Naturforsch. Teil B*, 31, 1295.

[104] Cameron, T. S. (1977). *Z. Naturforsch. Teil B*, 32, 1001.

[105] Li, N., Jiang, F., Chen, L., Li, X., Chen, Q. & Hong, M. (2011). *Chem. Commun.* 47, 2327.

[106] Liu, X., Ilankumaran, P., Guzei, I. A. & Verkade, J. G. (2000). *J. Org. Chem.* 65, 701.

[107] Bourne, S. A., Mbianda, X. Y., Modro, T. A., Nassimbeni, L. R. & Wan, H. (1998). *J. Chem. Soc., Perkin Trans.* 2, 83.

[108] Hsu, I.-N. & Craven, B. M. (1974). *Acta Cryst.* B30, 1299.

[109] Schmuck, C. & Wienand, W. (2003). *J. Am. Chem. Soc.* 125, 452.

[110] Fei, Z., Neda, I., Thönnessen, H., Jones, P. G. & Schmutzler, R. (1997). *Phosphorus, Sulfur, Silicon, Relat. Elem.* 131, 1.

[111] Gholivand, K., Shariatinia, Z., Yaghmaian, F. & Faramarzpour, H. (2006). *Bull. Chem. Soc. Jpn.* 79, 1604.

[112] Hutton, A. T., Modro, T. A., Niven, M. L. & Scaillet, S. (1986). *J. Chem. Soc., Perkin Trans.* 2, 17.

[113] Vasil'chenko, V. N., Mitkevich, V. V., Moiseenko, A. A., Khomenko, V. G. & Chernov, V. A. (1982). *Zh. Strukt. Khim.* 23, 107.

[114] Gholivand, K., Tadjarodi, A. & Ng, S. W. (2002). *Acta Cryst.* E58, o200.

[115] Kattuboina, A. & Li, G. (2008). *Tetrahedron Lett.* 49, 1573.

[116] Raissi Shabari, A., Pourayoubi, M., Taghizadeh, A., Ghoreishi, F. & Vahdani, B. (2011). *Acta Cryst.* E67, o2167.

[117] Rudd, M. D., Lindeman, S. V. & Husebye, S. (1996). *Acta Chem. Scand.* 50, 759.

[118] Hempel, A., Camerman, N., Mastropaolo, D. & Camerman, A. (1999). *Acta Cryst.* C55, 1173.

[119] Ghadimi, S., Pourayoubi, M. & Ebrahimi Valmoozi, A. A. (2009). *Z. Naturforsch. Teil B*, 64, 565.

[120] Sa, M. M., Silveira, G. P., Bortoluzzi, A. J. & Padwa, A. (2003). *Tetrahedron*, 59, 5441.

[121] Gholivand, K., Shariatinia, Z., Mahzouni, H. R. & Amiri, S. (2007). *Struct. Chem.* 18, 653.

[122] Bullen, G. J. & Dann, P. E. (1973). *Acta Cryst.* B29, 331.

[123] Engelhardt, U. & Franzmann, A. (1987). *Acta Cryst.* C43, 1313.

[124] Cadogan, J. I. G., Gould, R. O., Gould, S. E. B., Sadler, P. A., Swire, S. J. & Tait, B. S. (1975). *J. Chem. Soc., Perkin Trans.* 1, 2392.

[125] Attig, R. & Mootz, D. (1976). *Z. Anorg. Allg. Chem.* 419, 139.

[126] Ledger, J., Boomishankar, R. & Steiner, A. (2010). *Inorg. Chem.* 49, 3896.

[127] Fu, Z. & Liu, X. (2008). *Acta Cryst.* E64, o2171.

[128] Balazs, G., Drake, J. E., Silvestru, C. & Haiduc, I. (1999). *Inorg. Chim. Acta*, 287, 61.

[129] Gholivand, K., Dorosti, N., Ghaziany, F., Mirshahi, M. & Sarikhani, S. (2012). *Heteroatom Chem.* 23, 74

8

Phase Behavior and Crystal Structure of Binary Polycyclic Aromatic Compound Mixtures

Jinxia Fu[1,*], James W. Rice[2] and Eric M. Suuberg[2]
[1]Brown University Department of Chemistry, Providence, RI
[2]Brown University School of Engineering, Providence, RI
USA

1. Introduction

Polycyclic aromatic hydrocarbons (PAHs) are a class of compounds that consist of multiple fused aromatic rings. Concerns have been raised regarding PAHs due to their known health effects(Luthy et al., 1994; Sun et al., 2003). In addition PAHs, chlorinated and brominated polycyclic aromatic hydrocarbons (ClPAHs and BrPAHs) are of interest commercially and of concern for their environmental effects (Shiraishi et al., 1985; Haglund et al., 1987; Nilsson and Ostman, 1993; Koistinen et al., 1994a, b; Ishaq et al., 2003; Kitazawa et al., 2006; Horii et al., 2008; Horii et al., 2009; Ohura et al., 2009; Ni et al., 2010; Ohura et al., 2010). The thermodynamic properties of pure PAHs have been widely studied for more than 50 years (Szczepanik et al., 1963; Wakayama and Inokuchi, 1967; Murray and Pottie, 1974; De Kruif, 1980; Mackay et al., 1982; Bender et al., 1983; Sonnefeld et al., 1983; Hansen and Eckert, 1986; Sato et al., 1986; Hinckley et al., 1990; Nass et al., 1995; Oja and Suuberg, 1997; Ruzicka et al., 1998; Chickos and Acree, 1999; Shiu and Ma, 2000; Burks and Harmon, 2001; Lei et al., 2002; Mackay et al., 2006; Odabasi et al., 2006; Goldfarb and Suuberg, 2008b, a, c; Ma et al., 2010). However, PAHs and halogenated polycyclic aromatic hydrocarbons (HPAHs) often exist as solid and/or liquid mixtures. Therefore it is also important to understand the phase behavior and crystal structures of these PAH and HPAH mixtures.

Phase behavior involving solid-liquid equilibrium is the basis for crystallization in chemical and materials engineering. Binary mixture systems can have up to three degrees of freedom according to the Gibbs phase rule,

$$F=C-P+2 \tag{1}$$

where F is the degrees of freedom, C is the number of components, and P is the number of phases. Therefore, the equilibrium of binary systems is determined by three variables such as temperature, pressure, and composition, and this is of course increased by one compositional variable for each additional component.

More than half of the true binary organic mixture systems in the literature exhibit simple eutectic behavior (Matsuoka, 1991) (see Figure 1(A)), while about 10% of binary solid systems form solid solutions (Matsuoka, 1991) (see Figure 1(B)), in which the atoms or molecules of one of the components occupy sites in the crystal lattice of the other component without modifying its crystal structure. Additionally, about a quarter of these systems form intermolecular compounds (Matsuoka, 1991), such as monotectics (see Figure 1(C)). However, only limited research has been done on binary organic mixture systems, especially PAH binary mixture systems. Moreover, crystal morphology, i.e., polymorphs, racemates, and structural isomers, also affect the phase diagram and may induce non-ideal solid-liquid equilibrium.

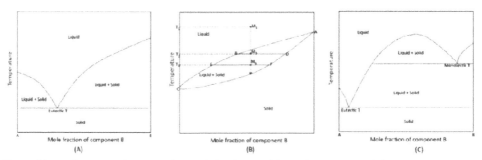

Fig. 1. Phase diagram of eutectic (A), solid solution (B), and monotectic (C) systems.

2. Eutectic systems

Figure 1(A) shows a phase diagram of a typical eutectic mixture system, which has a minimum melting temperature, i.e. a eutectic point. The eutectic point of a binary condensed mixture is defined as the temperature at which a solid mixture phase is in equilibrium with the liquid phase and a eutectic is generally considered to be a simple mechanical mixture of the solid and liquid (Rastogi and Bassi, 1964).

As in many other binary alloy mixtures, most PAH binary mixture systems exhibit eutectic behavior. Table 1 lists the eutectic point and eutectic concentration for about 50 binary PAH-containing mixture systems, in which at least one of the components is a PAH. The shape of the phase diagram for most of these binary mixture systems is similar to the phase diagram of anthrancene + pyrene mixture system (see Figure 2), except for a few systems, whose eutectic concentration is quite close to a pure component, such as in the naphthalene + chrysene system and phenanthrene + chrysene system.

For the studies preformed by this group on the anthracene + pyrene system (Rice et al., 2010), mixtures were prepared at various compositions by using a melt and quench-cool technique (Fu et al., 2010). Generally, the melting points and enthalpies of fusion of these PAH binary mixtures were found to often actually be independent of mixture preparation techniques. The liquidus and thaw points were determined according to the method proposed by Pounder an Masson (Pounder and Masson, 1934). The thaw temperature is the temperature at which the first droplet of liquid appears in a mixture-containing capillary. The liquidus temperature is the maximum temperature at which both solid crystals and liquid are observed to coexist. Above this temperature, there is only liquid phase present.

System	T_{fus1}/K	T_{fus2}/K	x_1	T_E/K
Naphthalene(1) + α-Naphthylamine(2) (Rastogi and Rama Varma, 1956)	353.5	323.2	0.360	301.3
Naphthalene(1) + α-Naphthol(2) (Rastogi and Rama Varma, 1956)	353.5	368.2	0.487	327.7
Naphthalene(1) + Phenanthrene(2) (Rastogi and Rama Varma, 1956; Rastogi and Bassi, 1964)	353.5	373.2	0.558	321.3
Naphthalene(1) + 2-methylnaphthalene(2) (Szczepanik et al., 1963)	353.5	307.6	0.362	298.7
Naphthalene(1) + Thionaphthene(2) (Szczepanik et al., 1963; Szczepanik and Ryszard, 1963)	353.5	305.2	0.063	302.4
Naphthalene(1) + Biphenyl(2) (Szczepanik et al., 1963; Szczepanik and Ryszard, 1963)	353.5	343.7	0.442	312.4
Naphthalene(1) + 2,6-dimethylnaphthalene(2)(Szczepanik and Ryszard, 1963)	353.5	383.2	0.665	333.7
Naphthalene(1) + 2,3-dimethylnaphthalene(2)(Szczepanik and Ryszard, 1963)	353.5	377.2	0.666	327.4
Naphthalene(1) + Acenaphthene(2) (Szczepanik et al., 1963; Szczepanik and Ryszard, 1963)	353.5	368.5	0.564	324.6
Naphthalene(1) + Fluorene(2) (Szczepanik et al., 1963; Szczepanik and Ryszard, 1963)	353.5	388.2	0.613	330.2
Naphthalene(1) + Phenanthrene (2) (Szczepanik et al., 1963; Szczepanik and Ryszard, 1963)	353.5	373.2	0.552	323.2
Naphthalene(1) + Fluoranthene (2) (Szczepanik et al., 1963; Szczepanik and Ryszard, 1963)	353.5	383.2	0.612	331
Naphthalene(1) + Pyrene (2) (Szczepanik et al., 1963; Szczepanik and Ryszard, 1963)	353.5	423.2	0.746	339.2
Naphthalene(1) + Chrysene(2) (Szczepanik et al., 1963; Szczepanik and Ryszard, 1963)	353.5	528.2	0.971	351.4
Biphenyl(1) + Fluorene(2) (Szczepanik et al., 1963; Szczepanik and Ryszard, 1963)	343.7	388.2	0.909	340.8
Biphenyl(1) + Acenaphthene(2) (Szczepanik et al., 1963; Szczepanik and Ryszard, 1963)	343.7	368.5	0.641	319.3

Diphenylene oxide(1) + Acenaphthene(2) (Szczepanik et al., 1963; Szczepanik and Ryszard, 1963)	359.2	368.5	0.578	326
Fluorene(1) + Acenaphthene(2) (Szczepanik et al., 1963; Szczepanik and Ryszard, 1963)	388.2	368.5	0.431	338.6
Fluorene(1) + 2,3,6-trimethylnaphthalene(2) (Szczepanik et al., 1963; Szczepanik and Ryszard, 1963)	388.2	375.2	0.658	361.6
Phenanthrene(1) + Biphenyl(2) (Szczepanik et al., 1963; Szczepanik and Ryszard, 1963)	373.2	343.7	0.691	324.8
Phenanthrene(1) + Acenaphthene(2) (Szczepanik et al., 1963; Szczepanik and Ryszard, 1963)	373.2	368.5	0.495	327.5
Phenanthrene(1) + 2,3,6-trimethylnaphthalene(2) (Szczepanik et al., 1963; Szczepanik and Ryszard, 1963)	373.2	375.2	0.704	331.5
Phenanthrene(1) + Fluorene(2) (Szczepanik et al., 1963; Szczepanik and Ryszard, 1963)	373.2	388.2	0.637	368.7
Phenanthrene(1) + Pyrene(2) (Szczepanik et al., 1963; Szczepanik and Ryszard, 1963)	373.2	423.2	0.747	354.7
Phenanthrene(1) + 3-methylphenanthrene(2) (Szczepanik et al., 1963; Szczepanik and Ryszard, 1963)	373.2	332	0.318	309.7
Phenanthrene(1) + 4,5-dimethylphenanthrene(2) (Szczepanik et al., 1963; Szczepanik and Ryszard, 1963)	373.2	388.2	0.621	342.6
Phenanthrene(1) + Fluoranthene (2) (Szczepanik et al., 1963; Szczepanik and Ryszard, 1963)	373.2	383.2	0.532	347.7
Phenanthrene(1) + Chrysene(2) (Szczepanik et al., 1963; Szczepanik and Ryszard, 1963)	373.2	528.2	0.957	369.2
Anthracene(1) + Carbazole(2) (Szczepanik et al., 1963; Szczepanik and Ryszard, 1963)	489.8	518	0.943	488.4
Anthracene(1) + 2-methylanthracene(2) (Szczepanik et al., 1963; Szczepanik and Ryszard, 1963)	489.8	472.2	0.108	471.5
Anthracene(1) + Chrysene(2) (Szczepanik et al., 1963; Szczepanik and Ryszard, 1963)	489.8	528.2	0.662	464.6

Anthracene(1) + Pyrene(2) (Szczepanik et al., 1963; Szczepanik and Ryszard, 1963)	489.8	423.2	0.221	404.6
Carbazole(1) + Fluoranthene(2) (Szczepanik et al., 1963; Szczepanik and Ryszard, 1963)	518	383.2	0.119	377.3
Carbazole(1) + Pyrene(2) (Szczepanik et al., 1963; Szczepanik and Ryszard, 1963)	518	423.2	0.154	409.1
Carbazole(1) + Chrysene(2) (Szczepanik et al., 1963; Szczepanik and Ryszard, 1963)	518	528.2	0.578	480.6
Fluoranthene(1) + Acenaphthene(2) (Szczepanik et al., 1963; Szczepanik and Ryszard, 1963)	383.2	368.5	0.433	336.9
Fluoranthene(1) + Fluorene(2) (Szczepanik et al., 1963; Szczepanik and Ryszard, 1963)	383.2	388.2	0.516	342.7
Fluoranthene(1) + 2-methylanthracene (2) (Szczepanik et al., 1963; Szczepanik and Ryszard, 1963)	383.2	472.2	0.794	368.7
Fluoranthene(1) + Pyrene(2) (Szczepanik et al., 1963; Szczepanik and Ryszard, 1963)	383.2	423.2	0.800	368.3
Fluoranthene(1) + Chrysene(2) (Szczepanik et al., 1963; Szczepanik and Ryszard, 1963)	383.2	528.2	0.952	379.6
Pyrene(1) + Chrysene(2) (Szczepanik et al., 1963; Szczepanik and Ryszard, 1963)	423.2	528.2	0.855	405.7
Acenaphthene (1) + 1,2-dimethylbenzene(2) (Szczepanik et al., 1963; Szczepanik and Ryszard, 1963)	368.5	298.7	0.055	295
Acenaphthene (1) + 1,2,4,5-tetramethylbenzene(2) (Szczepanik et al., 1963; Szczepanik and Ryszard, 1963)	368.5	352.3	0.423	323.6
Acenaphthene (1) + 2-methylnaphthalene(2) (Szczepanik et al., 1963; Szczepanik and Ryszard, 1963)	368.5	307.5	0.212	290.9
Acenaphthene (1) + 2,6-dimethylnaphthalene (2) (Szczepanik et al., 1963; Szczepanik and Ryszard, 1963)	368.5	383.2	0.598	339.7
Acenaphthene (1) + 2,7-dimethylnaphthalene (2) (Szczepanik et al., 1963; Szczepanik and Ryszard, 1963)	368.5	370.2	0.531	333.9

Acenaphthene (1) + Naphthalene (2) (Szczepanik et al., 1963; Szczepanik and Ryszard, 1963)	368.5	353.2	0.417	323.2
Acenaphthene (1) + Phenanthrene(2) (Szczepanik et al., 1963; Szczepanik and Ryszard, 1963)	368.5	372.5	0.492	329
Acenaphthene (1) + Fluorene(2) (Szczepanik et al., 1963; Szczepanik and Ryszard, 1963)	368.5	387.2	0.582	337.7
Acenaphthene (1) + Anthracene(2) (Szczepanik et al., 1963; Szczepanik and Ryszard, 1963)	368.5	489.7	0.914	361.2

Table 1. Melting temperatures of previously reported binary PAH eutectic systems

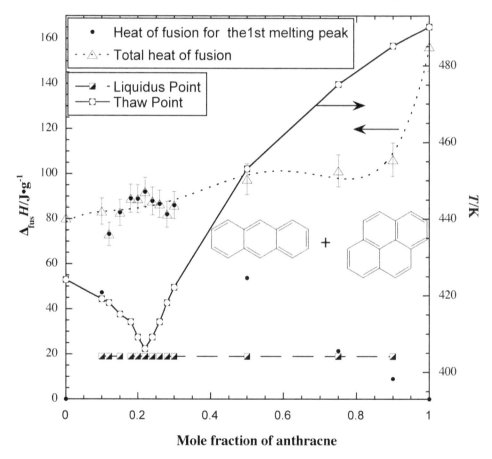

Fig. 2. Phase diagram and enthalpy of fusion of the anthracene (1) + pyrene (2) system (Rice et al., 2010).

The eutectic point for the anthracene (1) + pyrene (2) system occurs at 404 K at $x_1 = 0.22$ (see Figure 2). Only solid state exists below the thaw curve, i.e. eutectic temperature, and only liquid state exists above the liquidus curve. The areas between these two curves exhibit the coexistence of both solid and liquid phases.

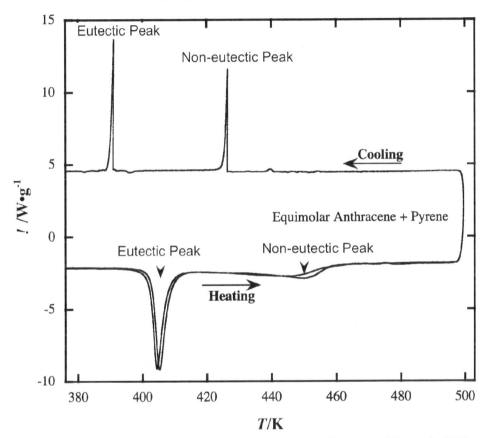

Fig. 3. Full DSC scan of an equimolar anthracene (1) + pyrene (2) mixture (Rice et al., 2010).

Figure 2 also displays the correlation between phase behavior and enthalpy of fusion, $\Delta_{fus}H$ for the system. The $\Delta_{fus}H$ observed for a DSC peak near the eutectic temperature of 404 K indicates the heat input for the initial melting of a eutectic solid phase to occur. The total $\Delta_{fus}H$ shown in Figure 2 is a summation of both endothermic phase transition peaks observed in the DSC scan, i.e. the eutectic phase melting and the non-eutectic phase melting (see Figure 3). It is worth noting that the total $\Delta_{fus}H$ is very similar to that of pure pyrene over a wide range of compositions and thus the $\Delta_{fus}H$ for both pure pyrene and the eutectic mixture are very similar. This means that when the mixture contains only a modest amount of anthracene, energetically it behaves quite similarly to pure pyrene, and this persists until the mixture is nearly pure anthracene (see Figure 2). There is a slight increase in fusion enthalpy when the mixtures are enriched in anthracene beyond the eutectic composition, but the shift is only modest as compared with the increase of fusion enthalpy to that of pure anthracene (see Figure 2). This indicates that the ability of anthracene to reach a lower energy crystalline configuration is significantly impeded by the presence of relatively small amounts of pyrene.

Additionally, Powder X-ray diffraction patterns for the same anthracene (1) + pyrene (2) system were also obtained. Figure 4 shows that the crystal structure of the eutectic mixture is similar to that of pyrene because peaks at 10.6, 11.6, 14.9, 16.3, 18.2, 23.3, 24.7 and 28.0 degree are all retained in the mixture diffraction pattern. This is consistent with the DSC result that implies that the $\Delta_{fus}H$ of the eutectic is very close to that of pure pyrene, and indicates that the crystal structures of the eutectic mixture and pure pyrene are similar. Likewise, Figure 4 shows that the crystal structure of a mixture at $x_1 = 0.90$ is comparable to that of pure anthracene.

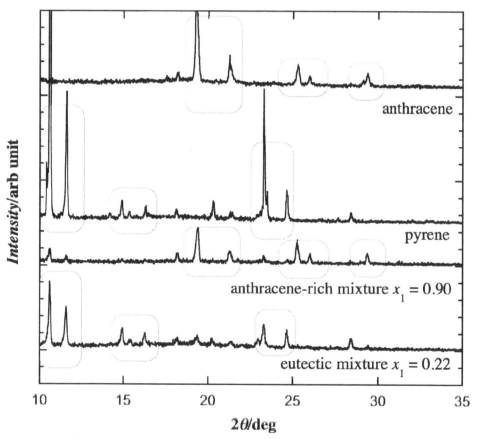

Fig. 4. X-ray diffraction patters of pure components and mixtures of anthracene (1) + pyrene (2) (Rice et al., 2010).

3. Monotectic systems

In contrast to eutectic systems, in which both components solidify below eutectic temperature, a monotectic reaction is characterized by the breakdown of a liquid into one solid and one liquid phase (Singh et al., 1985), i.e. one liquid phase decomposes into a solid phase and a liquid phase when the temperature is below the monotectic temperature. Figure 1(C) shows the phase diagram of a typical monotectic system. The monotectic composition is determined by the intersection of a liquidus line and a liquid miscibility gap (Singh et al., 1985). Generally, monotectic systems are less studied than eutectic systems.

Binary organic mixtures with PAHs can form monotectic systems. Table 2 lists the monotectic and eutectic point of a few monotectic forming PAH systems. Monotectic systems are characterized by monotectic, eutectic and upper consolute temperatures, though the upper consolute temperature is often not reported. The monotectic temperature, t_M, is the temperature at monotectic composition and the upper consolute temperature is the highest melting temperature of the mixture system, i.e. the critical point where the two liquid phases having identical composition become indistinguishable.

System	T_{fus1}/K	T_{fus2}/K	x_M	T_M/K	x_E	T_E/K
2,4-Dinitrophenol(1) + Naphthalene(2) (Singh et al., 2001; Singh et al., 2007)	378.2	353.2	0.316	357.7	0.838	344.2
Succinonitrile (1) + Pyrene(2) (Rai and Pandey, 2002)	330.2	423.2	0.025	416.5	0.744	328.5
Succinonitrile (1) + Phenanthrene (2) (Singh et al., 1985)	330.2	373.2	0.225	363.2[a]	0.975	~328.2
p-benzoquinone(1) + Pyrene(2) (Gupta and Singh, 2004)	388.2	423.2	0.324	392.2	0.792	376.2
m-dinitrobenzene(1)+ Pyrene(2) (Gupta and Singh, 2004)	362.2	423.2	0.301	363.2	0.702	361.2
m-nitrobenzoic acid(1) +Pyrene(2) (Gupta and Singh, 2004)	413.2	423.2	0.902	413.2	0.299	403.2

Table 2. Melting temperatures of previously reported binary PAH monotectic systems

Rai and Pandey studied the phase behavior of succinonitrile (1) + pyrene (2) mixture system (Rai and Pandey, 2002), which is a typical monotectic system (Figure 5). The enthalpy of fusion of pyrene, 17.65 kJ mole[-1] (Chickos and Acree, 1999), is much higher than that of succinonitrile, 3.7 kJ mole[-1] (Rai and Pandey, 2002). The monotectic point is 416.5 K (143.3°C) at $x_1=0.025$. The eutectic temperature is 328.5 K (55.4°C) at $x_1=0.744$ and the upper consolute temperature, t_C (465.2 K, 192.0°C), is 48.7 K above the monotectic point. When x_1 is between monotectic and eutectic composition, the two liquids, L_1 (rich in pyrene) and L_2 (rich in succinonitrile) are mutually immiscible. However, if the temperature is above the consolute temperature, there is complete miscibility in liquid state, i.e. only one liquid phase exists.

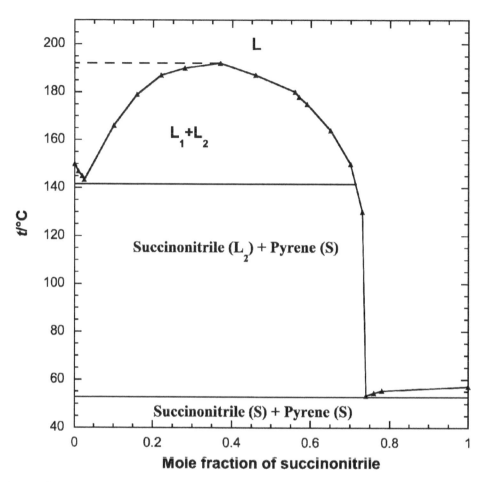

Fig. 5. Phase diagram of succinonitrile (1) + pyrene (2) mixture system (Rai and Pandey, 2002).

4. Solid solution

A solid solution is a solid mixture in which one or more atoms and/or molecules of one of the components occupies sites in the crystal lattice of the other component without significantly changing its crystal structure, even though the lattice parameter may vary. So this kind of system has a homogenous crystalline structure and is also called isomorphic system, because the components are completely miscible in both the liquid and solid phases. Figure 1(B) shows the phase behavior of a binary mixture system that forms a solid solution. In the diagram, the curve ABC and ADC are the liquidus and solidus curves, respectively. The area above ABC curve represents the region of homogeneous liquid solutions and the area below ADC curve represents the region of homogeneous solid solution. The area enclosed by ABCD is the region of liquid + solid solution. For instance, a mixture M_1 at temperature $T_1$is cooled to temperature T_2, the mixture M_2 becomes a mixture of liquid B and solid D. If M_2 is further cooled to temperature T_3, the liquid composition changes continuously from B to E along the liquidus curve, while the solid composition changes from D to F along the solidus curve. Additionally, the Hume-Rothery rules, named after William Hume-Rothery, are used to describe the conditions under which an element can dissolve in a metal and form a solid solution.

Szczepanik and Skalmowski (Szczepanik et al., 1963; Szczepanik and Ryszard, 1963) studied the phase behavior of over 60 PAH binary mixture systems, and demonstrate that PAH mixture systems also form solid solution, as shown by naphthalene + 1-methynaphthanlene, naphthalene + anthracene, phenanthrene + anthracene, phenanthrene + carbazole, anthracene + acridine, anthracene + fluoranthene, and chrysene + 1,2-benzanthracene systems. It is not known whether the the Hume-Rothery rules still work for PAH mixtures. However, it is worth noting that the number of such systems is small, compared with the number of eutectic-forming systems.

5. Systems with complex phase behavior

Due to the large molecular mass and complexity of the crystal structure of PAHs, i.e. polymorphism and racemate, the phase behavior of some of the PAH binary mixtures may be different from the above described three phase behaviors. Three complicated PAH binary mixture systems, i.e. anthracene + benzo[a]pyrene system (Rice and Suuberg, 2010), pyrene + 9,10-dibromoanthracene system (Fu et al., 2010), and anthracene + 2-bromoanthracene are introduced here.

5.1 Anthracene + benzo[a]pyrene system

Benzo[a]pyrene has a much larger molecular mass compared to pyrene, which leads to phase behavior in the anthracene (1) + benzo[a]pyrene (2) system (Rice and Suuberg, 2010) that is different from that of the anthracene + pyrene system. The phase diagram of anthracene (1) + benzo[a]pyrene (2) system (see Figure 6) indicates an eutectic-like mixture behavior. A eutectic-like phase is formed near $x_1 = 0.26$ between 414 and 420 K. There is however always a gap between the thaw curve and the lowest liquidus temperature, which is distinct from true eutectic behavior such as in Figure 1(A) or Figure 2. Therefore, mixtures of anthracene and benzo[a]pyrene form a single, amorphous, solid eutectic-like phase at $x_1 = 0.26$ that lacks any organized crystal structure and which melts throughout the 414 to 420 K temperature range. This region of phase transition, represented by the shaded region of Figure 6, is not rate dependant and is observed in both the DSC and melting temperature analysis for all combinations of anthracene + benzo[a]pyrene, providing evidence that this region represents the melting temperature range of a single, amorphous, solid phase. This conclusion is also supported by the X-ray diffraction results.

Powder X-ray diffraction studies were conducted to study the crystal structures of the anthracene (1) + benzo[a]pyrene (2) system (see Figure 7). The eutectic-like mixture lacks any organized crystal structure because the few peaks that exist in the X-ray pattern are not well defined and do not rise much above the baseline. Additionally, there is no real similarity between the eutectic mixture scan and those of the pure components. This result is consistent with the melting point studies that imply that the mixtures form a single, amorphous solid phase at the eutectic composition.

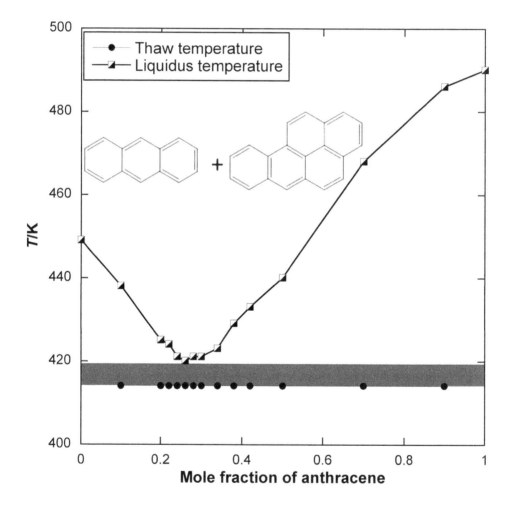

Fig. 6. Phase diagram of anthracene (1) + benzo[a]pyrene (2) system (Rice and Suuberg, 2010).

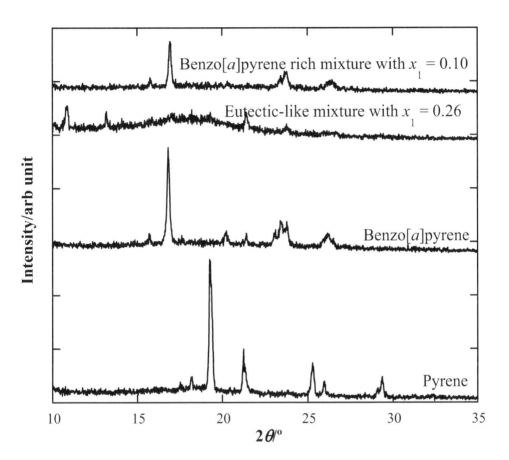

Fig. 7. X-ray diffraction patters of pure components and mixtures of anthracene (1) + benzo[a]pyrene (2) system (Rice and Suuberg, 2010).

5.2 Pyrene + 9,10-dibromoanthracene system

The influence of halogen substitution on the interaction energy between PAH molecules has also been investigated. Unlike the anthracene + pyrene mixture system, bromine substitution on anthracene induces a different kind of interaction in the pyrene (1) + 9,10-dibromoanthracene (2) mixture system, which also results in non-idealities in solid-liquid equilibrium (see Figure 8). The surface area and volume of the 9,10-dibromoanthracene molecule is much larger than that of pyrene.

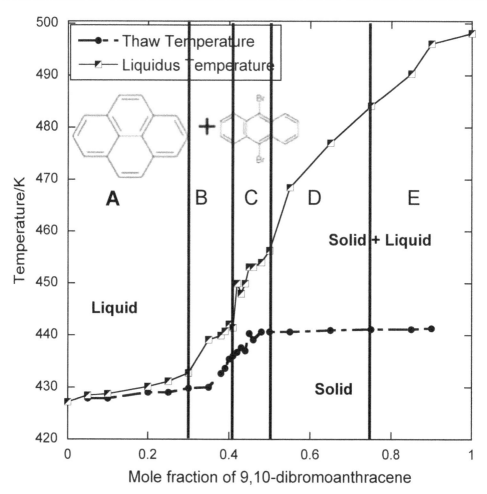

Fig. 8. Phase diagram of pyrene (1) + 9,10-dibromoanthracene (2) mixture system
(Fu et al., 2010).

The phase diagram of this system can be crudely divided into 5 regions. The mixtures with
relatively low mole fraction of 9,10-dibromoanthracene (< 0.30), in region A, form a pyrene
like phase. When the mole fraction of 9,10-dibromoanthracene is between 0.30-0.41, in
region B, the mixtures transition from a pyrene-like phase to two phases that both have low
melting temperatures. The divergence of the liquidus and thaw curve is 2-9 K. In region C,
mixtures containing about x_2 = 0.41-0.50 also show two-phase character and start to
transition to 9,10-dibromoanthracene behavior. Mixtures with x_2 = 0.50-0.75, in region D,
also have two phases with 9,10-dibromoanthracene like behavior and high melting
temperature. Only one of the phases evolves while the other gives a constant low melting
temperature (corresponding to the thaw point). In region E, a 9,10-dibromoanthracene like
phase is defined based upon the thermal behavior, shown below.

The full heating, cooling and reheating scan of a pyrene + 9,10-dibromoanthracene mixture at $x_2 = 0.48$ (in region C) is shown in Figure 9, where Φ is heat flow in the DSC. During the heating scan, two peaks appear at 428 K and 440 K, which indicates the two-phase character of the mixture. Two peaks are also observed in the cooling scan, in which the 9,10-dibromoanthracene like phase crystallizes first at 418 K, and then the pyrene like phase crystallizes at 410 K. The cooling scan also suggested two-phase behavior of the mixture just as did the melting behavior. When reheated, the phase transition enthalpies and associated temperatures matched those of the initial heating scan.

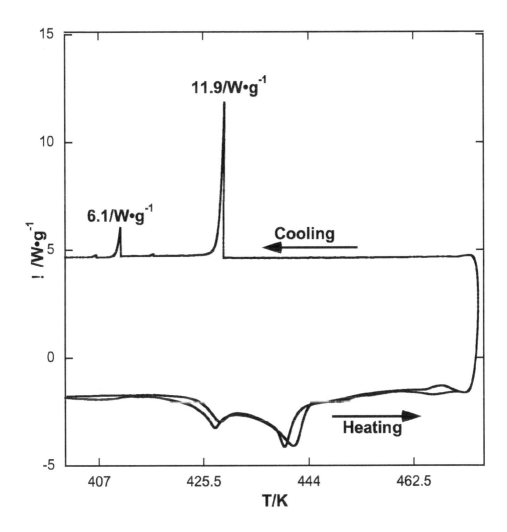

Fig. 9. Full DSC scan of a pyrene (1) + 9,10-dibromoanthracene (2) mixture at $x_2 = 0.48$ (Fu et al., 2010).

The temperature and enthalpy of crystallization (subcooled), shown in Figure 10, correspond to the results obtained from the phase diagram. Mixtures with a mole fraction of 9,10-dibromoanthraene 0.30-0.75, in regions B, C and D, have two-phase character, which is observed as two distinct phase-transition peaks during the cooling procedure. Region E showed two-phase melting behavior, but in the DSC experiments of Figure 10, the low temperature crystallization peak was absent. Likewise, region B showed two distinct melting peaks, whereas in the DSC experiment only a single peak was observed.

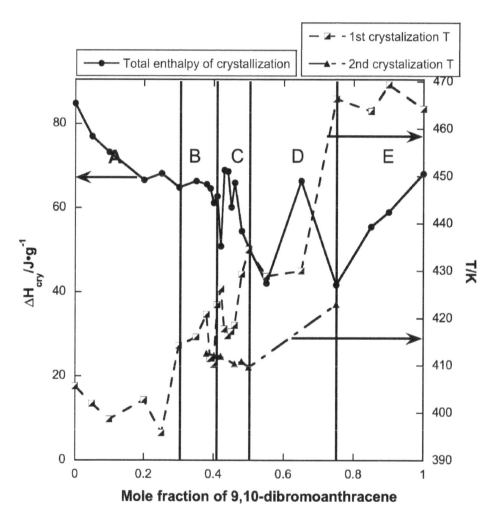

Fig. 10. Crystallization temperature and total enthalpy of crystallization of pyrene (1) + 9,10-dibromoanthracene (2) mixtures. 1st crystallization temperature is the higher temperature peak in the DSC cooling scan, and 2nd crystallization temperature is the lower temperature peak in the DSC cooling scan (Fu et al., 2010).

5.3 Anthracene + 2-bromoanthracene system

The influence of bromine substitution on thermochemical properties of PAH mixture systems was further investigated by studying the anthracene (1) + 2-bromoanthracene (2) system. The crystal structure is changed by addition one bromine atom on the aromatic ring. Moreover, the surface area and volume of 2-bromoanthracene is about 10% bigger than those of anthracene.

The solid-liquid equilibrium diagram of anthracene (1) + 2-bromoanthracene (2) system is shown in Figure 12. The diagram suggests the non-ideality of the anthracene + 2-bromoanthracene system. The melting temperature range (thaw to completion) of these mixtures at any given composition is observed to be 1.1 - 2.6 K. The reported solid-liquid equilibrium melting temperature is here taken as the thaw temperature, in Figure 12. The lowest solid-liquid equilibrium temperature for the system is 477.65 K at $x_1 = 0.74$, and the melting temperature range of this mixture is 1.8 K.

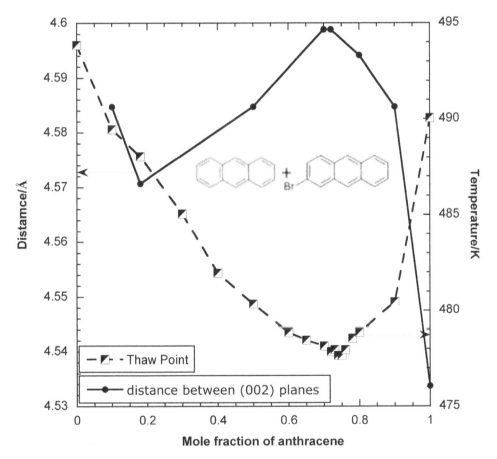

Fig. 12. Phase diagram and distance between (002) planes of anthracene (1) + 2-bromoanthracene (2) system.

5.3 Anthracene + 2-bromoanthracene system

The influence of bromine substitution on thermochemical properties of PAH mixture systems was further investigated by studying the anthracene (1) + 2-bromoanthracene (2) system. The crystal structure is changed by addition one bromine atom on the aromatic ring. Moreover, the surface area and volume of 2-bromoanthracene is about 10% bigger than those of anthracene.

The solid-liquid equilibrium diagram of anthracene (1) + 2-bromoanthracene (2) system is shown in Figure 12. The diagram suggests the non-ideality of the anthracene + 2-bromoanthracene system. The melting temperature range (thaw to completion) of these mixtures at any given composition is observed to be 1.1 - 2.6 K. The reported solid-liquid equilibrium melting temperature is here taken as the thaw temperature, in Figure 12. The lowest solid-liquid equilibrium temperature for the system is 477.65 K at $x_1 = 0.74$, and the melting temperature range of this mixture is 1.8 K.

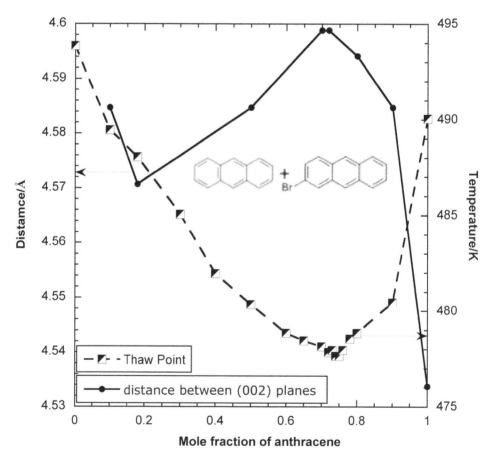

Fig. 12. Phase diagram and distance between (002) planes of anthracene (1) + 2-bromoanthracene (2) system.

The powder X-ray diffraction method was also used to study the crystal structures of pure anthracene, 2-bromoanthracene and their mixtures (see Figure 13). The lattice structure of anthracene crystals is monoclinic with a = 8.44 Å, b = 5.99 Å, c = 11.11 Å, β = 125.4° (Jo et al., 2006). The strong diffraction peak at 19.58° in pure anthracene corresponds to the (002) plane, and the spacing between the 002 planes is 4.53 Å. With the increase of the mole fraction of 2-bromoanthracene, x_2, in the mixture, the (002) plane spacing starts to shift to lower values. Moreover, a new diffraction peak occurs near 2θ = 17° with increasing x_2 in the mixture. This indicates that new mixture crystals are formed. The new peak appears at 2θ = 16.38° when x_1 = 0.70 roughly corresponding to the lowest solid-liquid equilibrium melting point. With increase of x_1, the peak position increases from 16.38° to 17.06° and disappears in pure anthracene. The diffraction data for mixtures with x_1 = 0.50 and 0.10 indicate relatively amorphous structures.

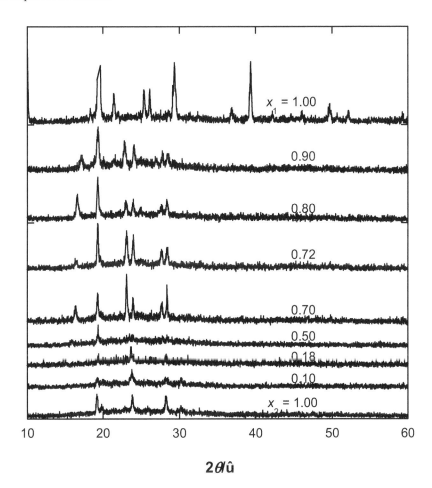

Fig. 13. X-ray diffraction patters of pure components and mixtures of anthracene (1) + 2-bromoanthracene(2).

The distance between 002 planes in the pure anthracene, pure 2-bromoanthracene and mixtures can be calculated by Bragg's law

$$n\lambda = 2d\sin\theta \tag{2}$$

where n is an integer, λ is the wavelength of the incident wave, d is the spacing between the planes in the atomic lattice, and θ is the angle between the incident ray and the scattering planes.

Figure 12 also shows changes of the distance between 002 planes in this system, which demonstrates that the spacings between 002 planes are stretched by adding 2-bromoanthracene into anthracene. The distance between 002 planes reaches a maximum when the mixture is near the lowest melting solid-liquid equilibrium point, which is in good agreement with the thermodynamic data in Figure 12, indicating the formation of the least stable solid state near the lowest solid-liquid equilibrium point. Interestingly, the mixture at $x_1 = 0.18$ gives a local minimum in the (002) plane spacing.

6. Conclusions

The phase behaviors of binary PAH-containing mixtures are complicated. Most of these mixture systems are eutectic systems, which have a behavior like the anthracene + pyrene system. Fewer binary PAH-containing mixtures can form monotectic and solid solution systems, such as succinonitrile + pyrene system and phenanthrene + anthracene system.

The phase behaviors of binary PAC mixtures are complicated and non-ideal. Mixtures with large PAHs, such as benzo[a]pyrene, can exhibit a gap between the thaw curve and liquidus curve. Halogen substitution (bromine substitution) also has significant effect on the thermochemical behaviors of binary PAC mixtures. Bromine substitution on anthracene results in non-ideal phase behavior in pyrene + 9,10-dibromoanthracene and anthracene + 2-bromoanthracene systems.

7. Acknowledgement

This project was supported by Grant Number P42 ES013660 from the National Institute of Environmental Health Sciences (NIEHS)/NIH, and the contents are solely the responsibility of the authors and do not necessarily represent the official views of the NIEHS/NIH.

8. References

Bender, R., Bieling, V., Maurer, G., 1983. The vapour pressures of solids: anthracene, hydroquinone, and resorcinol. The Journal of Chemical Thermodynamics 15, 585-594.

Burks, G.A., Harmon, T.C., 2001. Volatilization of solid-phase polycyclic aromatic hydrocarbons from model mixtures and lampblack-contaminated soils. J Chem Eng Data 46, 944-949.

Chickos, J.S., Acree, W.E., 1999. Estimating solid-liquid phase change enthalpies and entropies. J Phys Chem Ref Data 28, 1535-1673.

De Kruif, C., 1980. Enthalpies of sublimation and vapor-pressure of 11 polycyclic-hydrocarbons. J Chem Thermodyn 12, 243-248.

Fu, J., Rice, J.W., Suuberg, E.M., 2010. Phase behavior and vapor pressures of the pyrene + 9,10-dibromoanthracene system. Fluid Phase Equilibr 298, 219-224.

Goldfarb, J.L., Suuberg, E.M., 2008a. The effect of halogen hetero-atoms on the vapor pressures and thermodynamics of polycyclic aromatic compounds measured via the Knudsen effusion technique. J Chem Thermodyn 40, 460-466.

Goldfarb, J.L., Suuberg, E.M., 2008b. Vapor pressures and enthalpies of sublimation of ten polycyclic aromatic hydrocarbons determined via the Knudsen effusion method. J Chem Eng Data 53, 670-676.

Goldfarb, J.L., Suuberg, E.M., 2008c. Vapor pressures and thermodynamics of oxygen-containing polycyclic aromatic hydrocarbons measured using Knudsen effusion. Environ Toxicol Chem 27, 1244-1249.

Gupta, R.K., Singh, R.A., 2004. Thermochemical and microstructural studies on binary organic eutectics and complexes. Journal of Crystal Growth 267, 340-347.

Haglund, P., Alsberg, T., Bergman, A., Jansson, B., 1987. Analysis of Halogenated Polycyclic Aromatic-Hydrocarbons in Urban Air, Snow and Automobile Exhaust. Chemosphere 16, 2441-2450.

Hansen, P.C., Eckert, C.A., 1986. An Improved Transpiration Method for the Measurement of Very Low Vapor-Pressures. J Chem Eng Data 31, 1-3.

Hinckley, D.A., Bidleman, T.F., Foreman, W.T., Tuschall, J.R., 1990. Determination of Vapor-Pressures for Nonpolar and Semipolar Organic-Compounds from Gas-Chromatographic Retention Data. J Chem Eng Data 35, 232-237.

Horii, Y., Khim, J.S., Higley, E.B., Giesy, J.P., Ohura, T., Kannan, K., 2009. Relative Potencies of Individual Chlorinated and Brominated Polycyclic Aromatic Hydrocarbons for Induction of Aryl Hydrocarbon Receptor-Mediated Responses. Environ Sci Technol 43, 2159-2165.

Horii, Y., Ok, G., Ohura, T., Kannan, K., 2008. Occurrence and profiles of chlorinated and brominated polycyclic aromatic hydrocarbons in waste incinerators. Environ Sci Technol 42, 1904-1909.

Ishaq, R., Naf, C., Zebuhr, Y., Broman, D., Jarnberg, U., 2003. PCBs, PCNs, PCDD/Fs, PAHs and Cl-PAHs in air and water particulate samples - patterns and variations. Chemosphere 50, 1131-1150.

Jo, S., Yoshikawa, H., Fujii, A., Takenaga, M., 2006. Surface morphologies of anthracene single crystals grown from vapor phase. Appl Surf Sci 252, 3514-3519.

Kitazawa, A., Amagai, T., Ohura, T., 2006. Temporal trends and relationships of particulate chlorinated polycyclic aromatic hydrocarbons and their parent compounds in urban air. Environ Sci Technol 40, 4592-4598.

Koistinen, J., Paasivirta, J., Nevalainen, T., Lahtipera, M., 1994a. Chlorinated Fluorenes and Alkylfluorenes in Bleached Kraft Pulp and Pulp-Mill Discharges. Chemosphere 28, 2139-2150.

Koistinen, J., Paasivirta, J., Nevalainen, T., Lahtipera, M., 1994b. Chlorophenanthrenes, Alkylchlorophenanthrenes and Alkylchloronaphthalenes in Kraft Pulp-Mill Products and Discharges. Chemosphere 28, 1261-1277.

Lei, Y.D., Chankalal, R., Chan, A., Wania, F., 2002. Supercooled liquid vapor pressures of the polycyclic aromatic hydrocarbons. J Chem Eng Data 47, 801-806.

Luthy, R.G., Dzombak, D.A., Peters, C.A., Roy, S.B., Ramaswami, A., Nakles, D.V., Nott, B.R., 1994. Remediating Tar-Contaminated Soils at Manufactured-Gas Plant Sites. Environ Sci Technol 28, A266-A276.

Ma, Y.G., Lei, Y.D., Xiao, H., Wania, F., Wang, W.H., 2010. Critical Review and Recommended Values for the Physical-Chemical Property Data of 15 Polycyclic Aromatic Hydrocarbons at 25 degrees C. J Chem Eng Data 55, 819-825.

Mackay, D., Bobra, A., Chan, D.W., Shiu, W.Y., 1982. Vapor-Pressure Correlations for Low-Volatility Environmental Chemicals. Environ Sci Technol 16, 645-649.

Mackay, D., Shiu, W.Y., Ma, K.-C., Lee, S.C., 2006. Physical-Chemical Properties and Environmental Fate for Organic Chemicals. CRC Press.

Matsuoka, M.G., J. (ed); Davey, R.J (ed); Jones A. (ed), 1991. Advances in Industrial Crystallization. Butterworth-Heinemann, Oxford.

Murray, J.J., Pottie, R.F., 1974. The Vapor Pressures and Enthalpies of Sublimation of Five Polycyclic Aromatic Hydrocarbons. Canadian Journal of Chemistry 52, 557–563.

Nass, K., Lenoir, D., Kettrup, A., 1995. Calculation of the Thermodynamic Properties of Polycyclic Aromatic-Hydrocarbons by an Incremental Procedure. Angew Chem Int Edit 34, 1735-1736.

Ni, H.G., Zeng, H., Tao, S., Zeng, E.Y., 2010. Environmental and Human Exposure to Persistent Halogenated Compounds Derived from E-Waste in China. Environ Toxicol Chem 29, 1237-1247.

Nilsson, U.L., Ostman, C.E., 1993. Chlorinated Polycyclic Aromatic-Hydrocarbons - Method of Analysis and Their Occurrence in Urban Air. Environ Sci Technol 27, 1826-1831.

Odabasi, M., Cetin, E., Sofuoglu, A., 2006. Determination of octanol-air partition coefficients and supercooled liquid vapor pressures of PAHs as a function of temperature: Application to gas-particle partitioning in an urban atmosphere. Atmos Environ 40, 6615-6625.

Ohura, T., Morita, M., Kuruto-Niwa, R., Amagai, T., Sakakibara, H., Shimoi, K., 2010. Differential Action of Chlorinated Polycyclic Aromatic Hydrocarbons on Aryl Hydrocarbon Receptor-Mediated Signaling in Breast Cancer Cells. Environ Toxicol 25, 180-187.

Ohura, T., Sawada, K.I., Amagai, T., Shinomiya, M., 2009. Discovery of Novel Halogenated Polycyclic Aromatic Hydrocarbons in Urban Particulate Matters: Occurrence, Photostability, and AhR Activity. Environ Sci Technol 43, 2269-2275.

Oja, V., Suuberg, E.M., 1997. Development of a nonisothermal Knudsen effusion method and application to PAH and cellulose tar vapor pressure measurement. Anal Chem 69, 4619-4626.

Pounder, F.E., Masson, I., 1934. Thermal analysis, and its application to dinitrobenzenes. J. Chem. Soc., 1357-1360.

Rai, U.S., Pandey, P., 2002. Crystallization behaviour of metal-nonmetal monotectic alloys; Succinonitrile-pyrene system. Progress in Crystal Growth and Characterization of Materials 45, 59-64.

Rastogi, R.P., Bassi, P.S., 1964. Mechanism of Eutectic Crystallization. The Journal of Physical Chemistry 68, 2398-2406.

Rastogi, R.P., Rama Varma, K.T., 1956. Solid–liquid equilibria in solutions of non-electrolytes. Journal of the Chemical Society, 2097-2101.

Rice, J.W., Fu, J., Suuberg, E.M., 2010. Anthracene + Pyrene Solid Mixtures: Eutectic and Azeotropic Character. Journal of Chemical & Engineering Data, doi: 10.1021/je100208e.

Rice, J.W., Suuberg, E.M., 2010. Thermodynamic study of (anthracene + benzo[a]pyrene) solid mixtures. The Journal of Chemical Thermodynamics 42, 1356-1360.

Ruzicka, K., Mokbel, I., Majer, V., Ruzicka, V., Jose, J., Zabransky, M., 1998. Description of vapour-liquid and vapour-solid equilibria for a group of polycondensed compounds of petroleum interest. Fluid Phase Equilibria 148, 107-137.

Sato, N., Inomata, H., Arai, K., Saito, S., 1986. Measurement of Vapor-Pressures for Coal-Related Aromatic-Compounds by Gas Saturation Method. J Chem Eng Jpn 19, 145-147.

Shiraishi, H., Pilkington, N.H., Otsuki, A., Fuwa, K., 1985. Occurrence of Chlorinated Polynuclear Aromatic-Hydrocarbons in Tap Water. Environ Sci Technol 19, 585-590.

Shiu, W.Y., Ma, K.C., 2000. Temperature dependence of physical-chemical properties of selected chemicals of environmental interest. I. Mononuclear and polynuclear aromatic hydrocarbons. J Phys Chem Ref Data 29, 41-130.

Singh, N.B., Rai, U.S., Singh, O.P., 1985. Chemistry of eutectic and monotectic; phenanthrene-succinonitrile system. Journal of Crystal Growth 71, 353-360.

Singh, N.B., Srivastava, A., Singh, N.P., Gupta, A., 2007. Molecular interaction between naphthalene and 2,4-dinitrophenol in solid state. Molecular Crystals and Liquid Crystals 474, 43-+.

Singh, N.B., Srivastava, M.A., Singh, N.P., 2001. Solid-liquid equilibrium for 2,4-dinitrophenol plus naphthalene. J Chem Eng Data 46, 47-50.

Sonnefeld, W.J., Zoller, W.H., May, W.E., 1983. Dynamic Coupled-Column Liquid-Chromatographic Determination of Ambient-Temperature Vapor-Pressures of Polynuclear Aromatic-Hydrocarbons. Anal Chem 55, 275-280.

Sun, C.G., Snape, C.E., McRae, C., Fallick, A.E., 2003. Resolving coal and petroleum-derived polycyclic aromatic hydrocarbons (PAHs) in some contaminated land samples using compound-specific stable carbon isotope ratio measurements in conjunction with molecular fingerprints. Fuel 82, 2017-2023.

Szczepanik, Richard, Skalmowski, Wlodzimierz, 1963. Effects of crystal growth and volatilization of tar components on the aging of prepared tar. III. Solid-liquid phase relationship of the components of raw coal tar and prepared road tar. Bitumen, Teere, Asphalte, Peche 14, 506,508-512,514.

Szczepanik, Ryszard, 1963. Two- and multicomponent, solid-liquid systems formed by aromatic hydrocarbons, anthraquinone, and coal-tar fractions. Chem. Stosowana Ser. A 7, 621-660.

Wakayama, N., Inokuchi, H., 1967. Heats of sublimation of polycyclic aromatic hydrocarbons and their molecular packings. 40, 2267-2271.

9

The Diffusion Model of Grown-In Microdefects Formation During Crystallization of Dislocation-Free Silicon Single Crystals

V. I. Talanin and I. E. Talanin
Classic Private University
Ukraine

1. Introduction

Dislocation-free silicon single crystals are the basic material of microelectronics and nanoelectronics. Physical properties of semiconductor silicon are determined by the structural perfection of the crystals grown by the Czochralski and float-zone processes (Huff, 2002). In such crystals during their growth are formed grown-in microdefects.

Grown-in microdefects degrade the electronic properties of microdevices fabricated on silicon wafers. Optimizing the number and size of grown-in microdefects is crucial to improving processing yield of microelectronic devices. Many of the advances in integrated-circuit manufacturing achieved in recent years would not have been possible without parallel advances in silicon-crystal quality and defect engineering (Yang et al., 2009). The problem of defect formation in dislocation-free silicon single crystals during their growth is a fundamental problem of physics and chemistry of silicon. In particular it is the key to solving the problem engineering applications of silicon crystals. This is connected with the transformation grown-in microdefects during the technological treatment of silicon monocrystals.

Formation of grown-in microdefects occurs as a result of the interaction of point defects during crystal cooling. The distribution of grown-in microdefects in a growing crystal is influenced by its temperature field and the boundary conditions defined by its surfaces. Until recently it was assumed that the formation of grown-in microdefects is due to condensation of intrinsic point defects (Voronkov et. al., 2011). Recombination-diffusion model assumes fast recombination of intrinsic point defects at the initial moment of cooling the grown crystal. Fast recombination determines the type of dominant intrinsic point defects in the crystal. In this model was first used mathematical tool which allows you to associate the defect structure of crystal with distribution in the crystal thermal fields during the growth (Prostomolotov et al., 2011). It has been suggested that the fast recombination of intrinsic point defects near the crystallization front as a function of the growth parameter V_g/G (where V_g is the rate of crystal growth; G is the axial temperature gradient) leads to the formation of microvoids or interstitial dislocation loops (Voronkov, 2008). It is assumed that in the case $V_g/G < \xi_{crit}$ formed only interstitial A-microdefects as a result of aggregation of intrinsic interstitial silicon atoms. It is assumed that in the case $V_g/G > \xi_{crit}$ formed only microvoids as a result of aggregation of vacancies (Goethem et al., 2008; Kulkarni, 2008a). In this physical model, the interaction between the impurities and intrinsic point defects is not considered (Kulkarni et. al., 2004).

Recent versions of this model have suggested that part of the vacancies (v) in the temperature range 1683 ... 1373 K, due to the interaction with oxygen (O) and nitrogen (N) impurities, are bound into complexes of the vO, vO_2, and vN types (Kulkarni 2007; 2008b). After the formation of microvoids, the aforementioned complexes grow and take up vacancies. This model has ignored the growth of the complexes by means of the injection of intrinsic interstitial silicon atoms and the interaction of an impurity with intrinsic interstitial silicon atoms (Kulkarni 2007; 2008b).

In the general case recombination-diffusion model assumes that the process of defect formation in dislocation-free silicon single crystals occurs in four stages: (i) fast recombination of intrinsic point defects near the crystallization front; (ii) the formation in the narrow temperature range 1423...1223 K depending on the value of V_g/G microvoids or interstitial dislocation loops; (iii) the formation of oxygen clusters in the temperature range 1223...1023 K; (iv) growth of precipitates as a result of subsequent heat treatments.

Recombination-diffusion model is the physical basis for models of the dynamics of point defects. The mathematical model of point defect dynamics in silicon quantitatively explains the homogeneous mechanism of formation of microvoids and dislocation loops. It should be noted that, in the general case, the model of point defect dynamics includes three approximations: rigorous, simplified, and discrete–continuum approaches (Sinno, 1999; Dornberger et. al., 2001; Wang & Brown, 2001; Kulkarni et. al., 2004; Kulkarni, 2005; Prostomolotov & Verezub, 2009). The rigorous model requires the solution to integro-differential equations for point defect concentration fields, and the distribution of grown-in microdefects in this model is a function of the coordinates, the time, and the time of evolution of the size distribution of microdefects. A high consumption of time and cost for the performance of calculations required the development of a simplified model in which the average defect radius is approximated by the square root of the average defect area. This approximation is taken into account in the additional variable, which is proportional to the total area of the defect surface. The simplified model is effective for calculating the two-dimensional distribution of grown-in microdefects. Both models use the classical nucleation theory and suggest the calculation of the formation of stable nuclei and the kinetics of diffusion-limited growth of defects. The discrete–continuum approximation suggests a complex approach: the solution to discrete equations for the smallest defects and the solution to the Fokker–Planck equation for large-sized defects.

Recently, we proposed a new model for the formation grown-in microdefects. The physical model of the formation of grown-in microdefects assumes that the defect formation in dislocation-free Si single crystals upon cooling occurs in three stages: (i) the formation of impurity aggregates near the crystallization front, (ii) the formation and growth of impurity precipitates upon cooling from the crystallization temperature, and (iii) the formation of microvoids or dislocation loops (depending on the growth parameter V_g/G) – in a narrow temperature range of 1423...1223 K (V.I. Talanin & I.E. Talanin, 2006a; V.I. Talanin & I.E. Talanin, 2010b). This model on the experimentally and theoretically established fact the absence of recombination of intrinsic point defects near the crystallization front of the crystal is based (V.I. Talanin & I.E. Talanin, 2006a; V.I. Talanin & I.E. Talanin, 2007a).With the help of the diffusion model of formation grown-in microdefects was calculated process of high-temperature precipitation (V.I. Talanin & I.E. Talanin, 2010a). The processes of formation and growth of precipitates during cooling of the crystal is a controlling stage in the formation of the grown-in defect structure of dislocation-free silicon single crystals. At this stage, the formation and growth of oxygen and carbon precipitates occur in the temperature range from 1682 to 1423 K (V.I. Talanin & I.E. Talanin, 2010a).

The mathematical model of point defect dynamics can be adequately used on the basis of the physical model in which the impurity precipitation process occurs before the formation of microvoids or dislocation loops (V.I. Talanin & I.E. Talanin, 2010b). The model of point defect dynamics can be considered as component of the diffusion model for formation grown-in microdefects.

The aim of this paper is to present a diffusion model of formation grown-in microdefects in general and to discuss the possibility of its use as a tool for building the defect structure of dislocation-free silicon single crystal and device structures based on them.

2. Classification of grown-in microdefects

Currently, there are three classifications of grown-in microdefects: experimental classification, technological classification and physical classification.

Experimental classification of grown-in microdefects is based on the use of methods of selective etching, X-ray topography and transmission electron microscopy (Kock, 1970; Petroff & Kock, 1975; Foll & Kolbesen, 1975; Veselovskaya et al., 1977; Sitnikova et al., 1984; Sitnikova at al., 1985). A.J.R. de Kock entered the name of A-microdefects and B-microdefects, whereas E.G. Sheikhet entered the name C-microdefects, D-microdefects. We are entered the name (I+V)-microdefects (V.I.Talanin et al., 2002a, 2002b). These research allowed to establish the physical nature of A-microdefects, B-microdefects, C-microdefects, D-microdefects and (I+V)-microdefects. Experimental results indicated the identity of the processes of defect formation in crystals of FZ-Si and CZ-Si (Kock et al., 1979; V.I. Talanin & I.E. Talanin, 2003). This means that the classifications of grown-in microdefects in both types of crystals should also be identical (V.I. Talanin & I.E. Talanin, 2004).

Technological classification is used for large-scale crystals. The larger the diameter of the growing crystal, the lower growth rate, at which the same type of grown-in microdefects is formed. This occurs by reducing the axial temperature gradient in the crystal (Ammon et al., 1999). This leads to the appearance of a new type of grown-in microdefects (microvoids) and dislocation-free crystal growth in a narrow range of growth rates (Voronkov & Falster, 1998). In large crystals of interstitial dislocation loops and microvoids are considered as major grown-in microdefects in dislocation-free silicon crystals (Kulkarni et al., 2004).

Analysis of the experimental results of investigations of grown-in microdefects indicates that there are only three types of grown-in microdefects: precipitates of impurities ((I+V)-microdefects, D(C)-microdefects, B-microdefects), dislocation loops (A-microdefects) and microvoids (V.I. Talanin et. al., 2011b). We established that the basic elements of defect formation are primary oxygen-vacancy and carbon-interstitial agglomerates, which are formed at impurity centers near the crystallization front (V.I. Talanin & I.E. Talanin, 2006a).

An excess concentration of intrinsic point defects (vacancies or silicon self-interstitials) arises when the crystal is cooled under certain thermal conditions (Cho et al., 2006). This process leads to the formation of secondary grown-in microdefects (A-microdefects or microvoids) (V.I. Talanin & I.E. Talanin, 2004). We have proposed the physical classification of grown-in microdefects. It is based on the differences in the physical nature of the formation of primary and secondary grown-in microdefects (V.I. Talanin & I.E. Talanin, 2006a).

3. The diffusion model for formation of grown-in microdefects in dislocation-free silicon single crystals

We propose a new diffusion model of the formation and transformation of grown-in microdefects. It is based on the experimental studies of undoped dislocation-free Si single crystals grown by the floating zone and Czochralski methods. The diffusion model combines the physical model (the heterogeneous mechanism for the formation of grown-in microdefects), the physical classification of grown-in microdefects, and mathematical models of the formation of primary and secondary grown-in microdefects (Fig. 1).

Physical model based on the assumption about the absence of recombination intrinsic point defects at high temperatures. This assumption was confirmed in several experimental works (Talanin et al., 2002a; Talanin et al., 2002b; Talanin et al., 2003). In paper (V.I. Talanin & I.E. Talanin, 2007a) we first theoretically proved the absence of recombination of intrinsic point defects at high temperatures and fast recombination at low temperatures. The experimental data and the results obtained from thermodynamic calculations have demonstrated that the process of aggregation of point defects dominates over the process of recombination of intrinsic point defects. At high temperatures, the process of recombination makes an insignificant contribution to the process of aggregation. Consequently, vacancies and intrinsic interstitial atoms coexist in thermal equilibrium. As a result, intrinsic point defects of both types are simultaneously involved in the process of aggregation. The decomposition of a supersaturated solid solution of point defects occurs upon cooling through two mechanisms, namely, the vacancy and interstitial mechanisms, with the formation of oxygen-vacancy and carbon- interstitial agglomerates.

Absence of recombination intrinsic point defects at high temperatures allows us to propose the physical model of the formation grown-in microdefects. The basic concepts of the physics model for the formation of grown-in microdefects imply the following (V.I. Talanin & I.E. Talanin, 2006a): (i) the recombination of intrinsic point defects at high temperatures can be neglected; (ii) background carbon and oxygen impurities are involved in the defect formation as nucleation centers; (iii) the decay of the supersaturated solid solution of point defects when the crystal is cooled from the crystallization temperature occurs in two independent ways (branches): vacancy and interstitial; (iv) the defect formation is based on primary agglomerates formed as the crystal is cooled from the crystallization temperature due to the interaction between the impurities and intrinsic point defects; (v) when the crystal is cooled at temperatures below 1423 K, depending on the thermal growth conditions, secondary grown-in microdefects are formed due to the interaction between intrinsic point defects; (vi) the secondary grown-in microdefects are formed due to the coagulation (microvoids and A-microdefects) and deformation (A-microdefects) effects; (vii) the vacancy and interstitial branches of the heterogeneous mechanism have a symmetry, which implies simultaneous processes of defect formation during the decay of supersaturated solid solution of point defects; and (viii) the consequence of this symmetry is the formation of vacancy and interstitial grown-in microdefects of the same type and, correspondingly, the growth of dislocation-free Si single crystals in the same vacancy–interstitial mode (V.I. Talanin & I.E. Talanin, 2006b). It was revealed that the growth parameter $V_g/G = \xi_{crit}$ describes the conditions under which the (111) face appears on the crystallization front (V.I. Talanin & I.E. Talanin, 2006a). On the basis of physical model and concepts of primary and secondary grown-in microdefects we developed of physical classification of the grown-in microdefects (V.I. Talanin & I.E. Talanin, 2006a).

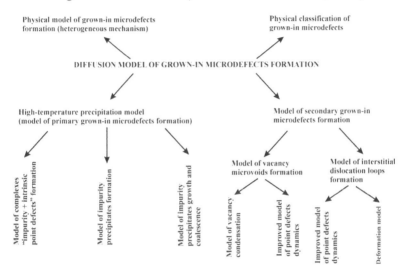

Fig. 1. Diffusion model of grown-in microdefects formation

A detailed description of the heterogeneous mechanism formation of grown-in microdefects and its correspondence to the results of experimental researches are presented in the articles (V.I. Talanin & I.E. Talanin, 2004; V.I. Talanin & I.E. Talanin, 2006a).

4. Diffusion kinetic of high-temperature precipitation

The calculation of the precipitation is carried out within the framework of the classical theory of nucleation, growth and coalescence of precipitates. For the calculation of formation and growth of precipitates are used analytic and approximate calculations. In the case of analytical calculations applied solution of differential equations of the dissociative diffusion (Talanin et al., 2007b, 2008). In the case of approximate calculations, the solution is sought in the form of systems of interconnected discrete differential equations of quasi-chemical reactions to describe the initial stages of nucleation of new phases and a similar system of continuous differential equations of the Fokker-Planck (V.I. Talanin & I.E. Talanin, 2010a).

4.1 Model of formation complex "impurity + intrinsic point defect"

The solution is sought within the model of dissociative diffusion–migration of impurities (Bulyarskii & Fistul', 1997). In this case, the difference from the decomposition phenomenon is that during diffusion (as a technological process), a diffusant is supplied to the sample from an external source, whereas in the case of decomposition it is produced by an internal source (lattice sites).

Vas'kin & Uskov are considered the problem of successive diffusion of a component A into a sample singly doped with a component B, taking into account the complex formation at the initial and boundary conditions (Vas'kin & Uskov, 1968). We are conducted similar consideration for our conditions (Talanin et al., 2007b; 2008). Under physical-model conditions (heterogeneous mechanism of grown-in microdefect formation), we assume that the component A is the background impurity (oxygen O or carbon C) and the component B is intrinsic point defects (vacancies V or interstitials I). For the vacancy and interstitial mechanisms, we consider, respectively, the oxygen+vacancy $(O+V)$ and carbon+interstitial $(C+I)$ interactions. The calculations performed in the framework of approximation of strong complex formation have demonstrated that the edge of the reaction front of the formation of a complex (i.e., the "oxygen+vacancy" and "carbon + self-interstitials" complex) is located at a distance of $\sim 3 \cdot 10^{-4}$ mm from the crystallization front (Talanin et al., 2007b). We have shown that complex formation occurs near the crystallization front. Detailed calculations are presented in the articles (Talanin et al., 2007b, 2008).

4.2 Model for the formation of precipitates

Let us consider a system of a growing undoped dislocation-free silicon single crystal. The concentrations of all point defects at the crystallization front are assumed to be equilibrium, and both the vacancies and the intrinsic interstitial silicon atoms are present in comparable concentrations. During cooling of the crystal after passing through the diffusion zone, an excessive (nonequilibrium) concentration of intrinsic point defects appears. Excess intrinsic point defects disappear on sinks whose role in this process is played by uncontrollable (background) impurities of oxygen and carbon (V.I. Talanin & I.E. Talanin, 2006a). In real silicon crystals, the concentrations of carbon and oxygen impurities are higher than the concentrations of the intrinsic point defects. The formation of complexes between the intrinsic point defects and impurities is governed, on the one hand, by the fact that both the intrinsic point defects and the impurities are sources of internal stresses in the lattice (elastic interaction) and, on the other hand, by the Coulomb interaction between them (provided the defects and the impurities are present in the charged state). The mathematical model under consideration allows for the elastic interaction and the absence of the recombination of intrinsic point defects in the high-temperature range (V.I. Talanin & I.E. Talanin, 2007a). The concentrations of intrinsic point defects $C_{i,v}(r,t)$ in the growing crystal satisfy the diffusion

equation $\dfrac{\partial C_{i,v}}{\partial t} = D_{i,v} \Delta \left(C_{i,v} - C_{ie,ve} \right)$ where r is the coordinate and t is the time. In the vicinity of the sinks (oxygen and carbon atoms), the concentration of intrinsic point defects $C_{ie,ve}$ is kept equilibrium, whereas the diffusion coefficients $D_{i,v}$ and the concentrations $C_{ie,ve}$ of intrinsic point defects decrease exponentially with decreasing temperature. Under these conditions, the formation of microvoids and interstitial dislocation loops is possible only at significant supersaturations of intrinsic point defects, which take place at a temperature $T = T_m - 300K$ (where T_m is the crystallization temperature). For the formation of precipitates in the high-temperature range $T \sim 1683 \ldots 1403$ K has been calculated using the model of dissociative diffusion (Talanin et al., 2007b; 2008). This approximation is valid at the initial stages of the formation of nuclei, when their sizes are small and the use of Fokker–Planck continuity differential equations is impossible. The calculations performed in the framework of this approximation have demonstrated that the edge of the reaction front of the formation of a complex is located at a distance of ~3·10⁻⁴ mm from the crystallization front. This spacing represents a diffusion layer in which an excessive concentration of intrinsic point defects appears.

We have considered the modern approach based on solving systems of coupled discrete differential equations of quasi-chemical reactions for the description of the initial stages of the formation of nuclei of new phases and a similar system of Fokker–Planck continuity differential equations.

In order to describe the kinetics of the simultaneous nucleation and growth (dissolution) of a new phase particles of several types in a supersaturated solid solution of an impurity in silicon was considered a system consisting of oxygen and carbon atoms, vacancies, and intrinsic interstitial silicon atoms. The interaction in this system during cooling of the crystal from 1683 K results in the formation of oxygen and carbon precipitates. In order to perform the computational experiments and to interpret their results was conducted a dimensional analysis of the kinetic equations and the conservation laws with the use of characteristic time constants and critical sizes of defects. This is made it possible to perform a comparative analysis of the joint evolution of oxygen and carbon precipitates and to optimize the computational algorithm for the numerical solution of the equations.

For example, for the case of a thin plane-parallel crystal plate of a large diameter, when the conditions in the plane parallel to the surface of the crystal can be considered to be uniform and the diffusion can be treated only along the normal to the surface (the z coordinate axis), the mass balance of point defects in the crystal is described by the system of diffusion equations for intrinsic interstitial silicon atoms, oxygen atoms, carbon atoms, and vacancies:

$$\frac{\partial C_o}{\partial t} = D_o \frac{\partial^2 C_o}{\partial z^2} - \frac{\partial C_o^{SiO_2}}{\partial t}$$
$$\frac{\partial C_c}{\partial t} = D_c \frac{\partial^2 C_c}{\partial z^2} - \frac{\partial C_c^{SiC}}{\partial t}$$
$$\frac{\partial C_i}{\partial t} = D_i \frac{\partial^2 C_i}{\partial z^2} + \frac{\partial C_i^{SiO_2}}{\partial t} - \frac{\partial C_i^{SiC}}{\partial t} \tag{1}$$
$$\frac{\partial C_v}{\partial t} = D_v \frac{\partial^2 C_v}{\partial z^2} - \frac{\partial C_v^{SiO_2}}{\partial t} + \frac{\partial C_v^{SiC}}{\partial t}$$

where C_o, C_c, C_i, C_v are the concentrations of oxygen, carbon, self-interstitials and vacancies respectively; D_o, D_c, D_i, D_v are the diffusion coefficients of oxygen, carbon, self-interstitials and vacancies respectively.

In the system of equations (1), we took into account that the oxygen precipitates serves as sinks for oxygen atoms and vacancies and as sources of interstitial silicon atoms. At the same time, the carbon precipitates, in turn, also serve as sinks for carbon atoms and interstitial silicon atoms and as sources for vacancies. Kinetic model of decomposition of solid solutions of oxygen and carbon impurities not only allows one to simulate the processes of precipitation during cooling of the as-grown silicon crystal to a temperature of 300 K but also adequately describes the available experimental data on the oxygen and carbon precipitation (V.I. Talanin & I.E. Talanin, 2011a).

The algorithm used for solving the problem of simulation of the simultaneous growth and dissolution of the oxygen and carbon precipitates due to the interaction of point defects during cooling of the crystal from the crystallization temperature is based on the monotonic explicit difference scheme of the first-order accuracy as applied to the Fokker–Planck equations.

Detailed calculations are presented in the articles (V.I. Talanin & I.E. Talanin, 2010a). These calculations demonstrate that intrinsic point defects (vacancies and intrinsic interstitial silicon atoms) exert a significant influence on the dynamics of mass exchange and mass transfer of point defects between the oxygen and carbon precipitates. The absorption of vacancies by the growing oxygen precipitates leads to the emission of silicon atoms into interstitial positions. The intrinsic interstitial silicon atoms, in turn, interact with the growing carbon precipitates, which, in the process of growth, supply vacancies for growing oxygen precipitates. This interaction leads to such a situation that, first, the growth of the precipitates is suppressed more weakly because of the slower increase in the supersaturation of the intrinsic point defects in the bulk of the growing crystal and, second, the critical radius of the formation of carbon precipitates increases more slowly, which favors a more rapid growth of the carbon precipitates. The higher rate of the evolution of the size distribution function for carbon precipitates can be associated with the higher mobility of interstitial silicon atoms as compared to vacancies in the high-temperature range. It can be assumed that the mutual formation and growth of oxygen and carbon precipitates result in a lower rate of the evolution of the size distribution function of the oxygen precipitates, regardless of their smaller critical size at the initial instant of time, owing to the effect of the carbon impurity.

4.3 Model of growth and coalescence of precipitates

In the classical theory of nucleation and growth of new-phase particles, the process of precipitation in a crystal is treated as a first-order phase transition and the kinetics of this process is divided into three stages: the formation of new-phase nuclei, the growth of clusters, and the coalescence stage. At the second stage of the precipitation process, clusters grow without a change in their number. At the third stage of the precipitation process, when the particles of the new phase are sufficiently large, the supersaturation is relatively low, new particles are not formed and the decisive role is played by the coalescence, which is accompanied by the dissolution of small-sized particles and the growth of large-sized particles. The condition providing for changeover to the coalescence stage is the ratio $u(t) = R(t) / R_{cr}(t) \approx 1$, where $R_{cr}(t)$ is the critical radius of the precipitate.

Detailed calculations stages of the growth and the coalescence are presented in the article (V.I. Talanin & I.E. Talanin, 2011a). The analysis was carried out under the assumption that precipitates grow at a fixed number of nucleation centers according to the diffusion mechanism of growth. The model corresponds to the precipitation uniform in the volume. An analysis of the results obtained and the data taken from (Talanin et al., 2007b; V.I.

Talanin & I.E. Talanin, 2010a) has demonstrated that the phase transition occurs according to the mechanism of nucleation and growth of a new phase so that these two processes are not separated in time and proceed in parallel.

The condition providing changeover to the stage of the coalescence is written in the form $R(t) \approx R_{cr}(t)$, which is satisfied for large-sized crystals at the temperature $T \approx 1423$ K. Taking into account the computational errors, this temperature for large-sized crystals corresponds to the initial point of the range of the formation of microvoids (at $V_g = 0.6$ mm/min). In this range, all impurities are bound and there arises a supersaturation with respect to vacancies, which is removed as a result of the formation of microvoids. With a change in the thermal conditions of the growth (for example, at $V_g = 0.3$ mm/min), there arises a supersaturation with respect to interstitial silicon atoms, which leads to the formation of dislocation loops. In this case, the condition $R(t) \approx R_{cr}(t)$ is satisfied at $T \approx 1418$ K. Consequently, the stage of the coalescence in large-sized silicon single crystals begins at temperatures close to the temperatures of the formation of clusters of intrinsic point defects (depending on the thermal growth conditions, these are microvoids or dislocation loops).

The absorption of vacancies by growing oxygen precipitates results in the emission of silicon atoms in interstitial sites. In turn, the intrinsic interstitial silicon atoms interact with growing carbon precipitates, which, in the course of their growth, supply vacancies for growing oxygen precipitates. This interplay between the processes leads to an accelerated changeover of the subsystems of oxygen and carbon precipitates to the stage of the coalescence as compared to the independent evolution of these two subsystems. The change in the thermal conditions for the growth of small-sized FZ-Si single crystals (high growth rates and axial temperature gradients) leads to the fact that the stage of the coalescence begins far in advance (at $T \approx T_m - 20K$). The results of theoretical calculations have demonstrated that a decrease in the concentrations of oxygen and carbon in small-sized single crystals leads to a further decrease in the time of occurrence of the growth stage of precipitates. The change in thermal conditions of crystal growth (in particular, an increase in the growth rate and in the axial temperature gradient in the crystal) substantially affects the stage of the growth of precipitates. In turn, the decrease in the time of occurrence of the growth stage of precipitates is associated, to a lesser extent, with the decrease in the concentration of impurities in crystals. Eventually, these factors are responsible for the decrease in the average size of the precipitates.

The kinetic model of growth and coalescence of oxygen and carbon precipitates in combination with the kinetic models describing their formation represents a unified model of the process of precipitation in dislocation-free silicon single crystals.

5. Diffusion kinetic of formation of the microvoids and dislocation loops

As mentioned earlier the defect formation processes in a semiconductor crystal, in general, and in silicon, in particular, have been described using the model of point defect dynamics; in this case, the crystal has been considered a dynamic system and real boundary conditions have been specified. However, the model of point defect dynamics has not been used for calculating the formation of interstitial dislocation loops and microvoids under the assumption that the recombination of intrinsic point defects is absent in the vicinity of the crystallization front. This fact is evidenced by experimental and theoretical investigations (V.I. Talanin & I.E. Talanin, 2006a, 2007a).

5.1 Kinetics of formation of microvoids

The experimentally determined temperature range of the formation of microvoids in crystals with a large diameter is 1403…1343 K (Kato et al., 1996; Itsumi, 2002). In this respect, the approximate calculations for the solution in terms of the model of point defect dynamics were performed at temperatures in the range 1403…1073 K. The computational model uses the classical theory of nucleation and formation of stable clusters and, in strict sense, represents the size distribution of clusters (microvoids) reasoning from the time process of their formation and previous history.

The calculations were carried out in the framework of the model of point defect dynamics, i.e., for the same crystals with the same parameters as in already the classical work on the simulation of microvoids and interstitial dislocation loops (A-microdefects) (Kulkarni et al., 2004). According to the analysis of the modern temperature fields used when growing crystals by the Czochralski method, the temperature gradient was taken to be $G = 2.5$ K/mm (Kulkarni et al., 2004). The simulation was performed for crystals 150 mm in diameter, which were grown at the rates V_g = 0.6 and 0.7 mm/min. These growth conditions correspond to the growth parameter $V_g/G > \xi_{crit}$.

Detailed calculations are presented in the articles (V.I. Talanin & I.E. Talanin, 2010b).Our results somewhat differ from those obtained in (Kulkarni et al., 2004). These differences are as follows: (i) the nucleation rate of microvoids at the initial stage of their formation is low and weakly increases with a decrease in the temperature and (ii) a sharp increase in the nucleation rate, which determines the nucleation temperature, occurs at a temperature T ~ 1333 K. These differences result from the fact that the recombination factor in our calculations was taken to be $k_{IV} = 0$. For $k_{IV} \neq 0$, consideration of the interaction between impurities and intrinsic point defects in the high-temperature range becomes impossible, which is accepted by the authors of the model of point defect dynamics (Kulkarni et al., 2004). In this case, in terms of the model, there arises a contradiction between the calculations using the mathematical model and the real physical system, which manifests itself in the ignoring of the precipitation process (Kulkarni et al., 2004).

5.2 Kinetics of formation of dislocation loops (A-microdefects)

The computational experiment was performed similarly to the calculations of the formation of microvoids. The simulation was performed for crystals 150 mm in diameter, which were grown at the rates V_g = 0.10 and 0.25 mm/min for the temperature gradient $G = 2.5$ K/mm. These growth conditions correspond to the growth parameter $V_g/G < \xi_{crit}$.

Detailed calculations are presented in the articles (V.I. Talanin & I.E. Talanin, 2010b). The temperature of the formation of A-microdefects corresponds to ~1153 K. An increase in the crystal growth rate weakly decreases the critical radius of A-microdefects and slightly affects the nucleation temperature. An increase in the crystal growth rate leads to an almost twofold decrease in the concentration of introduced defects.

The data of the computational experiment on the determination of the microvoid concentration correlate well with the experimentally observed results (10^4 ... 10^5 cm^{-3}) (Itsumi, 2002). For the A-microdefects, for which the concentration according to the experimental data is ~10^6...10^7 cm^{-3} (Petroff & Kock, 1975; Foll & Kolbesen, 1975), the discrepancy is as large as three orders of magnitude. This can be explained by the fact that, unlike microvoids, which are formed only through the coagulation mechanism, the formation of A-microdefects occurs according to both the coagulation mechanism and the mechanism of prismatic extrusion (deformation mechanism) (V.I. Talanin & I.E. Talanin, 2006a). The results of the calculations suggest that the main contribution to the formation of A-microdefects is made by the mechanism of prismatic extrusion when the formation of interstitial dislocation loops is associated with the relieving of stresses around the growing precipitate. Consequently, the impurity precipitation processes that proceed during cooling of the crystal from the crystallization temperature are fundamental (primary) in character and determine the overall defect formation process in the growth of dislocation-free silicon single crystals.

The calculations of the formation of microvoids and dislocation loops (A-microdefects) demonstrated that the above assumptions do not lead to substantial differences from the results of the previous calculations in terms of the model of point defect dynamics. This circumstance indicates that the mathematical model of point defect dynamics can be adequately used on the basis of the physical model in which the impurity precipitation process occurs before the formation of microvoids or interstitial dislocation loops. Moreover, the significant result of the calculations is the confirmation of the coagulation mechanism of the formation of microvoids and the deformation mechanism of the formation of interstitial dislocation loops. Therefore the model of the dynamics of point defects can be considered as component part of the diffusion model for formation grown-in microdefects.

5.3 Model of the vacancy coalescence

Model vacancy coalescence is a simplified model for the analysis of individual parameters of process of the formation microvoids. Detailed calculations are presented in the articles (V.I. Talanin & I.E. Talanin, 2010c). The fundamental interaction between impurities and intrinsic point defects upon crystal cooling under certain thermal conditions ($T < 1423$ K) leads to impurity depletion and the formation of a supersaturated solid solution of intrinsic point defects. The decay of this supersaturated solid solution causes the coagulation of intrinsic point defects in the form of microvoids.

An analysis of the experimental and calculated data within model of the vacancion coalescence in accordance with the heterogeneous diffusion model of the formation of grown-in microdefects revealed the following reasons for the occurrence of microvoids in dislocation-free silicon single crystals: (i) a sharp decrease in the concentration of background impurity that was not associated into impurity agglomerates (formed in the cooling range of 1683...1423 K); (ii) a large (over 80 mm) crystal diameter (in this case vacancies fail to drain from the central part of the crystal to the lateral surface); (iii) crystals of large diameter generally contain a ring of D-microdefects which forms due to the emergence of the (111) face on the crystallization front and which depletes the region inside with impurity atoms.

The growth parameter V_g / G describes the fundamental reasons related to the systematic nonuniform impurity distribution during crystal growth from a melt. Based on an analysis of the experimental results, one can suggest that the parameter V_g / G controls the growth because it describes the condition for the emergence of the (111) face at the crystallization front. Therefore, the impurity depletion inside the ring of D-microdefects upon crystal cooling at $T < 1423$ K is caused by two things: the impurity bonding during the formation of primary grown-in microdefects ((I+V)-microdefects) and the impurity drift to the (111) face, which is equivalent to the annular distribution of primary D-type grown-in microdefects. In this case, excess vacancies arise within the ring of D-microdefects to form a supersaturated solid solution with its subsequent decay and the formation of vacancy microvoids. In contrast, excess silicon self-interstitials arise beyond the D-ring to form a supersaturated solid solution with its subsequent decay and the formation of interstitial dislocation loops (A-microdefects) (V.I. Talanin & I.E. Talanin, 2010c).

The experimental classification of grown-in microdefects employs the terms such as A-microdefects, B-microdefects, D(C)-microdefects, (I+V)-microdefects and microvoids (V.I. Talanin & I.E. Talanin, 2006a). It was found that A-microdefects constitute interstitial-type dislocation loops, and B-microdefects, D(C)-microdefects, (I+V)-microdefects constitute precipitates of background oxygen and carbon impurities at different stages of their evolution (V.I. Talanin & I.E. Talanin, 2006a; V.I. Talanin & I.E. Talanin, 2011a)). At present, it is difficult to apply the experimental classification, since it is necessary to interpret the terms of every type of the grown-in microdefects for each publication. At the same time, from the physical point of view there are only three types of grown-in microdefects, i.e. impurity precipitates, dislocation loops and microvoids. Besides, when considering the formation of defects in silicon after processing (post-growth microdefects) the terms such as precipitates, dislocation loops and microvoids are also employed. Therefore, in order to harmonize a defect structure, we propose to switch to the physical classification of grown-in microdefects (V.I. Talanin & I.E. Talanin, 2004).

5.4 Kinetic model for the formation and growth of dislocation loops

Kinetics of high-temperature precipitation involves three stages: (i) the nucleation of a new phase, (ii) the growth stage and (iii) the stage coalescence. Precipitates originate from elastic interaction between point defects. They are, initially, present in coherent, elastic and deformable state, when lattice distortions close to the precipitate-matrix boundary are not large, and one atom of the precipitate corresponds to one atom of the matrix (Goldstein et al., 2011). Elastic deformations and any mechanical stress connected with them cause a transfer of excessive (deficient) substance from the precipitate or vice versa. Storage of elastic strain energy during the precipitate growth results in a loss of coherence by matrix. In this case it is impossible to establish one-to-one correspondence between atoms at different sides of the boundary. It results in structural relaxation of precipitates which occurs due to formation and movement of dislocation loops.

To simulate a stress state of the precipitate and the matrix surrounding it, it is sufficient to observe the precipitate which is simple spherical in shape. There can be found analytical solutions in respect of spherical precipitates (Kolesnikova & Romanov, 2004). Let us take the theoretical and experimental researches of stress relaxation at volume quantum dots as initial model (Chaldyshev et al., 2002; Kolesnikova & Romanov, 2004; Chaldyshev et al., 2005; Kolesnikova et al., 2007). According to these representations, as far as the precipitate grows, its elastic field induces the formation of a circular interstitial dislocation loop of mismatch. This process contributes to the decrease in total strain energy of the system. A growing precipitate displaces the matrix material in the crystal volume. Interstitial atoms form an interstitial dislocation loop near to the precipitate. At the same time, a mismatch dislocation loop is formed on the very precipitate (Kolesnikova et al., 2007). At the same time, the critical sizes of precipitates, at which formation of dislocations is energy favorable, have the same order as the critical size of dislocation loops (Kolesnikova et al., 2007).

In the volume of silicon the precipitate produces a stress field caused by mismatch between the lattice parameters of precipitate (a_1) and the surrounding matrix (a_2) (Kolesnikova et al., 2007). Then, the intrinsic deformation of the precipitate is defined as described bellow

$$\varepsilon = \frac{a_1 - a_2}{a_1} \tag{2}$$

In general, the precipitate intrinsic deformation in the matrix volume can be expressed as follows

$$\varepsilon^* = \begin{pmatrix} \varepsilon_{xx} & \varepsilon_{xy} & \varepsilon_{xz} \\ \varepsilon_{xy} & \varepsilon_{yy} & \varepsilon_{yz} \\ \varepsilon_{zx} & \varepsilon_{zy} & \varepsilon_{zz} \end{pmatrix} \delta\left(\Omega_{pr}\right) \tag{3}$$

where the diagonal terms constitute a dilatation mismatch between the precipitate and matrix lattices; the other terms are shear components; $\delta\left(\Omega_{pr}\right)$ is the Kronecker symbol. Elastic fields of precipitate (stresses σ_{ij} and deformation ε_{ij}) and field of full displacements are calculated taking into account their own deformation (3) and region of localization of the precipitate $\delta\left(\Omega_{pr}\right)$. The calculation of elastic fields of the precipitate is carried out by well-known scheme by using the elastic modules, Green's function of an elastic medium or its Fourier transform (Kolesnikova & Romanov, 2004).

Consider the simplest model of a spherical precipitate with equiaxed own deformation $\varepsilon_{ii}^* = \varepsilon, \varepsilon_{ij}^* = 0 (i \neq j; i, j, = x, y, z)$. The elastic strain energy of spheroidal defect with increasing radius of precipitate $\left(R_{pr}\right)$ increases as a cubic law (Kolesnikova et al., 2007):

$$E_{pr} = \frac{32 \cdot \pi}{45 \cdot (1 - \upsilon)} \cdot J \cdot \varepsilon^2 \cdot R_{pr}^3 \tag{4}$$

where J is the shear modulus; υ is the Poisson's ratio. From a certain critical radius R_{crit} takes effect mechanism for resetting the elastic energy of the precipitate. This mechanism leads to the formation of circular interstitial dislocation loop. Energy criterion of this mechanism is the condition $E^{initial} \geq E^{final}$, here $E^{initial}, E^{final}$ is the elastic energy of the system with the precipitate before and after relaxation, respectively (Kolesnikova et al., 2007).

In respect of a spherical precipitate with equiaxial intrinsic deformation, the calculation of elastic fields of the precipitate is substantially simplified. Let us assume that the intrinsic elastic strain energy of the precipitate before and after the formation of a dislocation loop of mismatch remains constant $E_{pr}^{initial} = E_{pr}^{final}$. Then the criterion of nucleation loop of misfit dislocation can be represented by the condition $0 \geq E_D + E_{prD}$, where E_D is the energy of loop of misfit dislocation; E_{prD} is the energy of interaction of precipitate with the dislocation loop (Kolesnikova et al., 2007).

To estimate believe that loop of misfit dislocation is the equatorial location on the spheroidal precipitate $R_D = R_{pr}$ the self-energy prismatic loop (Kolesnikova et al., 2007)

$$E_{loop} = \frac{J \cdot b^2 \cdot R_D}{2 \cdot (1 - v)} \cdot \left(\ln \frac{2 \cdot R_D}{f} - 2 \right)$$ (5)

where f is the radius of the core loop; b is the magnitude of the Burgers vector. The critical radius of precipitate for the formation of dislocation loop is determined from the expression (Kolesnikova et al., 2007)

$$R_{crit} = \frac{3b}{8\pi (1 + v)\varepsilon} \left(\ln \frac{1.08 \alpha R_{crit}}{b} \right)$$ (6)

where α is a constant contribution of the dislocation core. Expression (6) is approximate and can only be used to determine the value critical radius R_{crit}.

This paper (Bonafos et al., 1998) theoretically considers the increase kinetics for dislocation loops at the stages of loop growth and coalescence. It is assumed that, in general, the growth is either controlled by energy barrier when atom is captured by the loop, or by activation energy of interstitial atom diffusion. In conditions of cooling the crystal after being grown, we presume that the diffusion processes play a core role. The model (Burton & Speight, 1985) is further used in the calculations for evolution in size-dependant distribution of loops and for evolution in loop density.

The dislocation loops with a radius of $R > R_{crit}$ become bigger in size at the coalescence stage, while small dislocation loops with a radius of $R < R_{crit}$ will dissolve (Bonafos et al., 1998; Burton & Speight, 1985). The growth of dislocation loops during cooling after the growth of single crystal silicon occurs as due to dissolution of small loops with sizes less than critical, and as a result supersaturation for intrinsic interstitial silicon atoms. In this case, the crystal growth ratio is $V_g/G < \xi_{crit}$. When oversaturation of vacancies ($V_g/G > \xi_{crit}$) occurs, the interstitial dislocation loops start to dissolve. Increase in the radius of interstitial dislocation loop can be defined by the formula depending on the crystal cooling time (Burton & Speight, 1985):

$$R(t) = \sqrt{R_{crit}^2 + j \cdot D(t) \cdot t}$$ (7)

where $D(t)$ is the diffusion coefficient of intrinsic interstitial silicon atoms; t is the time cooling the crystal; j is the proportionality factor. The value of the cooling time of the crystal is determined from the dependence: $T(t) = \dfrac{T_m^2}{T_m + U \cdot t}$, where T_m is the crystallization temperature (melting) of silicon; $U = V_g \cdot G$ is the cooling rate of the crystal. The loop concentration depends on the crystal cooling time (Burton & Speight, 1985):

$$N(t) = \frac{M(t)}{1 + D(t) \cdot t / 2 \cdot R_{crit}^2} \qquad (8)$$

where $M(t)$ is the concentration of precipitates.

Initially, the precipitates act as stoppers for the dislocation loops restraining their distribution and generation. Then, the precipitates facilitate the formation of dislocation loops due to the action of Bardeen-Herring or Frank-Read sources (Gyseva et al., 1986). These processes lead to the formation and growth of complex dislocation loops. Formation and development of dislocation loops caused by the high-temperature precipitation of background impurities (oxygen and carbon). Growth and coalescence of dislocation loops are generally maintained due to the generation of growing precipitates instead of silicon self-interstitials, and as well to the dissolution of small dislocation loops.

If the parameter of crystal growth $V_g / G < \xi_{crit}$, for stress relaxation precipitate generates interstitial silicon atoms. If the parameter of crystal growth $V_g / G > \xi_{crit}$, for stress relaxation precipitate adsorbs vacancies. In this case is suppressed the formation of dislocation loops.

6. Construction of the defect structure of dislocation-free silicon single crystal and device structures on their base

Experimental studies require large material and time costs, while theoretical studies are carried out for single crystals with selected fixed parameters of their growth. It is necessary to develop a new method for studying the defect structure of silicon without these drawbacks. In the diffusion model of formation grown-in microdefects all the parameters of precipitates, microvoids and dislocation loops are determined through the thermal conditions of growth. Therefore, definition the type of defect structure and calculation of the formation of microdefects is conducted depending on the values of crystal growth rate, temperature gradients and cooling rate of the crystal. On this basis, we have developed a new method for studying the defect structure of silicon. This method allows to simulate a real experiment by the software (V.I. Talanin et. al., 2011b).

Electronic equivalent of an object for direct test on the computer are programs that converted the mathematical models and algorithms to the available computer language (C++). The program is written high-level language programming in C++ compiler Borland C++ Builder. Program complex consists of two consecutive parts: (i) the unit determination the type of defect structure and (ii) the unit of calculation and graphs.

At the stage of determining the type of the structure defect of software system works as follows. Initially is the choice of method of growing dislocation-free silicon single crystals (Czochralski method or the floating-zone method) and then is the choice of certain diameter of the crystal. The ratio of $V_{gcrit}/G = \xi$ theoretically and experimentally determined in a certain range of values (0.06 mm²/K·min ≤ ξ ≤ 0.3 mm²/K·min). Therefore, we choose the certain value ξ for the calculation. For a given diameter in a certain range of values are selected: the value of the axial temperature gradient in the center of the crystal (G_a) and the value of the minimum (V_{min}) and maximum (V_{max}) crystal growth rate. These values are determined from the analysis of experimental and theoretical data for different diameters of the crystal. Then produced choice of the axial temperature gradient at the edge of the crystal (G_e) in the range $G_e/G_a = 1.0 \ldots 2.5$. The reliability and accuracy of the computational experiments can be experimentally verified by means of selective etching crystal of the plane which passes through the center of the crystal and parallel to the direction of growth. In case of deviations can by using of the selection of parameters G_a, G_e, ξ achieve full compliance with theoretical and experimental data. In addition, this technique avoids the difficulties of experimental determination of the G_a and G_e, especially for large diameter crystals (V.I. Talanin et. al., 2011b). At the last stage of determining the type of defect structure on the resulting dependence of the critical growth rate $V_{gcrit}(r) = \xi \cdot G(r)$ is imposed value of real crystal growth rate $V = const$ (Fig. 2). This procedure allows determining the type of defect structure of a real crystal. Depending on the position of the line $V = const$ relative to the curve $V_{gcrit}(r) = \xi \cdot G(r)$ may be three areas of the defect structure. Calculation of these areas produced the block of calculation and graphs (Fig. 3).

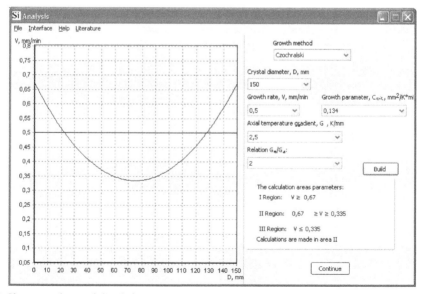

Fig. 2. Shape analysis of the defect structure

The Diffusion Model of Grown-In Microdefects Formation During Crystallization of Dislocation-Free Silicon Single Crystals

237

The first area of the calculation is characterized by high rates of crystal growth, when the above V-shaped distribution of precipitates formed only microvoids and precipitates. The second area of calculation is characterized by the average growth rate of the crystal, when a ring of precipitate in the plane perpendicular to the direction of growth crystal is formed. In this case inside the ring are formed precipitates and microvoids, outside the ring are formed precipitates and interstitial dislocation loops. The third area of calculation is characterized by low rates of crystal growth, when the below V-shaped distribution of precipitates are formed precipitates and interstitial dislocation loops. Calculation of the precipitates is carried out within the classical theory of nucleation, growth and coalescence of precipitates by means of the analytical and approximate calculations. Critical radius of precipitates, the distribution of precipitates in size, change in the average size of precipitates during the cooling of the crystal, and other parameters of the precipitation of carbon and oxygen are determined (Fig. 3). Mathematical models and calculation parameters are given in (V.I. Talanin et al., 2008; V.I. Talanin & I.E. Talanin, 2010a; 2011a). When calculating the vacancy microvoids initially are tested of conditions their formation, since microvoids are not formed at a cooling rate of the crystal $V_{cool} \geq 40K / min$ (Nakamura et al., 2002) and in crystals with a diameter less than 70 mm (V.I. Talanin et al., 2010c). Calculation of the microvoids and interstitial dislocation loops to determines for each of these types of defects such parameters as the critical radii and the concentrations (V.I. Talanin & I.E. Talanin, 2010b).

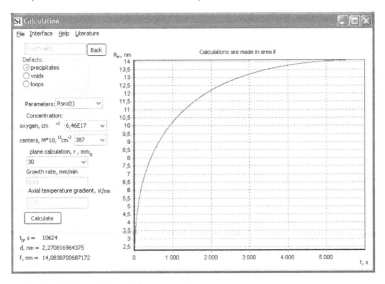

Fig. 3. Form of calculation of the defect structure

The software complex performs imitation of a real experiment and with the maximum precision reproduces the thermal characteristics of silicon single crystal growing. The software complex allows to determine the thermal conditions of crystal growth, to predict and control the defect structure of the crystal. Calculations formations of grown-in microdefects are in good agreement with the experimental results of research (V.I. Talanin et. al., 2011b).

The program complex is the first experience of a virtual experimental device for research the real structure of dislocation-free silicon single crystals. Currently, it can be used for the analysis and calculation of the defect structure of undoped single-crystal silicon. Depending on the thermal conditions of crystal growth can determine crucial parameters such as size and concentration of grown-in microdefects. Approach to the analysis and calculation of formation grown-in microdefects has an important advantage in simplicity, accessibility and sufficient adequacy of mathematical modeling in comparison with other methods. For its implementation does not require supercomputers, and can effectively use the experimental data, experience and intuition of physicists, materials scientists and technologists for the analytical calculation and design of the defect structure. The program complex is easy to implement on a personal computer in technology and research practices.

Disadvantages software system is determined deficiencies of the diffusion model of formation grown-in microdefects. These include: (i) one-dimensional model; (ii) failure to account for the width of the V-shaped distribution of precipitates; (iii) uncertainty in determining of thermal conditions of growth; (iv) the error of approximate numerical methods. Elimination of these deficiencies will increase the accuracy of the calculations.

The influence of other impurities (e.g., dopants, nitrogen, hydrogen, iron, and others) on the formation of the defect structure of silicon can be taken into account by using two approaches: a rigorous approach and a simplified approach. A rigorous approach requires accurate accounting of all the components in equations (1) mass balance of point defects in the crystal. In this case, the solution of the corresponding system of coupled equations of the Fokker-Planck equation can be considerably more difficult.

A simplified approach assumes the separation of impurities in the two groups. The first group contains impurities, which interact with vacancies. The second group contains impurities, which interact with self-interstitial atoms. This interaction for the first group of impurities is similar to the interaction of oxygen atoms with vacancies. This interaction for the second group of impurities is similar to carbon atoms interaction with self-interstitial atoms.. Therefore, in equations (1) the sum of the concentrations of the components of the first group of impurities is equivalent to the impurity concentration of oxygen, and oxygen diffusion coefficient is the sum of the diffusion coefficients of the system components. Are defined analogously the components of interaction in the second group of impurities. Then the system of equations reduced to the system of equations (1), followed by the task of determining the chemical nature of precipitates.

The task of construction the defect structure of dislocation-free silicon single crystal is an inverse problem the analysis and calculation of grown-in microdefects. In this case, you must first specify the type, size and concentration of grown-in microdefects. Parameters of grown-in microdefects are selected based on the requirements for defect structure of devices and integrated circuits. In the next stage are calculated parameters of crystal growth (the growth rate for a certain diameter of the crystal and temperature gradients), which provide presence of given defects of structure. We get that the defect structure determines thermal conditions of crystal growth. To automate the process of crystal growth need to carry out the development of software products based on the diffusion model in combination with known software products for modelling crystal growth.

Production technologies of devices based on silicon technology are connected with various impacts on the material. Heat treatments, ionizing radiations and mechanical effects have a critical impact on the initial defect structure of dislocation-free silicon single crystals. Technological impacts are lead to the transformation of grown-in microdefects. In the general case technological impacts are leads to: (i) the growth of initial grown-in microdefects; (ii) formation on grown-in microdefects of new defects (e.g., stacking faults); (iii) the formation of new defects. In contrast to grown-in microdefects these defects of crystal structure can be defined as postgrowth microdefects.

At the present time for the description of the formation of postgrowth microdefects are used a variety of models. Overview of the main models has been presented in (Sadamitsu et. al., 1993). The purpose of all models is to consider the formation and growth of defects as a result of technological impacts (e.g., by heating the crystal). A common deficiency of these models is the lack of consideration of the influence of grown-in microdefects, which are in the initial wafer.

Diffusion model of formation grown-in microdefects makes it possible to calculate of the defect structure of the initial silicon wafers. So from our point of view of theoretical analysis of the formation of postgrowth microdefects must be based on the diffusion model. It should be noted that taking into account all components of the general solution may be difficult. However, in some cases (e.g., heat treatment of silicon with a certain type of grown-in microdefects) can be solved in the near future. Building a general model for the formation of postgrowth microdefects will help optimize technological processes for production of devices. In this case, we can construct the defect structure of devices during their manufacture.

7. Conclusion

The diffusion model of the formation grown-in microdefects provides the unity and adequacy of physical and mathematical modeling. This model simulates of the defect structure of dislocation-free silicon single crystals of any diameters. The model of point defects dynamics can be considered as component of the diffusion model for formation grown-in microdefects. The diffusion model allowed to create software for personal computer. With the help of software can be conducted analytical researches which replace the expensive experimental researches.

Further development and modification of the software will lead to the development of information system of formation grown-in microdefects in dislocation-free silicon single crystals. The combinations of an information system with software for control the crystal growth will allow construct the defect structure of crystals during their growth. In turn, application a diffusion model of formation grown-in microdefects in the calculation of the formation of postgrowth microdefects allow to calculate the defect structure of silicon-based devices. In this case, it is possible to adequately construction the defect structure of silicon devices. We believe that the proposed in an article algorithm for the engineering of the defect structure of silicon can be used for other materials.

8. References

Ammon von, W.; Dornberger, E. & Hansson P.O. (1999). Bulk properties of very large diameter silicon single crystal. *Journal Crystal Growth*, Vol. 198-199, No. 1-4, pp. 390-398, ISSN 0022-0248.

Bonafos, C.; Mathiot, D. & Claverie A. (1998). Ostwald ripening of end-of-range defects in silicon. *Journal of Applied Physics*, Vol. 83, No. 6, pp. 3008-3018, ISSN 0021-8979.

Bulyarskii, S.V. & Fistul, V.I. (1997). *Thermodynamics and kinetics of interacting defects in semiconductors*, Nauka, ISBN 5-02-015164-5, Moscow, Russia.

Burton, B. & Speight M.V. (1985). The coarsening and annihilation kinetics of dislocation loops. *Philosophical Magazine A*, Vol. 53, No. 3, pp. 385-402, ISSN 0141-8610.

Chaldyshev, V.V.; Bert, N.A.; Romanov, A.E.; Suvorova, A.A.; Kolesnikova, A.L.; Preobrazhenskii, V.V.; Putyato, M.A.; Semyagin, B.R.; Werner, P.; Zakharov, N.D. & Claveria, A. (2002). Local stresses induced by nanoscale As-Sb clusters in GaAs matrix. *Applied Physics Letters*, Vol. 80, No. 3, pp. 377-381, ISSN 0003-6951.

Chaldyshev, V.V.; Kolesnikova, A.L.; Bert, N.A. & Romanov, A.E. (2005). Investigation of dislocation loops associated with As-Sb nanoclusters in GaAs. *Journal of Applied Physics*, Vol. 97, No. 2, pp. 024309-024319, ISSN 0021-8979.

Cho, H.-J.; Sim, B.-C. & Lee, J.Y. (2006). Asymmetric distributions of grown-in microdefects in Czochralski silicon. *Journal Crystal Growth*, Vol. 289, No. 2, pp. 458-463, ISSN 0022-0248.

Dornberger, E.; Ammon, von W.; Virbulis, J.; Hanna, B. & Sinno T. (2001). Modeling of transient point defect dynamics in Czochralski silicon crystal. *Journal Crystal Growth*, Vol. 230, No. 1-2, pp. 291-299, ISSN 0022-0248.

Föll, H. & Kolbesen B.O. (1975). Formation and nature of swirl defects in silicon. *Journal of Applied Physics*, Vol. 8, No. 3, pp. 319-331, ISSN 0021-8979.

Goethem, van N.; Potter, de A.; Bogaert, van den N. & Dupret, F. (2008). Dynamic prediction of point defects in Czochralski silicon growth. An attempt to reconcile experimental defect diffusion coefficients with the criterion V/G. *Journal of Physics and Chemistry of Solids*, Vol. 69, No. 2-3, pp. 320-324, ISSN 0022-3697.

Goldstein, R.V.; Mezhennyi, M.V.; Mil'vidskii, M.G.; Reznik, V.Ya.; Ustinov, K.B. & Shushpannikov, P.S. (2011). Experimental and theoretical investigation of formation of the oxygen-containing precipitate-dislocation loop system in silicon. *Physics of the Solid State*, Vol. 53, No. 3, pp. 527-538, ISSN 1063-7834.

Gyseva, N.B.; Sheikhet, E.G.; Shpeizman, V.V. & Shulpina I.L. (1986). Dislokazionnay aktivnost mikrodefektov v monokristalax kremnia. *Fizika tverdogo tela*, Vol. 28, No. 10, pp. 3192-3194, ISSN 0367-3294.

Huff, H.R. (2002). An electronics division retrospective (1952-2002) and future opportunities in the twenty-first century. *Journal of the Electrochemical Society*, Vol. 149, No. 5, pp. S35-S58, ISSN 0013-4651.

Itsumi, M. (2002). Octahedral void defects in Czochralski silicon. *Journal Crystal Growth*, Vol. 237-239, No. 3, pp. 1773-1778, ISSN 0022-0248.

Kato, M.; Yoshida, T.; Ikeda, Y. & Kitagawara, Y. (1996). Transmission electron microscope observation of "IR scattering defects" in as-grown Czochralski Si crystals. *Japanese Journal Applied Physics*, Vol. 35, No. 11, pp. 5597-5601, ISSN 0021-4922.

Kock de, A.J.R. (1970). Vacancy clusters in dislocation-free silicon. *Applied Physics Letters*, Vol. 16, No. 3, pp. 100-102, ISSN 0003-6951.

Kock de, A.J.R.; Stacy, W.T. & Wijgert van de, W.M. (1979). The effect of doping on microdefect formation in as-grown dislocation-free Czochralski silicon crystals. *Applied Physics Letters*, Vol. 34, No. 9, pp. 611-616, ISSN 0003-6951.

Kolesnikova, A.L. & Romanov, A.E. (2004). Misfit dislocation loops and critical parameters of quantum dots and wires. *Philosophical Magazine Letters*, Vol. 84, No. 3, pp. 501-506, ISSN 0950-0839.

Kolesnikova, A.L.; Romanov, A.E. & Chaldyshev V.V. (2007). Elastic-energy relaxation in heterostructures with strained nanoinclusions. *Physics of the Solid State*, Vol. 49, No. 4, pp. 667-674, ISSN 1063-7834.

Kulkarni, M.S.; Voronkov, V.V. & Falster, R. (2004). Quantification of defect dynamics in unsteady-state and steady-state Czochralski growth of monocrystalline silicon. *Journal Electrochemical Society*, Vol. 151. – No. 5, pp. G663-G669, ISSN 0013-4651.

Kulkarni, M.S. (2005). A selective review of the quantification of defect dynamics in growing Czochralski silicon crystals. *Ind. Eng. Chem. Res.*, Vol. 44, No. 16, pp. 6246-6263, ISSN 0888-5885.

Kulkarni, M.S. (2007). Defect dynamics in the presence of oxygen in growing Czochralski silicon crystals. *Journal Crystal Growth*, Vol. 303, No. 2, pp. 438-448, ISSN 0022-0248.

Kulkarni, M.S. (2008a). Lateral incorporation of vacancies in Czochralski silicon crystals. *Journal Crystal Growth*, Vol. 310, No. 13, pp. 3183-3191, ISSN 0022-0248.

Kulkarni, M.S. (2008b). Defect dynamics in the presence of nitrogen in growing Czochralski silicon crystals. *Journal Crystal Growth*, Vol. 310, No. 2, pp. 324-335, ISSN 0022-0248.

Nakamura, K.; Saishoji, T. & Tomioka, J. (2002). Grown-in defects in silicon crystals. *Journal Crystal Growth*, Vol. 237-239, No. 1-4, pp. 1678-1684, ISSN 0022-0248.

Petroff, P.M. & Kock de, A.J.R. (1975). Characterization of swirl defects in floating-zone silicon crystals. *Journal Crystal Growth*, Vol. 30, No. 1, pp. 117-124, ISSN 0022-0248.

Prostomolotov, A.I. & Verezub, N.A. (2009). Simplistic approach for 2D grown-in microdefects modeling. *Physica status solidi (c)*, Vol. 6, No. 8, pp. 1878-1881, ISSN 1610-1642.

Prostomolotov, A.I.; Verezub, N.A.; Mezhennii, M.V. & Reznik, V.Ua. (2011). Thermal optimization of CZ bulk growth and wafer annealing for crystalline dislocation-free silicon. *Journal of Crystal Growth*, Vol. 318, No. 1, pp. 187-192, ISSN 0022-0248.

Sadamitsu, S.; Umeno, S.; Koike, Y.; Hourai, M.; Sumita, S. & Shigematsu, T. (1993). Dislocations, precipitates and other defects in silicon crystals. *Japanese Journal Applied Physics*, Vol. 32, No. 9, pp. 3675-3679, ISSN 0021-4922.

Sinno, T. (1999). Modeling microdefect formation in Czochralski silicon. *Journal Electrochemical Society*, Vol. 146, No. 6, pp. 2300-2312, ISSN 0013-4651.

Sitnikova, A.A.; Sorokin, L.M.; Talanin, I.E.; Sheikhet, E.G. & Falkevich E.S. (1984). Electron-microscopic study of microdefects in silicon single crystals grown at high speed. *Physica Status Solidi (a)*, Vol. 81, No. 2, pp. 433-439, ISSN 1862-6300.

Sitnikova, A.A.; Sorokin, L.M.; Talanin, I.E.; Sheikhet, E.G. & Falkevich, E.S. (1985). Vacancy type microdefects in dislocation-free silicon single crystals. *Physica Status Solidi (a)*, Vol. 90, No. 1, pp. K31-K35, ISSN 1862-6300.

Talanin, V.I.; Talanin, I.E. & Levinson, D.I. (2002a). Physical model of paths of microdefects nucleation in dislocation-free single crystals float-zone silicon. *Cryst. Res. & Technol.*, Vol. 37, No. 9, pp. 983-1011, ISSN 0232-1300.

Talanin, V.I.; Talanin, I.E. & Levinson, D.I. (2002b). Physics of the formation of microdefects in dislocation-free monocrystals of float-zone silicon. *Semicond. Sci. Cryst. Res. & Technol.*, Vol. 17, No. 2, pp. 104-113, ISSN 0268-1242.

Talanin, V.I. & Talanin, I.E. (2003). Physical nature of grown-in microdefects in Czochralski-grown silicon and their transformation during various technological effects . *Physica Status Solidi (a)*, Vol. 200, No. 2, pp. 297-306, ISSN 1862-6300.

Talanin, V.I. & Talanin, I.E. (2004). Mechanism of formation and physical classification of the grown-in microdefects in semiconductor silicon. *Defect & Diffusion Forum*, Vol. 230-232, No. 1, pp. 177-198, ISSN 1012-0386.

Talanin, V.I. & Talanin, I.E. (2006a). Formation of grown-in microdefects in dislocation-free silicon monocrystals, In: *New research on semiconductors*, T.B. Elliot,(Ed.), 31-67, Nova Science Publishers, Inc., ISBN 1-59454-920-6, New York, USA.

Talanin, V.I. & Talanin, I.E. (2006b). On the formation of vacancy microdefects in dislocation-free silicon single crystals. *Ukranian Journal of Physics*, Vol. 51, No. 11-12, pp. 108-112, ISSN 0503-1265.

Talanin, V.I. & Talanin, I.E. (2007a). On the recombination of intrinsic point defects in dislocation-free silicon single crystals. *Physics of the Solid State*, Vol. 49, No. 3, pp. 467-470, ISSN 1063-7834.

Talanin, V.I.; Talanin, I.E. & Voronin, A.A. (2007b). About formation of grown-in microdefects in dislocation-free silicon single crystals. *Canadian Journal of Physics*, Vol. 85, No. 12, pp. 1459-1471, ISSN 1208-6045.

Talanin, V.I.; Talanin, I.E. & Voronin A.A. (2008). Modeling of the defect structure in dislocation-free silicon single crystals. *Crystallography Reports*, Vol. 53, № 7, pp. 1124-1132, ISSN 1063-7745.

Talanin, V.I. & Talanin, I.E. (2010a). Kinetic of high-temperature precipitation in dislocation-free silicon single crystals. *Physics of the Solid State*, Vol. 52, No. 10, pp. 2063-2069, ISSN 1063-7834.

Talanin, V.I. & Talanin, I.E. (2010b). Kinetics of formation of vacancy microvoids and interstitial dislocation loops in dislocation-free silicon single crystals. *Physics of Solid State*, Vol. 52, No. 9, pp. 1880-1886, ISSN 1063-7834.

Talanin, V.I. & Talanin, I.E. (2010c). Modeling of defect formation processes in dislocation-free silicon single crystals. *Crystallography Reports*, Vol. 55, No. 4, pp. 675-681, ISSN 1063-7745.

Talanin, V.I. & Talanin, I.E. (2011a). Kinetic model of growth and coalescence of oxygen and carbon precipitates during cooling of as-grown silicon crystals. *Physics of the Solid State*, Vol. 53, No. 1, pp. 119-126, ISSN 1063-7834.

Talanin, V.I.; Talanin, I.E. & Ustimenko N.Ph. (2011b). A new method for research of grown-in microdefects in dislocation-free silicon single crystals. *Journal of Crystallization Process & Technologys*, Vol.1, № 2, pp. 13-17, ISSN 2161-7678.

Yang, D.; Chen, J.; Ma, X. & Que, D. (2009). Impurity engineering of Czochralski silicon used for ultra large-scaled-integrated circuits. *Journal Crystal Growth*, Vol. 311, No. 3, pp. 837-841, ISSN 0022-0248.

Vas'kin, V.V. & Uskov V.A. (1968). Vlianie kompleksoobrazovania na diffyziu primesei v polyprovodnikax. *Fizika Tverdogo Tela*, Vol. 10, No. 6, pp. 1239-1241, ISSN 0367-3294.

Veselovskaya, N.V.; Sheikhet, E.G.; Neimark, K.N. & Falkevich, E.S. (1977). Defecty tipa klasterov v kremnii. *Proceedings of IV simposiuma Rost i legirovanie polyprovodnikovyx kristalov i plenok*, Vol. 2, pp. 284-288, Novosibirsk, USSR, June 1975.

Voronkov, V.V. & Falster, R. (1998). Vacancy-type microdefect formation in Czochralski silicon. *Journal Crystal Growth*, Vol. 194, No.1, pp. 76-88, ISSN 0022-0248.

Voronkov, V.V. (2008). Grown-in defects in silicon produced by agglomeration of vacancies and self-interstitial. *Journal Crystal Growth*, Vol. 310, No. 7-9, pp. 1307-1314, ISSN 0022-0248.

Voronkov, V.V.; Dai B. & Kulkarni M.S. (2011). Fundamentals and engineering of the Czochralski growth of semiconductor silicon crystals. *Comprehensive Semiconductor science and Technology*, Vol. 3, pp. 81-169, ISBN 978-0-444-53153-7.

Wang, Z. & Brown, R.A. (2001). Simulation of almost defect-free silicon crystal growth. *Journal Crystal Growth*, Vol. 231, No. 2, pp. 442-452, ISSN 0022-0248.

10

Synthesis and X-Ray Crystal Structure of α-Keggin-Type Aluminum-Substituted Polyoxotungstate

Chika Nozaki Kato[1], Yuki Makino[1], Mikio Yamasaki[2],
Yusuke Kataoka[3], Yasutaka Kitagawa[3] and Mitsutaka Okumura[3]
[1]Shizuoka University
[2]Rigaku Corporation
[3]Osaka University
Japan

1. Introduction

Aluminum and its derivatives such as alloys, oxides, organometallics, and inorganic compounds have attracted considerable attention because of their extreme versatility and unique range of properties, including acidity, hardness, and electroconductivity (Cotton & Wilkinson, 1988). Since the properties and activities of an aluminum species are strongly dependent on the structures of the aluminum sites, the syntheses of aluminum compounds with structurally well-defined aluminum sites are considerably significant for the development of novel and efficient aluminum-based materials. However, the use of these well-defined aluminum sites is slightly limited by the conditions resulting from the hydrolysis of the aluminum species by water (Djurdjevic et al., 2000; Baes & Mesmer, 1976; Orvig, 1993; Akitt, 1989).

Polyoxometalates have been of particular interest in the fields of catalytic chemistry, surface science, and materials science because their chemical properties such as redox potentials, acidities, and solubilities in various media can be finely tuned by choosing appropriate constituent elements and countercations (Pope, 1983; Pope & Müller, 1991, 1994). In particular, the coordination of metal ions to the vacant site(s) of lacunary polyoxometalates is one of the most effective techniques used for constructing efficient and well-defined active metal centers. Among various lacunary polyoxometalates, a series of Keggin-type phosphotungstates is one of the most useful types of lacunary polyoxometalates. Fig. 1 shows some examples of lacunary Keggin-type phosphotungstates, i.e., *mono*-lacunary α-Keggin [α-$PW_{11}O_{39}$]$^{7-}$ (Contant, 1987), *di*-lacunary γ-Keggin [γ-$PW_{10}O_{36}$]$^{7-}$ (Domaille, 1990; Knoth, 1981), and *tri*-lacunary α-Keggin [A-α-$PW_{9}O_{34}$]$^{9-}$ (Domaille, 1990) phosphotungstates. Knoth and co-workers first synthesized the Keggin derivative (Bu_4N)$_4$(H)$ClAlW_{11}PO_{39}$ by the reaction of *mono*-lacunary α-Keggin phosphotungstate with $AlCl_3$ in dichloroethane (Knoth et al., 1983). However, only a few aluminum-coordinated polyoxometalates (determined by X-ray crystallographic analysis) have been reported, e.g., a monomeric, *di*-aluminum-substituted γ-Keggin polyoxometalate $TBA_3H[\gamma\text{-}SiW_{10}O_{36}\{Al(OH_2)\}_2(\mu\text{-}$

OH)$_2$]·4H$_2$O (TBA = tetra-n-butylammonium) (Kikukawa et al., 2008), a monomeric, *mono*-aluminum-substituted α-Keggin polyoxometalate K$_6$H$_3$[ZnW$_{11}$O$_{40}$Al]·9.5H$_2$O (Yang et al., 1997), and a dimeric aluminum complex having *mono-* and *di*-aluminum sites sandwiched by *tri*-lacunary α-Keggin polyoxometalate K$_6$Na[(A-PW$_9$O$_{34}$)$_2${W(OH)(OH$_2$)}{Al(OH)(OH$_2$)} {Al(μ-OH)(OH$_2$)$_2$}$_2$]·19H$_2$O (Kato et al., 2010); these structures are shown in Fig. 2.

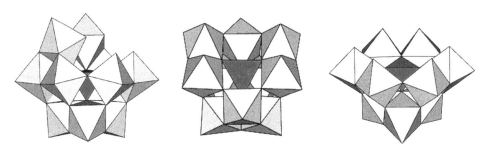

Fig. 1. Some examples of lacunary phosphotungstates. The polyhedral representations of *mono*-lacunary α-Keggin [α-PW$_{11}$O$_{39}$]$^{7-}$ (left), *di*-lacunary γ-Keggin [γ-PW$_{10}$O$_{36}$]$^{7-}$ (center), and *tri*-lacunary α-Keggin [A-α-PW$_9$O$_{34}$]$^{9-}$ (right) phosphotungstates. The WO$_6$ and internal PO$_4$ groups are represented by the white octahedra and red tetrahedron, respectively.

In this study, we successfully obtained a monomeric, α-Keggin *mono*-aluminum-substituted polyoxotungstate in the form of crystals (suitable for X-ray structure analysis) of [(n-C$_4$H$_9$)$_4$N]$_4$[α-PW$_{11}${Al(OH$_2$)}O$_{39}$] that were fully characterized by X-ray crystallography; elemental analysis; thermogravimetric/differential thermal analysis; Fourier transform infrared spectroscopy; and solution ^{31}P, ^{27}Al, and ^{183}W nuclear magnetic resonance spectroscopies. Although the X-ray crystallography of [α-PW$_{11}${Al(OH$_2$)}O$_{39}$]$^{4-}$ showed that the *mono*-aluminum-substituted site was not identified because of the high symmetry in the compound, the bonding mode (bond lengths and bond angles) were significantly influenced by the insertion of aluminum ions into the *mono*-vacant sites. In addition, density-functional-theory (DFT) calculations showed a unique coordination sphere around the *mono*-aluminum-substituted site in [α-PW$_{11}${Al(OH$_2$)}O$_{39}$]$^{4-}$; this was consistent with the X-ray crystal structure and spectroscopic results. In this paper, we report the complete details of the synthesis, molecular structure, and characterization of [(n-C$_4$H$_9$)$_4$N]$_4$[α-PW$_{11}$ {Al(OH$_2$)}O$_{39}$].

2. Experimental section

2.1 Materials

K$_7$[α-PW$_{11}$O$_{39}$]·11H$_2$O (Contant, 1987) and Cs$_7$[γ-PW$_{10}$O$_{36}$]·19H$_2$O (Domaille, 1990; Knoth, 1981) were prepared as described in the literature. The number of solvated water molecules was determined by thermogravimetric/differential thermal analyses. Acetonitrile-soluble, tetra-n-butylammonium salts of [α-PW$_{12}$O$_{40}$]$^{3-}$ and [α-PW$_{11}$O$_{39}$]$^{7-}$ were prepared by the addition of excess tetra-n-butylammonium bromide to the aqueous solutions of Na$_3$[α-PW$_{12}$O$_{40}$]·16H$_2$O (Rosenheim & Jaenicke, 1917) and K$_7$[α-PW$_{11}$O$_{39}$]·11H$_2$O. All the reagents and solvents were obtained and used as received from commercial sources. Al(NO$_3$)$_3$·9H$_2$O (Aldrich, 99.997% purity) was used in the synthesis. The X-ray crystal structure of

$[(CH_3)_2NH_2]_4[\alpha\text{-}PW_{11}Re^VO_{40}]$ (Kato et al., 2010) was resolved by SHELXS-97 (direct methods) and re-refined by SHELXL-97 (Sheldrick, 2008). The crystal data are as follows: $C_8H_{32}N_3O_4PReW_{11}$: M = 3063.87, *trigonal*, space group *R-3m*, a = 16.53(2) Å, c = 25.21(4) Å, V = 5963(12) Å^3, Z = 6, D_c = 5.119 g/cm^3, R_1 = 0.0559 ($I > 2\sigma(I)$) and wR_2 = 0.1513 (for all data). The four dimethylammonium ions could not be identified due to the disorder (Nomiya et al., 2001, 2002; Weakley & Finke, 1990; Lin et al., 1993). CCDC number 851154.

2.2 Instrumentation/analytical procedures

The elemental analysis was carried out by using Mikroanalytisches Labor Pascher (Remagen, Germany). The sample was dried overnight at room temperature under pressures of 10^{-3} – 10^{-4} Torr before analysis. Infrared spectra were recorded on a Parkin Elmer Spectrum100 FT-IR spectrometer in KBr disks at room temperature. Thermogravimetric (TG) and differential thermal analyses (DTA) data were obtained using a Rigaku Thermo Plus 2 series TG/DTA TG 8120. TG/DTA measurements were performed in air by constantly increasing the temperature from 20 to 500 °C at a rate of 4 °C per min. The ^{31}P nuclear magnetic resonance (NMR) (242.95 MHz) spectra in acetonitrile-d_3 solution were recorded in tubes (outer diameter: 5 mm) on a JEOL ECA-600 NMR spectrometer. The ^{31}P NMR spectra were referenced to an external standard of 85% H_3PO_4 in a sealed capillary. Negative chemical shifts were reported on the δ scale for resonance upfields of H_3PO_4 (δ 0). The ^{27}Al NMR (156.36 MHz) spectrum in acetonitrile-d_3 was recorded in tubes (outer diameter: 5 mm) on a JEOL ECA-600 NMR spectrometer. The ^{27}Al NMR spectrum was referenced to an external standard of saturated $AlCl_3$-D_2O solution (substitution method). Chemical shifts were reported as positive on the δ scale for resonance downfields of $AlCl_3$ (δ 0). The ^{183}W NMR (25.00 MHz) spectra were recorded in tubes (outer diameter: 10 mm) on a JEOL ECA-600 NMR spectrometer. The ^{183}W NMR spectra measured in acetonitrile-d_3 were referenced to an external standard of saturated Na_2WO_4-D_2O solution (substitution method).

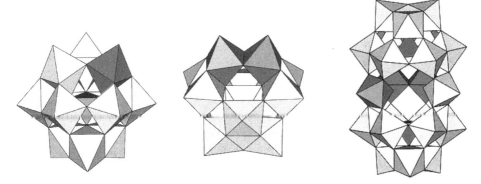

Fig. 2. The polyhedral representation of $K_6H_3[ZnW_{11}O_{40}Al]\cdot9.5H_2O$ (left), $TBA_3H[\gamma\text{-}SiW_{10}\text{-}O_{36}\{Al(OH_2)\}_2(\mu\text{-}OH)_2]\cdot4H_2O$ (TBA = tetra-*n*-butylammonium) (center), and $K_6Na[(A\text{-}PW_9\text{-}O_{34})_2\{W(OH)(OH_2)\}\{Al(OH)(OH_2)\}\{Al(\mu\text{-}OH)(OH_2)_2\}_2]\cdot19H_2O$ (right). The aluminum groups are represented by the blue octahedra. The WO_6 groups are represented by white octahedra. The internal ZnO_4, SiO_4, and PO_4 groups are represented by green, yellow, and red tetrahedra, respectively.

Chemical shifts were reported as negative for resonance upfields of Na_2WO_4 (δ 0). Potentiometric titration was carried out with 0.4 mol/L tetra-n-butylammonium hydroxide as a titrant under argon atmosphere (Weiner et al., 1996). The compound $[(n-C_4H_9)_4N]_4[\alpha-PW_{11}\{Al(OH_2)\}O_{39}]$ (0.018 mmol) was dissolved in acetonitrile (30 mL) at 25 °C and the solution was stirred for approximately 5 min. The titration data were obtained with a pH meter (Mettler Toledo). Data points were obtained in milivolt. A solution of tetra-n-butylammonium hydroxide (9.0 mmol/L) was syringed into the suspension in 0.25-equivalent intervals.

2.3 Synthesis of $[(n-C_4H_9)_4N]_4[\alpha-PW_{11}\{Al(OH_2)\}O_{39}]$

$Cs_7[\gamma-PW_{10}O_{36}]\cdot19H_2O$ (2.00 g; 0.538 mmol) was dissolved in water (600 mL) at 40 °C, and solid $Al(NO_3)_3\cdot9H_2O$ (0.250 g, 0.666 mmol) was added to the colorless clear solution. After stirring for 1 h at 40 °C, a solid $[(n-C_4H_9)_4N]_4Br$ (12.14 g; 37.7 mmol) was added to the solution, followed by stirring at 25 °C for 3 days. The white precipitate was collected on a glass frit (G4) and washed with water (ca. 1 L). At this stage, a crude product was obtained in a 1.662 g yield. The crude product (1.662 g) was dissolved in acetonitrile (10 mL), followed by filtering through a folded filter paper (Whatman #5). After the product was left standing for a week at 25 °C, colorless platelet crystals were formed. The obtained crystals weighted 0.752 g (the yield calculated considering that [mol of $[(n-C_4H_9)_4N]_4[\alpha-PW_{11}\{Al(OH_2)\}O_{39}]$/[mol of $Cs_7[\gamma-PW_{10}O_{36}]\cdot19H_2O]$ × 100 was 36.9%). The elemental analysis results were as follows: C, 20.73; H, 4.00; N, 1.58; P, 0.84; Al, 0.77; W, 54.6; Cs, <0.1%. The calculated values for $[(n-C_4H_9)_4N]_4[\alpha-PW_{11}\{Al(OH_2)\}O_{39}]$ = $C_{64}H_{146}AlN_4O_{40}PW_{11}$: C, 20.82; H, 3.99; N, 1.52; P, 0.84; Al, 0.73; W, 54.77; Cs, 0%. A weight loss of 2.16% was observed in the product during overnight drying at room temperature under 10^{-3}–10^{-4} Torr before the analysis, thereby suggesting the presence of two weakly solvated or adsorbed acetonitrile molecules (2.18%). TG/DTA under atmospheric conditions showed a weight loss of 31.0% with an exothermic peak at 337 °C was observed in the temperature range from 25 to 500 °C; our calculations indicated the presence of four $[(C_4H_9)_4N]^+$ ions, two acetonitrile molecules, and a water molecule (calcd. 28.4%). The results were as follows: IR soectroscopy results (KBr disk): 1078s, 964s, 887s, 818s, 749m, 702w, 518w cm^{-1}; ^{31}P NMR (25°C, acetonitrile-d_3): δ -12.5; ^{27}Al NMR (25 °C, acetonitrile-d_3): δ 16.1; ^{183}W NMR (25 °C, acetonitrile-d_3): δ -56.2 (2W), -93.1 (2W), -108.6 (2W), -115.8 (2W), -118.5 (1W), -153.9 (2W).

2.4 X-Ray crystallography

A colorless platelet crystal of $[(n-C_4H_9)_4N]_4[\alpha-PW_{11}\{Al(OH_2)\}O_{39}]$ (0.16 × 0.16 × 0.01 mm^3) was mounted on a MicroMount. All measurements were made on a Rigaku VariMax with a Saturn diffractometer using multi-layer mirror monochromated Mo Kα radiation (λ= 0.71075 Å) at 93 K. Data were collected and processed using CrystalClear for Windows, and structural analysis was performed using the CrystalStructure for Windows. The structure was solved by SHELXS-97 (direct methods) and refined by SHELXL-97 (Sheldrick, 2008). Since one aluminum atom was disordering over twelve tungsten sites in $[\alpha-PW_{11}\{Al(OH_2)\}O_{39}]^{4-}$, the occupancies for the aluminum and tungsten sites were fixed at 1/12 and 11/12 throughtout the refinement. Four tetra-n-butylammonium ions could not be modelled with disordered atoms. Accordingly, the residual electron density was removed using the SQUEEZE routine in PLATON (Spek, 2009).

2.5 Crystal data for [(n-C₄H₉)₄N]₄[α-PW₁₁{Al(OH₂)}O₃₉]

$C_{64}H_{146}AlN_4O_{40}PW_{11}$; M = 3692.17, *cubic*, space group *Im-3m* (#229), a = 17.665(2) Å, V = 5512.2(8) Å³, Z = 2, D_c = 2.224 g/cm³, μ(Mo-Kα) = 115.313 cm⁻¹. R_1 = 0.0220 ($I > 2\sigma(I)$) and wR_2 = 0.0554 (for all data). GOF = 1.093 (22662 total reflections, 652 unique reflections where $I > 2\sigma(I)$). CCDC number 851155.

2.6 Computational details

The optimal geometry of [α-PW₁₁{Al(OH₂)}O₃₉]⁴⁻ was computed by means of a DFT method. First, we optimized the crystal geometries and followed this up with single-point calculations with larger basis sets. All calculations were performed by a spin-restricted B3LYP on Gaussian09 program package (Frisch et al., 2009). The basis sets used for the geometry optimization were LANL2DZ for W atoms, 6-31+G* for P atoms and 6-31G* for H, O, and Al atoms. LANL2DZ and 6-31+G* were used for W and other atoms, respectively, for the single-point calculations. The geometry optimizations were started using the X-ray structure of [α-PW₁₂O₄₀]³⁻ as an initial geometry, and they were performed under the gas phase condition. The optimized geometries were confirmed to be true minima by frequency analyses. All atomic charges used in this text were obtained from Mulliken population analysis.

3. Results and discussion

3.1 Synthesis and molecular formula of [(n-C₄H₉)₄N]₄[α-PW₁₁{Al(OH₂)}O₃₉]

The tetra-*n*-butylammonium salt of [α-PW₁₁{Al(OH₂)}O₃₉]⁴⁻ was formed by the direct reaction of aluminum nitrate with [γ-PW₁₀O₃₆]⁷⁻ (the molar ratio of Al³⁺:[γ-PW₁₀O₃₆]⁷⁻ was ca. 1.0) in an aqueous solution at 40 °C under air, followed by the addition of excess tetra-*n*-butylammonium bromide. The crystallization was performed by slow-evaporation from acetonitrile at 25 °C. During the formation of [α-PW₁₁{Al(OH₂)}O₃₉]⁴⁻, the decomposition of a *di*-lacunary γ-Keggin polyoxotungstate, and isomerization of γ-isomer to α-isomer occurred in order to construct the *mono*-aluminum-substituted site in an α-Keggin structure. It was noted that the polyoxoanion [α-PW₁₁{Al(OH₂)}O₃₉]⁴⁻ was easily obtained by the stoichiometric reaction of aluminum nitrate with a *mono*-lacunary α-Keggin polyoxotungstate, [α-PW₁₁O₃₉]⁷⁻, in an aqueous solution; however, a single species of [α-PW₁₁{Al(OH₂)}O₃₉]⁴⁻ could not be obtained as a tetra-*n*-butylammonium salt by using [α-PW₁₁O₃₉]⁷⁻ as a starting polyoxoanion.[1] Thus, single crystals that were suitable for X-ray crystallography could be obtained for the crystallization of the tetra-*n*-butylammonium salt of [α-PW₁₁{Al(OH₂)}O₃₉]⁴⁻ synthesized by using a *di*-lacunary γ-Keggin polyoxotungstate.

---

[1] The ³¹P NMR spectrum in acetonitrile-*d₃* of the tetra-*n*-butylammonium salt of [α-PW₁₁{Al(OH₂)}O₃₉]⁴⁻ prepared by the stoichiometic reaction of [α-PW₁₁O₃₉]⁷⁻ with Al(NO₃)₃·9H₂O in an aqueous solution showed two signals at -12.35 ppm and -12.48 ppm. The signal at -12.48 ppm was assigned to the internal phosphorus atom in [α-PW₁₁{Al(OH₂)}O₃₉]⁴⁻, whereas the signal at -12.35 ppm could not be identified; however, the signal was not due to the proton isomer, as reported for [(CH₃)₂NH₂]₁₀[Hf(PW₁₁O₃₉)₂]·8H₂O (Hou et al., 2007).

The sample for the elemental analysis was dried overnight at room temperature under a vacuum of 10^{-3} – 10^{-4} Torr. The elemental results for C, H, N, P, Al, and W were in good agreement with the calculated values for the formula without any absorbed or solvated molecules for $[(n\text{-}C_4H_9)_4N]_4[\alpha\text{-}PW_{11}\{Al(OH_2)\}O_{39}]$.

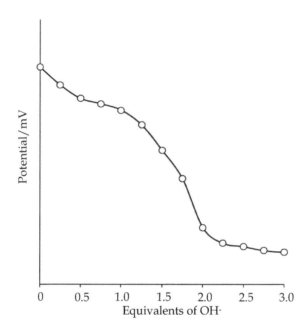

Fig. 3. Profile for the potentiometric titration of $[(n\text{-}C_4H_9)_4N]_4[\alpha\text{-}PW_{11}\{Al(OH_2)\}O_{39}]$ with tetra-n-butylammonium hydroxide as a titrant.

The Cs analysis revealed no contamination of cesium ions from $Cs_7[\gamma\text{-}PW_{10}O_{36}]\cdot19H_2O$. The weight loss observed during the course of drying before the analysis was 2.16% for $[(n\text{-}C_4H_9)_4N]_4[\alpha\text{-}PW_{11}\{Al(OH_2)\}O_{39}]$; this corresponded to two weakly solvated or adsorbed acetonitrile molecules. On the other hand, in the TG/DTA measurement performed under atmospheric conditions, a weight loss of 31.0% observed in the temperature range from 25 to 500 °C corresponded to four tetra-n-butylammonium ions, two acetonitrile molecules, and a water molecule.

From the potentiometric titration, a break point at 2.0 equivalents of added base was observed, as shown in Fig. 3. The titration profile revealed that $[(n\text{-}C_4H_9)_4N]_4[\alpha\text{-}PW_{11}\{Al(OH_2)\}O_{39}]$ had two titratable protons dissociated from the Al-OH$_2$ group. This result was consistent with the elemental analysis result.

3.2 The molecular structure of [(n-C$_4$H$_9$)$_4$N]$_4$[α-PW$_{11}${Al(OH$_2$)}O$_{39}$]

The molecular structure of [α-PW$_{11}${Al(OH$_2$)}O$_{39}$]$^{4-}$ as determined by X-ray crystallography is shown in Figs. 4 and 5. The bond lengths and bond angles are summarized in appendix. The molecular structure of [α-PW$_{11}${Al(OH$_2$)}O$_{39}$]$^{4-}$ was identical to that of a monomeric, α-Keggin polyoxotungstate [α-PW$_{12}$O$_{40}$]$^{3-}$ (Neiwert et al., 2002; Busbongthong & Ozeki, 2009). Due to the high symmetry space group, the eleven tungsten(VI) atoms were disordered and the *mono*-aluminum-substituted site was not identified, as observed for [W$_9$ReO$_{32}$]$^{5-}$ (Ortéga et al., 1997), [α-PW$_{11}$Re^{V}O$_{40}$]$^{5-}$ (Kato et al., 2010), [{SiW$_{11}$O$_{39}$Cu(H$_2$O)}{Cu$_2$(ac)-(phen)$_2$(H$_2$O)}]$^{14-}$ (phen = phenanthroline, ac = acetate) (Reinoso et al., 2006), (ANIH)$_5$[PCu(H$_2$O)W$_{11}$O$_{39}$](ANI)·8H$_2$O (ANI = aniline, ANIH$^+$ = anilinium ion) (Fukaya et al., 2011), Cs$_5$[PMn(H$_2$O)W$_{11}$O$_{39}$]·4H$_2$O (Patel et al., 2011), and Cs$_5$[PNi(H$_2$O)W$_{11}$O$_{39}$]·2H$_2$O (T. J. R. Weakley, 1987). However, the bond lengths of [(n-C$_4$H$_9$)$_4$N]$_4$[α-PW$_{11}${Al(OH$_2$)}O$_{39}$] were clearly influenced by the insertion of aluminum ion into the vacant site as compared with those of [CH$_3$NH$_3$]$_3$[PW$_{12}$O$_{40}$]·2H$_2$O, [(CH$_3$)$_2$NH$_2$]$_3$[PW$_{12}$O$_{40}$], and [(CH$_3$)$_3$NH]$_3$-[PW$_{12}$O$_{40}$] (Busbongthong & Ozeki, 2009) (Table 1). Thus, the lengths of the oxygen atoms belonging to the central PO$_4$ tetrahedron (O$_a$) are longer than those of the three alkylammonium salts of [PW$_{12}$O$_{40}$]$^{3-}$; whereas, the lengths of the bridging oxygen atoms between corner-sharing MO$_6$ (M = W and Al) octahedra (O$_c$) and bridging oxygen atoms between edge-sharing MO$_6$ octahedra (O$_e$) are shorter than those of [PW$_{12}$O$_{40}$]$^{3-}$. For comparisons, the bond lengths of *mono*-metal-substituted α-Keggin phosphotungstates, e.g., [(CH$_3$)$_2$NH$_2$]$_4$[α-PW$_{11}$Re^{V}O$_{40}$], (ANIH)$_5$[PCu(H$_2$O)W$_{11}$O$_{39}$](ANI)·8H$_2$O (ANI = aniline,

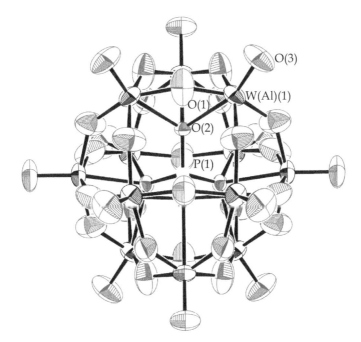

Fig. 4. The molecular structure (ORTEP drawing) of [α-PW$_{11}${Al(OH$_2$)}O$_{39}$]$^{4-}$.

	$[(n\text{-}C_4H_9)_4N]_4[\alpha\text{-}PW_{11}\{Al(OH_2)\}O_{39}]$
W(Al)-O_a	2.466 (2.466)
W(Al)-O_c	1.883 (1.883)
W(Al)-O_e	1.883 (1.883)
W(Al)-O_t	1.667 (1.667)
P-O	1.5206 (1.5206)
	$[CH_3NH_3]_3[\alpha\text{-}PW_{12}O_{40}]\cdot 2H_2O$
W-O_a	2.4077 - 2.4606 (2.4398)
W-O_c	1.8766 - 1.9407 (1.9076)
W-O_e	1.8808 - 1.9448 (1.9166)
W-O_t	1.6818 - 1.7068 (1.6951)
P-O	1.5286 - 1.5377 (1.5324)
	$[(CH_3)_2NH_2]_3[\alpha\text{-}PW_{12}O_{40}]$
W-O_a	2.4273 - 2.4568 (2.4430)
W-O_c	1.9044 - 1.9164 (1.9103)
W-O_e	1.9029 - 1.9234 (1.9158)
W-O_t	1.7000 - 1.7038 (1.7026)
P-O	1.5220 - 1.5348 (1.5313)
	$[(CH_3)_3NH]_3[\alpha\text{-}PW_{12}O_{40}]$
W-O_a	2.4313 – 2.4497 (2.4313)
W-O_c	1.8840 – 1.9286 (1.9127)
W-O_e	1.8996 – 1.9437 (1.9186)
W-O_t	1.6890 – 1.6970 (1.6933)
P-O	1.5296 – 1.5355 (1.5340)

Table 1. Ranges and mean bond distances (Å) for $[(n\text{-}C_4H_9)_4N]_4[\alpha\text{-}PW_{11}\{Al(OH_2)\}O_{39}]$, and the three alkylammonium salts of $[PW_{12}O_{40}]^{3-}$. The terms O_a, O_c, O_e, and O_t are explained in Fig. 5. The mean values are provided in parentheses.

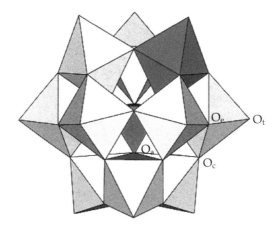

Fig. 5. The polyhedral representation of $[\alpha\text{-}PW_{11}\{Al(OH_2)\}O_{39}]^{4-}$. In the polyhedral representation, the AlO_6 and WO_6 groups are represented by blue and white octahedra, respectively. The internal PO_4 group is represented by the red tetrahedron. Further, O_a,

oxygen atoms belonging to the central PO_4 tetrahedron; O_c, bridging oxygen atoms between corner-sharing MO_6 (M = Al and W) octahedra; O_e, bridging oxygen atoms between edge-sharing MO_6 octahedra (M = Al and W); O_t, terminal oxygen atoms.

$ANIH^+$ = anilinium ion), $Cs_5[PMn(H_2O)W_{11}O_{39}]\cdot4H_2O$, and $Cs_5[PNi(H_2O)W_{11}O_{39}]\cdot2H_2O$ as determined by X-ray crystallography are summarized in Table 2. Although a simple comparison was difficult to draw, the following trends were observed: The W-O_a bond lengths of $[PCu(H_2O)W_{11}O_{39}]^{5-}$, $[PMn(H_2O)W_{11}O_{39}]^{5-}$, and $[PNi(H_2O)W_{11}O_{39}]^{5-}$ were significantly longer than those of $[\alpha-PW_{12}O_{40}]^{3-}$ and $[\alpha-PW_{11}Re^VO_{40}]^{4-}$, as observed for $[\alpha-PW_{11}\{Al(OH_2)\}O_{39}]^{4-}$ due to the presence of a water molecule coordinated to the mono-metal-substituted sites. The W(M)-O_c and W(M)-O_e (M = Re, Cu, Mn, and Ni) bond lengths of the four polyoxoanions mentioned in Table 2 were similar to those of $[\alpha-PW_{12}O_{40}]^{3-}$, whereas, the bond lengths of $[\alpha-PW_{11}\{Al(OH_2)\}O_{39}]^{4-}$ were clearly shorter than those of $[\alpha-PW_{12}O_{40}]^{3-}$.

	$[(CH_3)_2NH_2]_4[\alpha-PW_{11}Re^VO_{40}]$
W(Re)-O_a	2.418 – 2.441 (2.432)
W(Re)-O_c	1.896 – 1.914 (1.906)
W(Re)-O_e	1.895 – 1.922 (1.907)
W(Re)-O_t	1.647 – 1.694 (1.680)
P-O	1.538 – 1.540 (1.539)
	$(ANIH)_5[PCu(H_2O)W_{11}O_{39}](ANI)\cdot8H_2O$
W(Cu)-O_a	2.4784 – 2.5044 (2.4916)
W(Cu)-O_c	1.8946 – 1.9277 (1.9077)
W(Cu)-O_e	1.8946 – 1.9277 (1.9077)
W(Cu)-O_t	1.7163 – 1.7220 (1.7178)
P-O	1.4925 – 1.5078 (1.4965)
	$Cs_5[PMn(H_2O)W_{11}O_{39}]\cdot4H_2O$
W(Mn)-O_a	2.4220 – 2.5520 (2.4874)
W(Mn)-O_c	1.9223 – 1.8698 (1.9051)
W(Mn)-O_e	1.8689 – 1.9620 (1.9079)
W(Mn)-O_t	1.6678 – 1.752 (1.6889)
P-O	1.4902 – 1.602 (1.5265)
	$Cs_5[PNi(H_2O)W_{11}O_{39}]\cdot2H_2O$
W(Ni)-O_a	2.4013 – 2.5152 (2.4792)
W(Ni)-O_c	1.8628 – 1.9430 (1.8974)
W(Ni)-O_e	1.8633 – 1.9421 (1.8964)
W(Ni)-O_t	1.6714 – 1.7354 (1.7010)
P-O	1.5150 – 1.5256 (1.5209)

Table 2. Ranges and mean bond distances (Å) for four mono-metal-substituted α-Keggin phosphotungstates. The terms O_a and O_t are explained in Fig. 5. The terms O_c and O_e indicate bridging oxygen atoms between corner- and edge-sharing MO_6 (M = W, Re, Cu, Mn, Ni) octahedra. The mean values are provided in parentheses.

To investigate the coordination sphere around the mono-aluminum-substituted site in $[\alpha-PW_{11}\{Al(OH_2)\}O_{39}]^{4-}$, the optimized geometry was computed by means of a DFT method, as shown in Figs. 6 and 7. The ranges and mean bond distances, and the Millken charges for the DFT-optimized $[\alpha\text{-PW}_{11}\{\text{Al(OH}_2)\}\text{O}_{39}]^{4-}$ are summarized in Tables 3 and 4. It was noted that the *mono*-aluminum-substituted site was uniquely concave downward, which caused the extension of the P-O bond linkaged to the aluminum atom (1.5654 Å), whereas the Al-O bond linkaged to the internal phosphorus atom was shortened due to the insertion of the Al^{3+} ion that has a smaller ionic radius (0.675 Å) than that of W^{6+} (0.74 Å) into the *mono*-vacant site (Shannon, 1976). The lengths of Al-O bonds at the corner- and edge-sharing Al-O-W bondings were shorter than those of W-O bonds at the corner- and edge-sharing W-O-W bondings, which caused shortening of the average W(Al)-O bond lengths, as observed by X-ray crystallography.

The Mulliken charges of all oxygen atoms linkaged to aluminum atoms in $[\alpha\text{-PW}_{11}\{\text{Al(OH}_2)\}\text{O}_{39}]^{4-}$ were more positive than those linkaged to tungsten atoms in $[\alpha\text{-PW}_{12}\text{O}_{40}]^{3-}$; whereas the charges of oxygen atoms linkaged to tungsten atoms in $[\alpha\text{-PW}_{11}\{\text{Al(OH}_2)\}\text{O}_{39}]^{4-}$ were similar to those in $[\alpha\text{-PW}_{12}\text{O}_{40}]^{3-}$. In addition, the atomic charge of the phosphorus atom in $[\alpha\text{-PW}_{11}\{\text{Al(OH}_2)\}\text{O}_{39}]^{4-}$ was more negative than that in $[\alpha\text{-PW}_{12}\text{O}_{40}]^{3-}$. In the case of *mono*-vanadium(V)-substituted Keggin silicotungstate $[\text{SiW}_{11}\text{VO}_{40}]^{5-}$, the net charge associated with the inner tetrahedron was very similar to that supported by SiO_4 in $[\text{SiW}_{12}\text{O}_{40}]^{4-}$ (Maestre et al., 2001). Thus, the difference in the charge on the internal phosphorus atom for $[\alpha\text{-PW}_{11}\{\text{Al(OH}_2)\}\text{O}_{39}]^{4-}$ and $[\alpha\text{-PW}_{12}\text{O}_{40}]^{3-}$ might be due to the gravitation of aluminum atoms towards the internal PO_4 group.

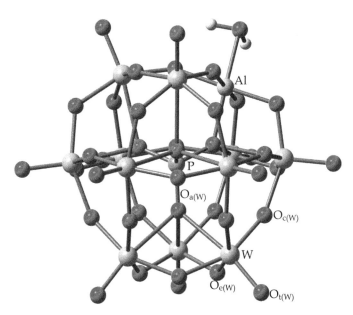

Fig. 6. The DFT-optimized geometry of $[\alpha\text{-PW}_{11}\{\text{Al(OH}_2)\}\text{O}_{39}]^{4-}$. The phosphorus, oxygen, aluminum, tungsten, and hydrogen atoms are represented by orange, red, pink, blue, and white balls, respectively.

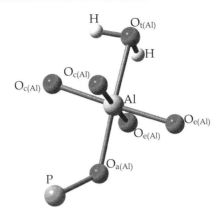

Fig. 7. The coordination sphere around the *mono*-aluminum-substituted site in DFT-optimized $[\alpha\text{-PW}_{11}\{\text{Al(OH}_2)\}\text{O}_{39}]^{4-}$.

	$[\alpha\text{-PW}_{11}\{\text{Al(OH}_2)\}\text{O}_{39}]^{4-}$	$[\alpha\text{-PW}_{12}\text{O}_{40}]^{3-}$
W-O$_a$	2.4422 – 2.5140 (2.4702)	2.4568 – 2.4579 (2.4574)
W-O$_c$	1.8311 – 1.9828 (1.9206)	1.9202 – 1.9216 (1.9209)
W-O$_e$	1.8373 – 1.9918 (1.9267)	1.9262 – 1.9276 (1.9267)
W-O$_t$	1.7196 – 1.7246 (1.7210)	1.7103 – 1.7106 (1.7105)
P-O	1.5450 – 1.5654 (1.5517)	1.5530 – 1.5535 (1.5533)
Al-O$_a$	1.9487 (1.9487)	–
Al-O$_c$	1.8519, 1.8955 (1.8737)	–
Al-O$_e$	1.8723, 1.9215 (1.8969)	–
Al-OH$_2$	2.0983 (2.0983)	–

Table 3. Ranges and mean bond distances (Å) for $[\alpha\text{-PW}_{11}\{\text{Al(OH}_2)\}\text{O}_{39}]^{4-}$ and $[\alpha\text{-PW}_{12}\text{O}_{40}]^{3-}$ optimized by DFT calculations. The terms O$_a$, O$_c$, O$_e$, and O$_t$ are explained in Fig. 5. The average values are provided in parentheses.

	$[\alpha\text{-PW}_{11}\{\text{Al(OH}_2)\}\text{O}_{39}]^{4-}$	$[\alpha\text{-PW}_{12}\text{O}_{40}]^{3-}$
O$_{a\,(W)}$	-0.7356 – -0.8445 (-0.7734)	-0.8951 – -0.8990 (-0.8968)
O$_{c\,(W)}$	-1.226 – -1.345 (-1.317)	-1.353 – -1.355 (-1.353)
O$_{e\,(W)}$	-1.030 – -1.160 (-1.074)	-1.085 – -1.087 (-1.086)
O$_{t\,(W)}$	-0.6757 – -0.6991 (-0.6882)	-0.6273 – -0.6277 (-0.6275)
P	7.255 (7.255)	9.256 (9.256)
W	2.101 – 2.343 (2.257)	2.343 – 2.346 (2.345)
O$_{a(Al)}$	-0.1495 (-0.1495)	–
O$_{c(Al)}$	-0.3332, -0.5920 (-0.4626)	–
O$_{e(Al)}$	-0.4910, -0.7848 (-0.6379)	–
O$_{t(Al)}$	-0.5553 (-0.5553)	–
Al	-0.5307 (-0.5307)	–
H	0.5754, 0.5796 (0.5775)	–

Table 4. Mulliken charges computed for $[\alpha\text{-PW}_{11}\{\text{Al(OH}_2)\}\text{O}_{39}]^{4-}$ and $[\alpha\text{-PW}_{12}\text{O}_{40}]^{3-}$. The terms O$_{a(M)}$, O$_{c(M)}$, O$_{e(M)}$, and O$_{t(M)}$ (M = Al and W) are explained in Figs. 6 and 7. The average values are provided in parentheses.

3.2 Spectroscopic data for [(n-C₄H₉)₄N]₄[α-PW₁₁{Al(OH₂)}O₃₉]

The FTIR spectra measured as a KBr disk of $[(n\text{-}C_4H_9)_4N]_4[\alpha\text{-}PW_{11}\{Al(OH_2)\}O_{39}]$, $K_7[\alpha\text{-}PW_{11}O_{39}]\cdot 11H_2O$, $Cs_7[\gamma\text{-}PW_{10}O_{36}]\cdot 19H_2O$, and $Na_3[\alpha\text{-}PW_{12}O_{40}]\cdot 16H_2O$ are shown in Fig. 8. For

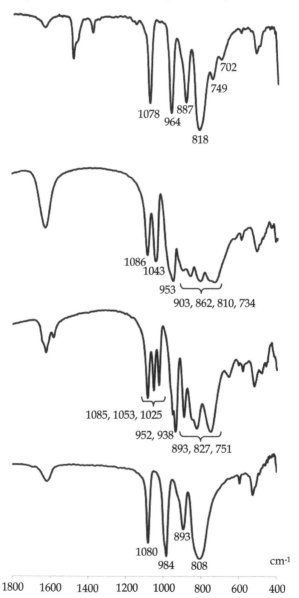

Fig. 8. FTIR spectra (as KBr disks) in the range of 1800 – 400 cm⁻¹ for $[(n\text{-}C_4H_9)_4N]_4[\alpha\text{-}PW_{11}\{Al(OH_2)\}O_{39}]$ (top), $K_7[\alpha\text{-}PW_{11}O_{39}]\cdot 11H_2O$ (the second top), $Cs_7[\gamma\text{-}PW_{10}O_{36}]\cdot 19H_2O$ (the third top), and $Na_3[\alpha\text{-}PW_{12}O_{40}]\cdot 16H_2O$ (bottom)

$[(n\text{-}C_4H_9)_4N]_4[\alpha\text{-}PW_{11}\{Al(OH_2)\}O_{39}]$, the P-O band was observed at 1078 cm^{-1}, and the W-O bands were observed at 964, 887, 818, 749, and 702 cm^{-1}, these were different from those of $K_7[\alpha\text{-}PW_{11}O_{39}]\cdot11H_2O$ (1086, 1043, 953, 903, 862, 810, and 734 cm^{-1}) and $Cs_7[\gamma\text{-}PW_{10}O_{36}]\cdot19H_2O$ (1085, 1053, 1025, 952, 938, 893, 827, and 751 cm^{-1}) (Rocchiccioli-Deltcheff et al., 1983; Thouvenot et al., 1984). This result suggested that the aluminum atom was coordinated into the vacant site in the polyoxometalate. It should be noted that the bands observed for $[(n\text{-}C_4H_9)_4N]_4[\alpha\text{-}PW_{11}\{Al(OH_2)\}O_{39}]$ were significantly different from those of $Na_3[\alpha\text{-}PW_{12}O_{40}]\cdot16H_2O$ (1080, 984, 893, and 808 cm^{-1}). This was consistent with the results observed by X-ray crystallography and DFT calculations, as mentioned above.

The ^{31}P NMR spectrum of $[(n\text{-}C_4H_9)_4N]_4[\alpha\text{-}PW_{11}\{Al(OH_2)\}O_{39}]$ in acetonitrile-d_3 at ~25 °C was a clear single line spectrum at -12.5 ppm due to the internal phosphorus atom, thereby confirming the compound's purity and homogeneity, as shown in Fig. 9. The signal exhibited a shift from the signals of tetra-n-butylammonium salts of $[\alpha\text{-}PW_{12}O_{40}]^{3-}$ (δ -14.6) and $[\alpha\text{-}PW_{11}O_{39}]^{7-}$ (δ -12.0), suggesting the insertion of aluminum ion into the vacant site.

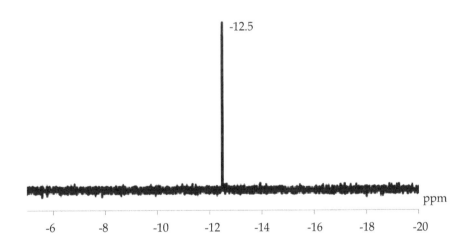

Fig. 9. ^{31}P NMR spectrum in acetonitrile-d_3 of $[(n\text{-}C_4H_9)_4N]_4[\alpha\text{-}PW_{11}\{Al(OH_2)\}O_{39}]$.

The ^{27}Al NMR spectrum (Fig. 10) of $[(n\text{-}C_4H_9)_4N]_4[\alpha\text{-}PW_{11}\{Al(OH_2)\}O_{39}]$ in acetonitrile-d_3 at ~25 °C showed a broad signal at 16.1 ppm due to the $mono$-aluminum-substituted site in $[\alpha\text{-}PW_{11}\{Al(OH_2)\}O_{39}]^{4-}$.

The ^{183}W NMR spectrum (Fig. 11) of $[(n\text{-}C_4H_9)_4N]_4[\alpha\text{-}PW_{11}\{Al(OH_2)\}O_{39}]$ in acetonitrile-d_3 at ~25 °C was a six-line spectrum of (δ -56.2, -93.1, -108.6, -115.8, -118.5, -153.9) with 2:2:2:2:1:2 intensities, which were in accordance with the presence of eleven tungsten atoms with Cs symmetry. These spectral data were completely consistent with the X-ray structure and the optimized structure, suggesting that the solid structure was maintained in the solution.

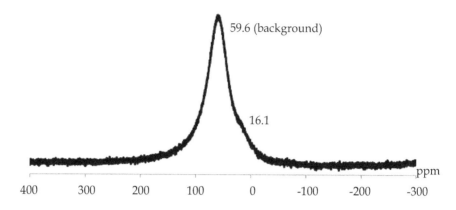

Fig. 10. ^{27}Al NMR spectrum in acetonitrile-d_3 of [(n-C$_4$H$_9$)$_4$N]$_4$[α-PW$_{11}${Al(OH$_2$)}O$_{39}$].

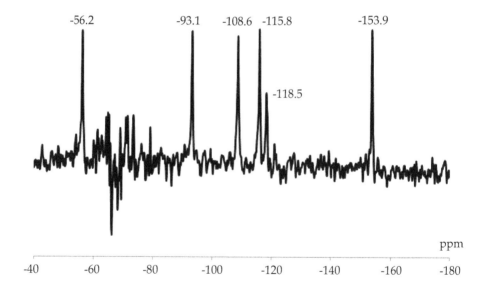

Fig. 11. ^{183}W NMR spectrum in acetonitrile-d_3 of [(n-C$_4$H$_9$)$_4$N]$_4$[α-PW$_{11}${Al(OH$_2$)}O$_{39}$].

4. Conclusion

The synthesis of a monomeric, *mono*-aluminum-substituted α-Keggin polyoxometalate is described in this study. We successfully obtained single crystals of acetonitrile-soluble tetra-n-butylammonium salt [(n-C$_4$H$_9$)$_4$N]$_4$[α-PW$_{11}${Al(OH$_2$)}O$_{39}$] by reacting aluminum nitrate with a *di*-lacunary γ-Keggin phosphotungstate. The characterization of [(n-C$_4$H$_9$)$_4$N]$_4$[α-PW$_{11}${Al(OH$_2$)}O$_{39}$] was accomplished by X-ray crystallography, elemental analysis, thermogravimetric/differential thermal analysis, Fourier transform infrared spectra, and solution ^{31}P, ^{27}Al, and ^{183}W nuclear magnetic resonance spectroscopy. The single-crystal X-ray structure analysis, revealed as $[(n\text{-}C_4H_9)_4N]_4[\alpha\text{-}PW_{11}\{Al(OH_2)\}O_{39}]$, was a monomeric, α-Keggin structure, and the *mono*-aluminum-substituted site could not be identified due to the high symmetry in the product. In contrast, the DFT-optimized geometry of $[\alpha\text{-}PW_{11}\{Al(OH_2)\}O_{39}]^{4-}$ showed that the *mono*-aluminum-substituted site was uniquely concave downward, which caused the extension of the P-O bond linkaged to the aluminum atom, whereas the Al-O bond linkaged to the phosphorus atom was shortened. This structural difference strongly influenced the bonding mode (bond lengths and bond angles) as determined by X-ray crystallography. In addition, the Mulliken charges clearly exhibited the effect caused by the insertion of aluminum atoms into the *mono*-vacant sites.

5. Acknowledgment

This work was supported by a Grant-in-Aid for Scientific Research on Innovative Areas (No. 21200055) of the Ministry of Education, Culture, Sports, Science and Technology, Japan. Y. Kataoka acknowledges the JSPS Research Fellowship for Young Scientist. Y. Kitagawa also has been supported by Grant-in-Aid for Scientific Research on Innovative Areas ("Coordination Programming" area 2170, No. 22108515) from the Ministry of Education, Culture, Sports, Science and Technology (MEXT). This research was partially carried out using equipment at the Center for Instrumental Analysis, Shizuoka University.

6. Appendix

Bond lengths (Å) of $[(n\text{-}C_4H_9)_4N]_4[\alpha\text{-}PW_{11}\{Al(OH_2)\}O_{39}]$: W(1)-O(1) 1.883(4); W(1)-O(1)[1] 1.883(4); W(1)-O(1)[2] 1.883(4); W(1)-O(1)[3] 1.883(4); W(1)-O(2) 2.465(5); W(1)-O(2)[4] 2.465(5); W(1)-O(3) 1.667(4); P(1)-O(2) 1.522(5); P(1)-O(2)[5] 1.522(5); P(1)-O(2)[6] 1.522(5); P(1)-O(2)[7] 1.522(5); P(1)-O(2)[4] 1.522(5); P(1)-O(2)[8] 1.522(5); P(1)-O(2)[9] 1.522(5); P(1)-O(2)[10] 1.522(5); Al(1)-O(1) 1.883(4); Al(1)-O(1)[1] 1.883(4); Al(1)-O(1)[2] 1.883(4); Al(1)-O(1)[3] 1.883(4); Al(1)-O(3) 1.667(4). Symmetry operators: (1) X,Z,Y (2) Z,Y,-X+1 (3) Z,-X+1,Y (4) Y,Z,-X+1 (5) Y,Z,X (6) Z,X,Y (7) X,Y,-Z+1 (8) Z,X,-Y+1 (9) -Z+1,X,-Y+1 (10) -Y+1,-Z+1,-X+1.

Bond angles (°) of $[(n\text{-}C_4H_9)_4N]_4[\alpha\text{-}PW_{11}\{Al(OH_2)\}O_{39}]$: O(1)-W(1)-O(1)[1] 87.5(2); O(1)-W(1)-O(1)[2] 87.08(18); O(1)-W(1)-O(1)[3] 154.8(2); O(1)-W(1)-O(2) 63.32(19); O(1)-W(1)-O(2)[4] 92.40(18); O(1)-W(1)-O(3) 102.58(17); O(1)[1]-W(1)-O(1)[2] 154.8(2) O(1)[1]-W(1)-O(1)[3] 87.08(18); O(1)[1]-W(1)-O(2) 63.32(19); O(1)[1]-W(1)-O(2)[4] 92.40(18); O(1)[1]-W(1)-O(3) 102.58(17); O(1)[2]-W(1)-O(1)[3] 87.5(2); O(1)[2]-W(1)-O(2) 92.40(18); O(1)[2]-W(1)-O(2)[4] 63.32(19); O(1)[2]-W(1)-O(3) 102.58(17); O(1)[3]-W(1)-O(2) 92.40(18); O(1)[3]-W(1)-O(2)[4] 63.32(19); O(1)[3]-W(1)-O(3) 102.58(17); O(2)-W(1)-O(2)[4] 41.76(15); O(2)-W(1)-O(3) 159.12(11); O(2)[4]-W(1)-O(3) 159.12(11); O(2)-P(1)-O(2)[5] 109.5(3); O(2)-P(1)-O(2)[6] 109.5(3); O(2)-P(1)-O(2)[7] 70.5(3); O(2)-P(1)-O(2)[4] 70.5(3); O(2)-P(1)-O(2)[8] 180.0(4); O(2)-P(1)-O(2)[9] 109.5(3); O(2)-P(1)-O(2)[10] 70.5(3); O(2)[5]-P(1)-O(2)[6] 109.5(3); O(2)[5]-P(1)-O(2)[7] 70.5(3); O(2)[5]-P(1)-O(2)[4] 70.5(3); O(2)[5]-P(1)-O(2)[8] 70.5(3); O(2)[5]-P(1)-O(2)[9] 109.5(3); O(2)[5]-P(1)-O(2)[10] 180.0(4); O(2)[6]-P(1)-O(2)[7] 180.0(4); O(2)[6]-P(1)-O(2)[4] 70.5(3); O(2)[6]-P(1)-O(2)[8] 70.5(3); O(2)[6]-P(1)-O(2)[9] 109.5(3); O(2)[6]-P(1)-O(2)[10] 70.5(3); O(2)[7]-P(1)-O(2)[4] 109.5(3); O(2)[7]-P(1)-O(2)[8] 109.5(3); O(2)[7]-P(1)-O(2)[9] 70.5(3); O(2)[7]-P(1)-O(2)[10] 109.5(3); O(2)[4]-P(1)-O(2)[8] 109.5(3); O(2)[4]-P(1)-O(2)[9] 180.0(4); O(2)[4]-P(1)-O(2)[10] 109.5(3); O(2)[8]-P(1)-O(2)[9] 70.5(3); O(2)[8]-P(1)-O(2)[10] 109.5(3); O(2)[9]-P(1)-O(2)[10] 70.5(3); O(1)-Al(1)-O(1)[1] 87.5(2); O(1)-

Al(1)-O(1)[2] 87.08(18); O(1)-Al(1)–O(1)[3] 154.8(2); O(1)-Al(1)-O(3) 102.58(17); O(1)[1]-Al(1)-O(1)[2] 154.8(2); O(1)[1]-Al(1)-O(1)[3] 87.08(18); O(1)[1]-Al(1)-O(3) 102.58(17); O(1)[2]-Al(1)-O(1)[3] 87.5(2); O(1)[2]-Al(1)-O(3) 102.58(17); O(1)[3]-Al(1)-O(3) 102.58(17); W(1)-O(1)-W(1)[11] 140.7(3); W(1)-O(1)-Al(1)[11] 140.7(3); W(1)[11]-O(1)-Al(1) 140.7(3); Al(1)-O(1)-Al(1)[11] 140.7(3); W(1)-O(2)-W(1)[11] 91.97(16); W(1)-O(2)-W(1)[12] 91.97(16); W(1)-O(2)-P(1) 123.9(3); W(1)-O(2)-O(2)[7] 131.4(3); W(1)-O(2)-O(2)[4] 69.1(3); W(1)-O(2)-O(2)[10] 131.4(3); W(1)[11]-O(2)-W(1)[12] 91.97(16); W(1)[11]-O(2)-P(1) 123.9(3); W(1)[11]-O(2)-O(2)[7] 69.1(3) W(1)[11]-O(2)-O(2)[4] 131.4(3); W(1)[11]-O(2)- O(2)[10] 131.4(3); W(1)[12]-O(2)-P(1) 123.9(3); W(1)[12]-O(2)-O(2)[7] 131.4(3); W(1)[12]-O(2)-O(2)[4] 131.4(3); W(1)[12]-O(2)-O(2)[10] 69.1(3); P(1)-O(2)-O(2)[7] 54.7(3); P(1)-O(2)-O(2)[4] 54.7(3); P(1)-O(2)-O(2)[10] 54.7(3); O(2)[7]-O(2)-O(2)[4] 90.0(3); O(2)[7]-O(2)-O(2)[10] 90.0(3); O(2)[4]-O(2)-O(2)[10] 90.0(3). Symmetry operators: (1) X,Z,Y (2) Z,Y,-X+1 (3) Z,-X+1,Y (4) Y,Z,-X+1 (5) Y,Z,X (6) Z,X,Y (7) X,Y,-Z+1 (8) Z,X,-Y+1 (9) -Z+1,X,-Y+1 (10) -Y+1,-Z+1,-X+1 (11) -Y+1,Z,-X+1 (12) -Z+1,-X+1,Y.

7. References

Akitt, J. W. (1988). Multinuclear Studies of Aluminum Compounds. *Prog. Nucl. Magn. Res. Spectr.*,Vol.21, No.1-2, pp. 1–149

Baes, C. F. Jr. & Mesmer, R. E. (1976). *The Hydrolysis of Cations*, pp. 112–123, John Wiley, New York, 1976

Busbongthong, S. & Ozeki, T. (2009). Structural Relationships among Methyl-, Dimethyl-, and Trimethylammonium Phosphdodecatungstates. *Bull. Chem. Soc. Jpn.*, Vol.82, No.11, pp. 1393–1397

Contant, R. (1987). Relation between Tungstophosphates Related to the Phosphorus Tungsten Oxide Anion ($PW_{12}O_{40}{}^{3-}$). Synthesis and Properties of a New Lacunary Potassium Polytungstophosphate ($K_{10}P_2W_{20}O_{70}\cdot24H_2O$). *Can. J. Chem.*, Vol.65, No.3, pp. 568–573

Cotton, F. A. & Wilkinson, G. (1988). *Advanced Inorganic Chemistry, Fifth Edition*, John Wiley & Sons, New York

Djurdjevic P.; Jelic, R. & Dzajevic, D. (2000). The Effect of Surface Active Substances on Hydrolysis of Aluminum(III) Ion. *Main Metal Chemistry*, Vol.23, No.8, pp. 409–421

Domaille, P. J. (1990). Vanadium(V) Substituted Dodecatungstophosphates. *Inorg. Synth.*, Vol.27, pp. 96–104

Frisch, M. J.; Trucks, G. W.; Schlegel, H. B.; Scuseria, G. E.; Robb, M. A.; Cheeseman, J. R.; Scalmani, G.; Barone, V.; Mennucci, B.; Petersson, G. A.; Nakatsuji, H.; Caricato, M.; Li, X.; Hratchian, H. P.; Izmaylov, A. F.; Bloino, J.; Zheng, G.; Sonnenberg, J. L.; Hada, M.; Ehara, M.; Toyota, K.; Fukuda, R.; Hasegawa, J.; Ishida, M.; Nakajima, T.; Honda, Y.; Kitao, O.; Nakai, H.; Vreven, T.; Montgomery, Jr., J. A.; Peralta, J. E.; Ogliaro, F.; Bearpark, M.; Heyd, J. J.; Brothers, E.; Kudin, K. N.; Staroverov, V. N.; Kobayashi, R.; Normand, J.; Raghavachari, K.; Rendell, A.; Burant, J. C.; Iyengar, S. S.; Tomasi, J.; Cossi, M.; Rega, N.; Millam, N. J.; Klene, M.; Knox, J. E.; Cross, J. B.; Bakken, V.; Adamo, C.; Jaramillo, J.; Gomperts, R.; Stratmann, R. E.; Yazyev, O.; Austin, A. J.; Cammi, R.; Pomelli, C.; Ochterski, J. W.; Martin, R. L.; Morokuma, K.; Zakrzewski, V. G.; Voth, G. A.; Salvador, P.; Dannenberg, J. J.; Dapprich, S.; Daniels, A. D.; Farkas, Ö.; Foresman, J. B.; Ortiz, J. V.; Cioslowski, J. & Fox, D. J. (2009). *Gaussian 09, Revision B.1*, Gaussian, Inc., Wallingford CT

Fukaya, K.; Srifa, A.; Ishikawa, E. & Naruke, H. (2010). Synthesis and Structural Characterization of Polyoxometalates Incorporating with Anilinium Cations and Facile Preparation of Hybrid Film. *J. Mol. Struc.* Vol.979, pp. 221–226

Hou, Y.; Fang, X. & Hill, C. L. (2007). Breaking Symmetry: Spontaneous Resolution of a Polyoxometalate. *Chem. Eur. J.* Vol.13, pp. 9442–9447

Kato, C. N.; Hara, K.; Kato, M.; Amano, H.; Sato, K.; Kataoka, Y. & Mori, W. (2010). EDTA-Reduction of Water to Molecular Hydrogen Catalyzed by Visible-Light-Response TiO_2-Based Materials Sensitized by Dawson- and Keggin-Type Rhenium(V)-Containing Polyoxotungstates. *Materials,* Vol.3, pp. 897–917

Kato, C. N.; Katayama, Y.; Nagami, M.; Kato, M. & Yamasaki, M. (2010). A Sandwich-type Aluminium Complex Composed of Tri-lacunary Keggin-type Polyoxotungstate: Synthesis and X-Ray Crystal Structure of $[(A-PW_9O_{34})_2\{W(OH)(OH_2)\}\{Al(OH)(OH_2)\}\{Al(\mu-OH)(OH_2)_2\}_2]^{7-}$. *Dalton Trans.,* Vol.39, pp. 11469–11474

Kikukawa, Y.; Yamaguchi, S.; Nakagawa, Y.; Uehara, K.; Uchida, S.; Yamaguchi, K. & Mizuno, N. (2008). Synthesis of a Dialuminum-Substituted Silicotungstate and the Diasteroselective Cyclization of Citronellal Derivatives. *J. Am. Chem. Soc.,* Vol.130, No.47, 15872–15878

Knoth, W. H.; Domaille, P. J. & Roe, D. C. (1983). Halometal Derivatives of $W_{12}PO_{40}{}^{3-}$ and Related [183]W NMR Studies. *Inorg. Chem.* Vol.22, 198–201

Knoth, W. H. & Harlow, R. L. (1981). New Tungstophosphates: $Cs_6W_5P_2O_{23}$, $Cs_7W_{10}PO_{36}$, and $Cs_7Na_2W_{10}PO_{37}$. *J. Am. Chem. Soc.,* Vol.103. No.7, pp. 1865–1867

Lin, Y.; Weakley, T. J. R.; Rapko, B. & Finke, R. G. (1993). Polyoxoanions Derived from Tungstosilicatie ($A-\beta-SiW_9O_{34}{}^{10-}$): Synthesis, Single-crystal Structural Determination, and Solution Structural Characterization by Tungsten-183 NMR and IR of Titanotungstosilicate ($A-\beta-Si_2W_{18}Ti_6O_{77}{}^{14-}$). *Inorg. Chem.,* Vol.32, No.23, pp. 5095–5101

Maestre, J. M.; Lopez, X.; Bo, C.; Poblet, J.-M. & Casan-Pastor N. (2001). Electronic and Magnetic Properties of α-Keggin Anions: A DFT Study of $[XM_{12}O_{40}]^{n-}$, (M = W, Mo; X = Al[III], Si[VI], P[V], Fe[III], Co[II], Co[III]) and $[SiM_{11}VO_{40}]^{m-}$ (M = Mo and W). *J. Am. Chem. Soc.,* Vol.123, pp. 3749–3758

Neiwert, W. A.; Cowan, J. J.; Hardcastle, K. I.; Hill, C. L. & Weinstock, I. A. (2002). Stability and Structure in α- and β-Keggin Heteropolytungstates, $[X^{n+}W_{12}O_{40}]^{(8-n)-}$, X = *p*-Block Cation. *Inorg. Chem.,* Vol.41, 6950–6952

Nomiya, K.; Takahashi, M.; Ohsawa, K. & Widegren, J. A. (2001). Synthesis and Characterization of Tri-titanium(IV)-1,2,3-substituted α-Keggin Polyoxotungstates with Heteroatoms P and Si. Crystal Structure of the Dimeric, Ti-O-Ti Bridged Anhydride Form $K_{10}H_2[\alpha,\alpha-P_2W_{18}Ti_6O_{77}]\cdot17H_2O$ and Confirmation of Dimeric Forms in Aqueous Solution by Ultracentrifugation Molecular Weight Measurements. *J. Chem. Soc., Dalton Trans.* No.19, pp. 2872-2878

Nomiya, K.; Takahashi, M. Widegren, J. A.; Aizawa, T.; Sakai, Y. & Kasuga, N. C. (2002). Synthesis and pH-Variable Ultracentrifugation Molecular Weight Measurements of the Dimeric, Ti-O-Ti Bridged Anhydride Form of a Novel Di-Ti[IV]-substituted α-Keggin Polyoxotungstate. Molecular Structure of the $[(\alpha-1,2-PW_{10}Ti_2O_{39})_2]^{10-}$ Polyoxoanion. *J. Chem. Soc., Dalton Trans.* No.19, pp. 3679–3685

Ortéga, F.; Pope, M. T. & Evans, H.T., Jr. (1997). Tungstorhenate Heteropolyanions. 2. Synthesis and Characterization of Enneatungstorhebates(V), -(VI) and -(VII). *Inorg. Chem.,* Vol.36, No.10, pp. 2166–2169

Orvig, C. (1993). The Aqueous Coordination Chemistry of Aluminum In: *Coordination Chemistry of Aluminum,* G.H. Robinson, (Ed.), 85–121, VCH, Weinheim

Patel, K.; Shringarpure, P. & Patel, A. (2011). One-step Synthesis of a Keggin-type Manganese(II)-substitited Phosphotungstate: Structrural and Spectroscopic

Characterization and Non-solvent Liquid Phase Oxidation of Styrene. *Transition Met. Chem.*, Vol.36, pp. 171–177

Pope, M. T. (1983). *Heteropoly and Isopoly Oxometalates*, Springer-Verlag, Berlin

Pope, M. T. & Müller, A. (1991). Chemistry of Polyoxometallates. Actual Variation on an Old Theme with Interdisciplinary References. *Angew. Chem. Int. Ed. Engl.*, Vol.30, No.1, pp. 34–48

Pope, M. T. & Müller, A. (Eds.), (1994). *Polyoxometalates: From Platonic Solids to Anti-Retroviral Activity*, Kluwer Academic Publishers, Dordrecht, The Netherlands

Reinoso, S.; Vitoria, P.; Felices, L. S.; Lezama, L. & Gutiérrez-Zorrilla, J. M. (2006). Analysis of Weak Interactions in the Crystal Packing of Inorganic Metalorganic Hybrids Based on Keggin Polyoxometalates and Dinuclear Copper(II)-Acetate Complexes. *Inorg. Chem.*, Vol.45, pp. 108–118

Rocchiccioli-Deltcheff, C.; Fournier, M.; Franck, R. & Thouvenot, R. (1983). Vibrational Investigations of Polyoxometalates. 2. Evidence for Anion-Anion Interactions in Molybdenum(VI) and Tungsten(VI) Compounds Related to the Keggin Structure. *Inorg. Chem.*, Vol.22, pp. 207–216

Rosenheim, A. & Jaenicke, J. Z. (1917). Iso- and Heteropoly Acids. XV. Heteropoly tungstates and Some Heteropoly Molybdates. *Anorg. Allg. Chem.*, Vol.101, pp. 235–275

Shannon, R. D. (1976). Revised Effective Ionic Radii and Systematic Studies of Interatomic Distances in Halides and Chalcogenides. *Acta Crystallogr., Sect. A*, Vol.A32, pp. 751–767

Sheldrick, G. M. (2008). A Short History of SHELX. *Acta Crystallogr., Sect. A*, Vol.A46, No.1, pp. 112–122

Spek, A. L. (2009). Structure Validation in Chemical Crystallography. *Acta Crystallogr., Sect. D*, Vol.D65, No.2, pp. 148–155

Thouvenot, R.; Fournier, M.; Franck, R. & Rocchiccioli-Deltcheff, C. (1984). Vibrational Investigations of Polyoxometalates. 3. Isomerism in Molybdenum(VI) and Tungsten(VI) Compounds Related to the Keggin Structure. *Inorg. Chem.*, Vol.23, pp. 598–605

Weakley, T. J. R. (1987). Crystal Structure of Cesium Aquanickelo(II)undecatungstophsphate Dihydrate. J. Cryst. Spectro. Res., Vol.17, No.3, pp. 383–391

Weakley, T. J. R. & Finke, R. G. (1990). Single-crystal X-Ray Structures of the Polyoxotungstate Salts $K_{8.3}Na_{1.7}[Cu_4(H_2O)_2(PW_9O_{34})_2]\cdot24H_2O$ and $Na_{14}Cu[Cu_4(H_2O)_2(P_2W_{15}O_{56})_2]\cdot53H_2O$. *Inorg. Chem.*, Vol.29, No.6, pp. 1235–1241

Weiner, H.; Aiken III, J. D. & Finke, R. G. (1996). Polyoxometalate Catalyst Precursors. Improved Synthesis, H^+-Titration Procedure, and Evidence for ^{31}P NMR as a Highly Sensitive Support-Site Indicator for the Prototype Polyoxoanion-Organometallic-support System $[(n-C_4H_9)_4N]_9P_2W_{15}Nb_3O_{62}$. *Inorg. Chem.*, Vol.35, pp. 7905–7913

Yang, Q. H.; Zhou, D. F.; Dai, H. C.; Liu, J. F.; Xing, Y.; Lin, Y. H. & Jia, H. Q. (1997). Synthesis, Structure and Properties of Undecatungstozincate Containing 3A Elements. *Polyhedron*, Vol.16, No.23, 3985–3989

11

Preparation of Carvedilol Spherical Crystals Having Solid Dispersion Structure by the Emulsion Solvent Diffusion Method and Evaluation of Its *in vitro* Characteristics

Amit R. Tapas, Pravin S. Kawtikwar and Dinesh M. Sakarkar
Sudhakarrao Naik Institute of Pharmacy, Pusad, Dist Yavatmal, Maharashtra
India

1. Introduction

Solid dispersion is one of the most efficient techniques to improve the dissolution rate of poorly water-soluble drugs, leading to an improvement in the relative bioavailability of their formulations. At present, the solvent method and the melting method are widely used in the preparation of solid dispersions. In general, subsequent grinding, sieving, mixing and granulation are necessary to produce the different desired formulations.

The spherical agglomeration technique has been used as an efficient particle preparation technique developed by Kawashima in the 1980s (Kawashima et al., 1994). Initially, spherical agglomeration technique was used to improve powder flowability, packability, and compressibility (Usha et al. 2008; Yadav and Yadav, 2008; Bodmeier and Paeratakul et al., 1989). Then polymers were introduced in this system to modify their release (Di Martino et al., 1999). Currently, this technique is used more frequently for the solid dispersion preparation of water-insoluble drugs in order to improve their solubility, dissolution rate and simplify the manufacturing process (Cui et al., 2003, Tapas et al. 2009, 2010). Spherical crystallization has been developed by Yoshiaki Kawashima and co-workers as a novel particulate design technique to improve processibility such as mixing, filling, tableting characteristics and dissolution rate of pharmaceuticals (Kawashima et al., 1974, 1976, 1981, 1982, 1983, 1984, 1985, 1989, 1991, 1994, 1995, 2002, 2003). The resultant crystals can be designated as spherical agglomerates (Kulkarni and Nagavi, 2002). Spherical crystallization is an effective alternative to improve dissolution rate of drugs (Sano et al., 1992). Now days functional drug devices such as microspheres, microcapsules, microballoons and biodegradable nanospheres were developed using the emulsion solvent diffusion techniques involving the introduction of a functional polymer into the system (Di Martino et al., 1999; Marshall and York, 1991; Garekani and Ford, 1999). This can be achieved by various methods such as

1. Spehrical Agglomeration (SA)
2. Quasi Emulsion Solvent Diffusion (QESD)
3. Ammonia Diffusion System (ADS)
4. Neutralization (NT)

Out of which first two are the most common methods in practice.

In the spherical crystallization process, crystal formation, growth and agglomeration occur simultaneously within the same system. In this method, a third solvent called the bridging liquid is added in a smaller amount to purposely induce and promote the formation of agglomerates. Crystals are agglomerated during the crystallization process and large spherical agglomerates are produced. A near saturated solution of the drug in a good solvent is poured into a poor solvent. The poor and good solvents are freely miscible and the "affinity" between the solvents is stronger than the affinity between drug and good solvent, leading to precipitation of crystals immediately. Under agitation, the bridging liquid (the wetting agent) is added, which is immiscible with the poor solvent and preferentially wet the precipitated crystals. As a result of interfacial tension effects and capillary forces, the bridging liquid acts to adhere the crystals to one another and facilitates them to agglomerate (Fig. 1).

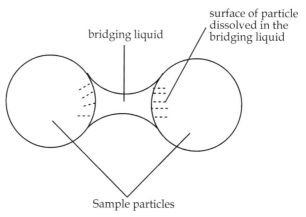

Fig. 1. Two sample particles joined together by a liquid bridge

In spherical agglomeration method, when a drug solution (in good solvent) was poured into a poor solvent under agitation, the drug crystals were formed immediately and agglomerated with a bridging liquid dispersed in the poor solvent, because the bridging liquid has a preference for wetting the drug crystals.

Quasi emulsion solvent diffusion method is also known as transient emulsion method. Firstly the drug was dissolved in a mixed solvent of good solvent and bridging liquid. Because of the increased interfacial tension between the two solvents, the solution is dispersed into the poor solvent producing emulsion (quasi) droplets, even though the pure solvents are miscible. The good solvent diffuses gradually out of the emulsion droplets into the surrounding poor solvent phase, and the poor solvent diffuses into the droplets by which the drug crystallizes inside the droplets. The method is considered to be simpler than the SA method, but it can be difficult to find a suitable additive to keep the system emulsified and to improve the diffusion of the poor solute into the dispersed phase. Especially hydrophilic/hydrophobic additives are used to improve the diffusion remarkably. In this method the shape and the structure of the agglomerate depend strongly on the good solvent to poor solvent ratio and the temperature difference between the two

Preparation of Carvedilol Spherical Crystals Having Solid Dispersion Structure by the Emulsion Solvent Diffusion Method and Evaluation of Its in vitro Characteristics

263

solvents when the drug solution was introduced into the poor solvent under certain temperature and stirring, the drug solution was dispersed immediately to form quasi o/w emulsion droplets, the emulsion droplets were gradually solidified, forming spherical agglomerates along with the diffusion of the good solvent from the droplets into the poor solvent.

Spherical agglomeration has got more importance than other methods because it is easy to operate and the selection of the solvents is easier than in the other methods. Quasi emulsion solvent diffusion method has the second importance.

An ammonia diffusion system is applicable to amphoteric drug substances. In this method, the mixture of three partially immiscible solvent i.e. acetone, ammonia water, dichloromethane was used as crystallization system. In this system ammonia water acted as bridging liquid as well as good solvent, acetone as the water miscible but poor solvent, thus drug precipitated by solvent change without forming ammonium salt. Water immiscible solvent such as hydrocarbon or halogenated hydrocarbons e.g. dichloromethane induced liberation of ammonia water.

In neutralization method sodium hydroxide acts as a good solvent and hydrochloric acid as a poor solvent or vice-versa. These solutions were added to each other in order to get neutralization. The bridging liquid was added drop wise under agitation to form spherical agglomerates.

In this study special attention was given to improving the solubility and dissolution rate of poorly water soluble drug carvedilol using quasi emulsion solvent diffusion method. Carvedilol (CAR) (fig. 2), (±)-1-(carbazol-4yloxy)-3-[[2-(o-methoxyphenoxy)ethyl]amino]-2-propanol is an α_1, β_1 and β_2 adrenergic receptor antagonist (Sweetman, 2002). It is used to treat mild to moderate essential hypertension, mild to severe heart failure, and patients with systolic dysfunction after myocardial infraction. Carvedilol is practically insoluble in water and exhibits pH dependent solubility. Its solubility is <1μg/ml above pH 9.0, 23 μg/ml at pH 7, and about 100 μg/ml at pH 5 at room temperature. It's extremely low solubility at alkaline pH levels may prevent the drug from being available for absorption in the small intestine and colon, thus making it poor candidate for an extended release dosage form. In the present study, to overcome the problems related to solubility, dissolution rate, flowability, and compressibility, the microspheres having solid dispersions structure of Carvedilol were prepared by emulsion solvent diffusion method by using a poloxamer (polxamer F68 and poloxamer F127) as a hydrophilic polymer.

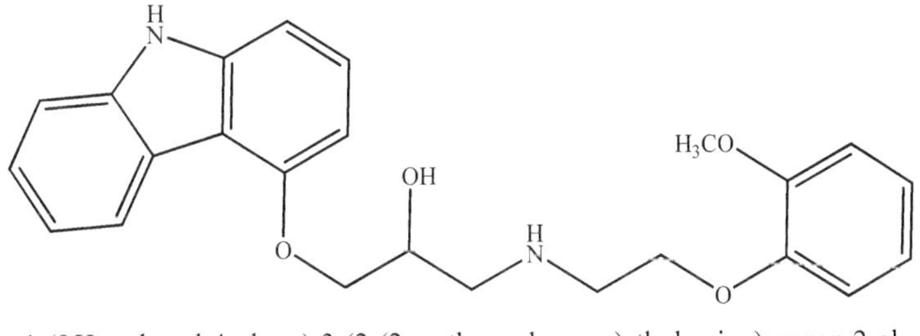

1-(9H-carbazol-4-yloxy)-3-(2-(2-methoxyphenoxy)ethylamino)propan-2-ol

Fig. 2. Chemical structure of Carvedilol (CAR)

2. Materials and methods

2.1 Materials

Carvedilol was supplied by Dr. Reddy's Laboratory, Hyderabad, India as a gift sample. Poloxamer F68 and F127 were supplied by Lupin Research Park, Pune, India. All other chemicals used were of analytical grade.

2.2 Methods

Carvedilol (1.0 g) with poloxamer was dissolved in good solvent methanol (12.0 mL). The bridging liquid dichloromethane (2.0 mL) was added to it. The resulting solution was then poured dropwise in to the poor solvent distilled water (100 mL) containing Aerosil 200 Pharma (0.1 g). The mixture was stirred continuously for a period of 0.5 h using a controlled speed mechanical stirrer (Remi motors, India) at 1000 rpm. As the good solvent diffused into the poor solvent, droplets gradually solidified. Finally the coprecipitated microspheres of the drug-polymer were filtered through Whatman filter paper (No.1) and dried in desiccator at room temperature. The amount of poloxamers was altered to get desired microspheres. The composition is given in Table 1.

Composition/Parameters	CP681	CP682	CP1271	CP1272
CAR (g)	1.0	1.0	1.0	1.0
Methanol (ml)	12.0	12.0	12.0	12.0
DCM (ml)	2.0	2.0	2.0	2.0
Water (ml)	100	100	100	100
Poloxamer F68 (g)	1.5	3.0	--	--
Poloxamer F127 (g)	--	--	1.5	3.0
Aerosil 200 pharma (g)	0.1	0.1	0.1	0.1
Stirring speed (rpm)	1000	1000	1000	1000

Table 1. Composition of spherical crystals

2.2.1 Drug content study

The drug content study of agglomerates was determined by dissolving 100 mg of crystals in 3 ml methanol and diluting further with distilled water (100 ml) followed by measuring the absorbance of appropriately diluted solution spectrophotometrically (PharmaSpec UV-1700, UV-Vis spectrophotometer, Shimadzu) at 286 nm.

2.2.2 Fourier Transform Infrared Spectroscopy (FTIR)

The FTIR spectra of powder CAR, and their agglomerates were recorded on an FTIR spectrophotometer (JASCO, FTIR V-430 Plus).

2.2.3 Differential Scanning Calorimetry (DSC)

DSC analysis was performed using a DSC 823 calorimeter (Mettler Toledo model) operated by STARe software. Samples of CAR and its agglomerates were sealed in an aluminium crucible and heated at the rate of 10 °C/min up to 300 °C under a nitrogen atmosphere (40 ml/min).

2.2.4 Powder X-ray diffraction studies

Powder X-ray diffraction patterns (XRD) of the CAR and its spherical agglomerates were monitored with an x-ray diffractometer (Philips Analytical XRD) using copper as x-ray target, a voltage of 40 KV, a current of 25 mA and with 2.28970 Å wavelength. The samples were analyzed over 2θ range of 10.01-99.99⁰ with scanning step size of 0.02⁰ (2θ) and scan step time of 0.8 second.

2.2.5 Scanning electron microscopy

The surface morphology of the agglomerates was accessed by SEM. The crystals were splutter coated with gold before scanning.

2.2.6 Micromeritic properties

The size of agglomerates was determined by microscopic method using stage and eyepiece micrometers. The shape of the agglomerates was observed under an optical microscope (×60 magnification) attached to a computer. Flowability of untreated carvedilol and agglomerates was assessed by determination of angle of repose, Carr's index (CI) and Hausner's ratio (HR) (Wells, 2002). Angle of repose was determined by fixed funnel method (Martin et al., 2002). The mean of three determinations was reported. The CI and HR were calculated from the loose and tapped densities. Tapped density was determined by tapping the samples into a 10 ml measuring cylinder. The CI and HR were calculated according to the following equation 1 and 2.

$$\text{C.I.} = \frac{\text{Tapped density} - \text{Bulk density}}{\text{Tapped density}} \times 100 \qquad (1)$$

$$\text{H.R.} = \frac{\text{Tapped density}}{\text{Bulk density}} \qquad (2)$$

2.2.7 Solubility studies

A quantity of crystals (about 100 mg) was shaken with 10 mL distilled water in stoppered conical flask at incubator shaker for 24 h at room temperature. The solution was then passed through a whatmann filter paper (No. 42) and amount of drug dissolved was analyzed spectrophotometrically.

2.2.8 Dissolution rate studies

The dissolution rate studies of carvedilol alone and its spherical agglomerates were performed in triplicate in a dissolution apparatus (Electrolab, India) using the paddle method (USP Type II). Dissolution studies were carried out using 900 ml of 0.1N HCl (pH 1.2) at 37 ± 0.5 ⁰C at 50 rpm as per US FDA guidelines (U.S. Food and drug administration [USFDA], 2010 and Bhutani et al., 2007.). 12.5 mg of carvedilol or its equivalent amount of spherical agglomerates were added to 900 ml of 0.1N HCl (pH 1.2). Samples (5 ml) were withdrawn at time intervals of 10, 20, 30, and 60 min. The volume of dissolution medium was adjusted to 900 ml by replacing each 5 ml aliquot withdrawn with 5 ml of fresh 0.1N HCl (pH 1.2). The solution was immediately filtered, suitably diluted and the concentrations of carvedilol in samples were determined spectrophotometrically at 286 nm. The results obtained from the dissolution studies were statistically validated using ANOVA.

2.2.9 Dissolution efficiency studies

The dissolution efficiency (DE) of the batches was calculated by the method mentioned by Khan (Khan, 1975). It is defined as the area under the dissolution curve between time points t_1 and t_2 expressed as a percentage of the curve at maximum dissolution, y100, over the same time period or the area under the dissolution curve up to a certain time, t, (measured using trapezoidal rule) expressed as a percentage of the area of the rectangle described by 100% dissolution in the same time. DE_{60} values were calculated from dissolution data and used to evaluate the dissolution rate (Anderson et al., 1998).

$$\text{Dissolution Efficiency} = \frac{\int_0^t ydt}{y100(t_2 - t_1)} \times 100 \tag{3}$$

3. Result and discussion

3.1 Preparation of spherical crystals

A typical spherical crystallization system involved a good solvent, a poor solvent for a drug and a bridging liquid. The selection of these solvent depends on miscibility of the solvents and solubility of drug in individual solvents. Accordingly acetone, dichloromethane, water were selected as a good solvent, bridging liquid, and poor solvent, respectively. CAR is soluble in methanol, but poorly soluble in water. Also it is soluble in dichloromethane which is immiscible in water. Hence, this solvent system was used in the present study. When drug polymer solution was poured into the poor solvent under agitation at selected temperature, the drug polymer solution became immediately semitransparent due to the presence of small sized emulsion droplets. Gradually emulsion droplets solidified along with diffusion of the good solvents, as bridging liquid dichloromethane was commixed with good solvent, when the good solvent in the droplets diffused into the poor solvent, the residual dichloromethane in the droplets bridged the Aerosil, coprecipitated drug, and polymer to form spherical crystals. The Aerosil acts as a dispering agent and mass compactor, because coacervation droplets formed from the drug-polymer droplets during the solidifying period were sticky and readily coalesced, while the introduction of Aerosil efficiently prevented coalescence and produced compact spherical crystals.

3.2 Optimization of process variables for preparation of spherical crystals

To optimize the Carvedilol spherical crystallization by methanol, water, dichloromethane solvent system following parameters considered amount and mode of addition of bridging liquid, stirring speed and temperature. (Table 2).

S.No.	Parameter	Variables	Observation
1	Conc. of Bridging Liquid (Dichloromethane)	1 ml	No agglomeration
		2 ml	Agglomeration
		3 ml	No agglomeration
2	Agitation speed	800 rpm	Spherical but large
		1000 rpm	Spherical
		1200 rpm	Irregular shape and small
3	Agitation time	15 min	Incomplete agglomerates
		30 min	Spherical agglomerates
4	Mode of addition of bridging liquid	Whole at a time	Crystals of irregular geometry
		Drop wise	Spherical agglomerates

Table 2. Parameters affecting spherical agglomeration

3.3 Drug content study

Percent drug content was found to be in the range of 92.12±1.60 to 94.4±2.37 (Table 3).

3.4 Micromeritic properties

Pure CAR could not pass through the funnel during the angle of repose experiment which could be due to the irregular shape and high fineness of the powder, which posed hurdles in the uniform flow from the funnel. It exhibited poor flowability and packability as indicated by Hausner ratio (1.52) and Carr's Index (34.37%). All agglomerates showed excellent flowability and packability (Angle of repose: 23-28⁰; Carr's Index: 15-18%; Hausner ratio: 1.13-1.17) when compared to pure CAR. The improved flowability of agglomerates may be due to good sphericity and larger size of agglomerates. During the tapping process, smaller agglomerates might have infiltrated into the voids between larger particles, which could result improved packability (lower CI). The results of micromeritic properties are shown in Table 3.

S.No.	Samples	Drug Content	Carr's Index	Hausner Ratio	Angle of Repose	Particle Size (μm)	Aqueous Solubility (μgmL^{-1})
1.	CAR	100.0±0.0	34.37±1.79	1.52±1.26	---	71.55±5.37	21.0 ± 0.51
2.	CP681	94.4±2.37	17.53±1.49[b]	1.15±2.67[b]	27.12±2.43	112.7±31.26[b]	32.20 ± 1.32
3.	CP682	92.31±2.02	15.64±1.40[b]	1.13±2.21[b]	25.36±0.67	136.43±18.57[b]	39.04 ± 1.65[b]
4.	CP1271	94.19±1.15	18.63±2.90[b]	1.17±1.82[b]	28.42±0.23	88.92±25.38[b]	58.08 ± 1.28[b]
5.	CP1272	92.12±1.60	17.24±1.87[b]	1.14±1.53[b]	23.05±0.89	102.03±20.11[b]	64.25 ± 1.31[b]

[a]Mean ± SD, n = 3; [b]Significantly different compared to pure CAR ($p<0.05$).

Table 3. Micromeritics, Drug Content, Particle size, aqueous solubility data of carvedilol and its spherical crystals

3.5 FTIR, DSC and powder X-ray studies

The FTIR spectra of CAR as well as its spherical crytals are presented in Figure 3. FTIR of CAR showed a characteristic peaks at 3343.96 (N-H str. Aromatic Amines), 3062.41 (C-H str.

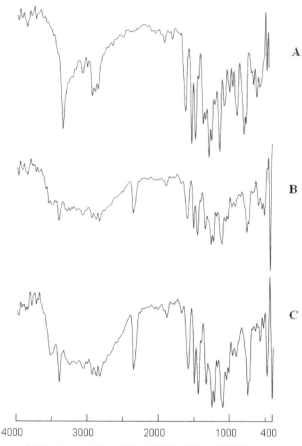

Fig. 3. FTIR spectra of (A) – Carvedilol, (B) – CP682, (C) – CP1272

Aromatic Hydrocarbon), 2923.56 (C-H str. in –CH$_3$/ –CH$_2$), 1592.91 (C=C str. Aromatic), 1253.5, 1214.93, 1099.23 (C-O str. in Ar C=C-O-C) cm^{-1}. There was no considerable change in the IR peaks of the spherical agglomerates when compaired with pure CAR, which revealed that no chemical interaction had occurred between drug and polymer during agglomeration process.

Figure 4 shows the DSC thermogram of pure CAR and its spherical crystals. DSC thermogram of CAR showed endothermic peak at 120.47^{0}C, which represented melting of carvedilol. There was negligible change in the melting point endotherms of prepared spherical crystals compared to pure drug (CP682 = 115.47^{0}C, CP1272 = 118.67^{0}C). The endotherms at 57.05^{0}C and 59.07^{0}C ascribed to the melting of Poloxamer F68 and Poloxamer F127 respectively. This observation further supports the IR spectroscopy results, which indicated the absence of any interactions between the drug and additives used in the preparation. However, there was a decrease, although very small, in the melting point of the drug in the spherical crystals compared to that of pure carvedilol. This indicates the little amorphization of carvedilol when prepared in the form of spherical crystals.

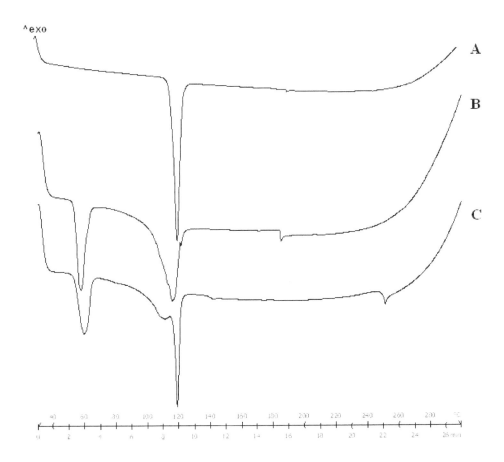

Fig. 4. DSC thermogram of (A) – Carvedilol, (B) – CP682, (C) – CP1272.

The XRD patterns of CAR shown in figure 5. The intense peaks at 2θ of 26.16⁰, 27.48⁰, 36.47⁰ and 39.34⁰ with peak intensities (counts) 310, 256, 228 and 135 respectively obtained from CAR confirmed the crystalline form of CAR. The PXRD patterns of CAR spherical crystals could be distinguished from the pure CAR. Peaks at around 8.4, 17, 22⁰ 2θ confirms the change in crystal arrangements of CAR in its spherical crystal form.

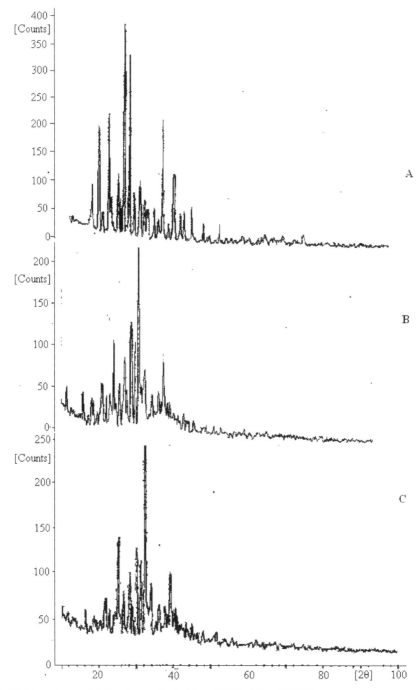

Fig. 5. XRD patterns of (A) – Carvedilol, (B) – CP682, (C) – CP1272.

Preparation of Carvedilol Spherical Crystals Having Solid Dispersion Structure by the Emulsion Solvent Diffusion
Method and Evaluation of Its in vitro Characteristics

271

3.6 Scanning electron microscopy

The results of surface morphology studies are shown in Figure 6. The SEM results revealed the spherical structure of agglomerates. The surface morphology studies also revealed that the agglomerates were formed by very small crystals, which were closely compacted into spherical form. These photo-micrographs show that the prepared agglomerates were spherical in shape which enabled them to flow very easily.

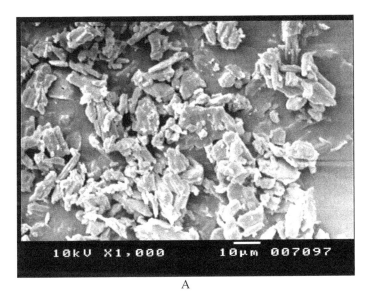

A

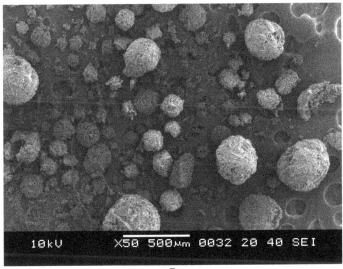

B

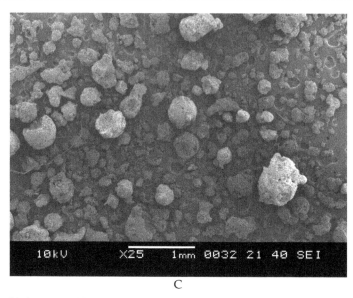

C

Fig. 6. SEM of (A) – Carvedilol, (B) – CP682, (C) – CP1272

3.7 Dissolution rate studies

The dissolution curves of pure carvedilol and its spherical crystals in 0.1 N HCl (pH 1.2) are shown in fig. 7. The release rate profiles were expressed as the percentage drug released vs. time. Table 4 shows % drug dissolved in 1h (DP_{60}) and dissolution efficiency values at 30 min (DE_{30}) for carvedilol and its spherical crystals. These values are tested statistically through one way ANOVA and are found significantly different ($p<0.05$) from pure carvedilol. As indicated carvedilol was dissolved more than 80% from spherical crystals CP681, CP1271 and CP1272 after 1h and more than 90% from spherical crystal CP682 while the pure CAR powder was just dissolved 34.37% at comparable time. The results revealed that the spherical crystals caused significant increase ($P<0.05$) in drug release compared to the pure drug. Enhancement in dissolution rate of spherical agglomerates as compared to pure drug may be due the presence of hydrophilic polymer, Poloxamer. The mechanism

Sample	Carvedilol Release after 1h	Dissolution Efficiency at 30 min
CAR	34.37±0.19	20.56±0.09
CP681	83.42±0.25[b]	66.12±0.31[b]
CP682	95.52±0.25[b]	70.63±1.38[b]
CP1271	82.26±0.96[b]	65.90±0.14[b]
CP1272	84.42±0.07[b]	67.39±0.39[b]

[a]Mean ± SD, n = 3
[b]Significantly different compared to carvedilol ($p < 0.05$).

Table 4. Drug Release and Dissolution Efficiency[a]

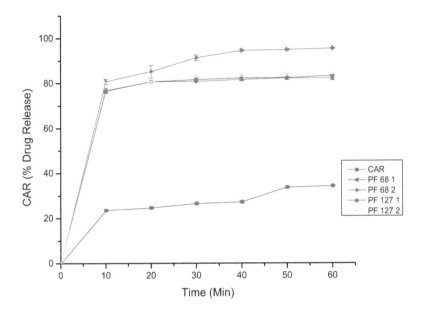

Fig. 7. Dissolution profile of CAR and its agglomerates. (Mean ±SD, n = 3.)

behind the greater solubility and dissolution of CAR from its agglomerated form may resemble the solid dispersion mechanism despite the larger particle size of agglomerates. This effect may be due to improved wettability of the surface of agglomerates by the adsorption of poloxamer onto the surfaces of crystals. These results confirm that the dissolution rate of carvedilol was increased in form of spherical crystals when compared to its pure form.

4. Conclusion

CAR-poloxamer spherical crystals were prepared successfully by ESD method. The resultant crystals have the desired micromeritic properties, such as flowability and packability. In the present investigation Poloxamer F68 and Poloxamer F127 has significantly improved dissolution rate of carvedilol. However in vivo bioavailability studies are required to ensure whether, the results obtain in this investigation can be extrapolated to the in vivo conditions.

5. Acknowledgments

The authors gratefully acknowledge Dr. Reddy's Laboratory, Hyderabad, India for the gift sample of Carvedilol. The authors are thankful to AISSMS college of Pharmacy, Pune, India for providing FTIR and DSC facilities. Also the authors would like to thank Shivaji University, Kolhapur, India for providing PXRD facility.

6. References

Anderson, N.H., Bauer, M., Boussac, N., Khan-Malek, R., Munden, P., Sardaro, M. (1998). An evaluation of fit factors and dissolution efficiency for the comparison of in vitro dissolution profiles. *Journal of Pharmaceutical and Biomedical Analysis,* 17, 811–822.

Bhutani, S., Hiremath, S.N., Swamy, P.V., Raju, S.A. (2001). Preparation and evaluation of inclusion complexes of carvedilol. *Journal of Scientific and Industrial Research,* 66, 830-834.

Bodmeier R., Paeratakul R. (1989). Spherical agglomerates of water-insoluble drugs. *Journal of Pharmaceutical Sciences,* 78, 964-967.

Cui F., Yang M., Jiang Y., Cun D., Lin W., Fan Y., Kawashima Y (2003). Design of sustained-release nitrendipine microspheres having solid dispersion structure by quasi-emulsion solvent diffusion method. *Journal of ControlledRelease,* 91, 375-384.

Di Martino, P., Barthelemy, C., Piva, F., Joiris, E., Palmieri, G.F., Martelli, S. (1999). Improved dissolution behavior of fenbufen by spherical crystallization. *Drug Delivery and Industrial Pharmacy,* 25, 1073-1081.

Garekani, H.A., Ford, J.L., Rubinstein, M.H., Rajabi-Siahboomi, A.R. (1999). Formation and Compression Properties of Prismatic Polyhedral and Thin Plate-like Crystals of Paracetamol. *International Journal of Pharmaceutics,* 187, 77-89.

Kawashima Y, Aoki S, Takenaka H. Spherical agglomeration of aminophylline crystals during reaction in liquid by the spherical crystallization technique. *Chem. Pharm. Bull.* 1982; 30, 1900-2.

Kawashima Y, Capes CE. Experimental study of the kinetics of spherical agglomeration in as stirred vessel. *Powder Technol.* 1974; 10, 85-92.

Kawashima Y, Capes CE. Further studies of the kinetics of spherical agglomeration in a stirred vessel. *Powder Technol.* 1976; 13, 279-288.

Kawashima Y, Cui F, Takeuchi H, Niwa T, Hino T, Kiuchi K. Improvements in flowability and compressibility of pharmaceutical crystals for direct tabletting by spherical crystallization with a 2 solvent system. *Powder Technol.* 1994; 78, 151-157.

Kawashima Y, Cui F, Takeuchi H, Niwa T, Hino T, Kiuchi K. Improved static compression behaviors and tablettabilities of spherically agglomerated crystals produced by the spherical crystallization technique with a two-solvent system. *Pharma. Res.* 1995; 12, 1040-4.

Kawashima Y, Cui F, Takeuchi H, Niwa T, Hino T, Kiuchi K. Parameters determining the agglomeration behavior and the micromeritic properties of spherically agglomerated crystals prepared by the spherical crystallization technique with miscible solvent systems. *Int. J. Pharm.* 1995; 119, 139-47.

Kawashima Y, Furukawa K, Takenaka H. The physicochemical parameters determining the size of agglomerate prepared by the wet spherical agglomeration technique. *Powder Technol.* 1981; 30, 211-16.

Kawashima Y, Imai M, Takeuchi H, Yamamoto H, Kamiya K, Hino T. Improved flowability and compactibility of spherically agglomerated crystals of ascorbic acid for direct tabletting designed by spherical crystallization process. *Powder Technol.* 2003; 130, 283-289.

Kawashima Y, Imai M, Takeuchi H, Yamamoto H, and Kamiya, K, Development of Agglomerated Crystals of Ascorbic acid by the Spherical Crystallization Technique for Direct Tabletting, and Evaluation of their Compactibilities, *Kona.* 2002; 20, 251-261.

Kawashima Y, Lin SY, Ogawa M, Handa T, Takenaka H. Preparations of agglomerated
crystals of polymorphic mixtures and a new complex of indomethacin-epirizole by
the spherical crystallization technique. *J. Pharm. Sci.* 1985; 74, 1152-6.

Kawashima Y, Morishima K, Takeuchi H, Niwa T, Hino T, Kawashima Y. Crystal design for
direct tabletting and coating by the spherical crystallization technique. *AIChE
Symposium Series.* 1991; 284, 26-32.

Kawashima Y, Naito M, Lin SY, Takenaka H. An experimental study of the kinetics of the
spherical crystallization of sodium theophylline monohydrate. *Powder Technol.*
1983; 34, 255-60.

Kawashima Y, Okumura M, Takenaka H. Spherical crystallization: direct spherical
agglomeration of salicylic acid crystals during crystallization. *Science.* 1982;
4550(216), 1127-8.

Kawashima Y, Okumura M, Takenaka H. The effects of temperature on the spherical
crystallization of salicylic acid. *Powder Technol.* 1984; 39, 41-47.

Kawashima Y, Takeuchi H, Niwa T, Hino T, Yamakoshi M, Kihara K. Preparation of
spherically agglomerated crystals of an antibacterial drug for direct tabletting by a
novel spherical crystallization technique. *Congr. Int. Technol. Pharm.* 1989; 5, 228-34.

Kulkarni, P.K. and Nagavi B.G. (2002). Spherical crystallization. *Indian Journal of
Pharmaceutical. Education,* 36, 66-71.

Khan, K.A. (1975). The concept of dissolution efficiency. *Journal Pharmacy and Pharmacology,*
27, 48-49.

Marshall, P.V., York, P. (1991). Compaction Properties of Nitrofurantoin Samples
Crystallised from Different Solvents. *International Journal of Pharmaceutics,* 67, 59-65.

Martin, A., Bustamante, P., Chun, A. (2002). Micromeritics, In: *Physical Pharmacy- physical
chemical principles in the pharmaceutical sciences,* 4th ed., pp. 423-452, Lippincott
Williams amd Wilkins, Baltimore.

Sano, A., Kuriki T., Kawashima Y., Takeuchi H., Hino T., and Niwa T. (1992). Particle design
of tolbutamide by spherical crystallization technique. V. Improvement of
dissolution and bioavailability of direct compressed tablets prepared using
tolbutamide agglomerated crystals. *Chemical and Pharmaceutical. Bulletin,* 40, 3030-
3035.

Sweetman, S.C. (Ed(s).). (2002). *Martindale The Complete Drug Reference.* 33rd ed.,
Pharmaceutical Press, London, 2002, pp. 855-856.

Tapas, A.R., Kawtikwar, P.S., Sakarkar, D.M. (2009). Enhanced dissolution rate of felodipine
using spherical agglomeration with Inutec SP1 by quasi emulsion solvent diffusion
method. *Research in Pharmaceutical Sciences,* 4, 77-84.

Tapas, A.R., Kawtikwar, P.S., Sakarkar, D.M. (2010). Spherically agglomerated solid
dispersions of valsartan to improve solubility, dissolution rate and micromeritic
properties. *International Journal of Drug Delivery,* 2, 304-313.

Usha, A.N., Mutalik, S., Reddy, M.S., Rajith, A.K., Kushtagi, P., Udupa, N. (2008).
Preparation and in vitro, preclinical and clinical studies of aceclofenac spherical
agglomerates. *European Journal of Pharmaceutics and Biopharmaceutics,* 70, 674-683.

U.S. Food and drug administration [Internet]. Dissolution methods for drug products; 2010.
Available from:
http://www.accessdata.fda.gov/scripts/cder/dissolution/dsp_SearchResults_Dis
solutions.cfm?PrintAll=1.

Wells, J. (2002). Pharmaceutical preformulation, the physicochemical properties of drug substances, In: *Pharmaceutics- the science of dosage form design*, M.E. Aulton (Ed), pp. 113-138, Churchill Livingstone, London.

Permissions

The contributors of this book come from diverse backgrounds, making this book a truly international effort. This book will bring forth new frontiers with its revolutionizing research information and detailed analysis of the nascent developments around the world.

We would like to thank Prof. Yitzhak Mastai, for lending his expertise to make the book truly unique. He has played a crucial role in the development of this book. Without his invaluable contribution this book wouldn't have been possible. He has made vital efforts to compile up to date information on the varied aspects of this subject to make this book a valuable addition to the collection of many professionals and students.

This book was conceptualized with the vision of imparting up-to-date information and advanced data in this field. To ensure the same, a matchless editorial board was set up. Every individual on the board went through rigorous rounds of assessment to prove their worth. After which they invested a large part of their time researching and compiling the most relevant data for our readers. Conferences and sessions were held from time to time between the editorial board and the contributing authors to present the data in the most comprehensible form. The editorial team has worked tirelessly to provide valuable and valid information to help people across the globe.

Every chapter published in this book has been scrutinized by our experts. Their significance has been extensively debated. The topics covered herein carry significant findings which will fuel the growth of the discipline. They may even be implemented as practical applications or may be referred to as a beginning point for another development. Chapters in this book were first published by InTech; hereby published with permission under the Creative Commons Attribution License or equivalent.

The editorial board has been involved in producing this book since its inception. They have spent rigorous hours researching and exploring the diverse topics which have resulted in the successful publishing of this book. They have passed on their knowledge of decades through this book. To expedite this challenging task, the publisher supported the team at every step. A small team of assistant editors was also appointed to further simplify the editing procedure and attain best results for the readers.

Our editorial team has been hand-picked from every corner of the world. Their multi-ethnicity adds dynamic inputs to the discussions which result in innovative outcomes. These outcomes are then further discussed with the researchers and contributors who give their valuable feedback and opinion regarding the same. The feedback is then collaborated with the researches and they are edited in a comprehensive manner to aid the understanding of the subject.

Apart from the editorial board, the designing team has also invested a significant amount of their time in understanding the subject and creating the most relevant covers. They scrutinized every image to scout for the most suitable representation of the subject and create an appropriate cover for the book.

The publishing team has been involved in this book since its early stages. They were actively engaged in every process, be it collecting the data, connecting with the contributors or procuring relevant information. The team has been an ardent support to the editorial, designing and production team. Their endless efforts to recruit the best for this project, has resulted in the accomplishment of this book. They are a veteran in the field of academics and their pool of knowledge is as vast as their experience in printing. Their expertise and guidance has proved useful at every step. Their uncompromising quality standards have made this book an exceptional effort. Their encouragement from time to time has been an inspiration for everyone.

The publisher and the editorial board hope that this book will prove to be a valuable piece of knowledge for researchers, students, practitioners and scholars across the globe.

List of Contributors

Joanna Jaworska
Adam Mickiewicz University, Poland

Masaumi Nakahara
Japan Atomic Energy Agency, Nuclear Fuel Cycle Engineering Laboratories, Japan

Tianlong Deng
Tianjin Key Laboratory of Marine Resources and Chemistry, College of Marine Science and Engineering, Tianjin University of Science and Technology, TEDA, Tianjin, People Republic of China
CAS Key Laboratory of Salt Lake Resources and Chemistry, Institute of Salt Lakes, Chinese Academy of Sciences, Xining, Qinghai, People Republic of China

Zanqun Liu
School of Civil Engineering, Central South University, Changsha, Hunan, P.R China
Magnel Laboratory for Concrete Research, Department of Structural Engineering, Ghent University, Ghent, Belgium
National Engineering Laboratory for High Speed Railway Construction, Changsha, Hunan, P.R China

Dehua Deng and Zhiwu Yu
School of Civil Engineering, Central South University, Changsha, Hunan, P.R China
National Engineering Laboratory for High Speed Railway Construction, Changsha, Hunan, P.R China

Geert De Schutter
Magnel Laboratory for Concrete Research, Department of Structural Engineering, Ghent University, Ghent, Belgium

Ana Ecija, Karmele Vidal, Aitor Larrañaga, Luis Ortega-San-Martín and María Isabel Arriortua
Universidad del País Vasco/Euskal Herriko Unibertsitatea (UPV/EHU), Facultad de Ciencia y Tecnología, Dpto. Mineralogía y Petrología, Leioa, Spain
Pontificia Universidad Católica del Perú (PUCP), Dpto. Ciencias, Sección Químicas, Lima, Peru

Tomasz Wróbel
Silesian University of Technology, Foundry Department, Poland

Mehrdad Pourayoubi and Atekch Tarahhomi
Department of Chemistry, Ferdowsi University of Mashhad, Mashhad, Iran

Fahimeh Sabbaghi
Department of Chemistry, Zanjan Branch, Islamic Azad University, Zanjan, Iran

Vladimir Divjakovic
Department of Physics, Faculty of Sciences, University of Novi Sad, Novi Sad, Serbia

Jinxia Fu
Brown University Department of Chemistry, Providence, RI, USA

james W. Rice and Eric M. Suuberg
Brown University School of Engineering, Providence, RI, USA

V. I. Talanin and I. E. Talanin
Classic Private University, Ukraine

Chika Nozaki Kato and Yuki Makino
Shizuoka University, Japan

Mikio Yamasaki
Rigaku Corporation, Japan

Yusuke Kataoka, Yasutaka Kitagawa and Mitsutaka Okumura
Osaka University, Japan

Amit R. Tapas, Pravin S. Kawtikwar and Dinesh M. Sakarkar
Sudhakarrao Naik Institute of Pharmacy, Pusad, Dist Yavatmal, Maharashtra, India

Printed in the USA
CPSIA information can be obtained
at www.ICGtesting.com
JSHW011454221024
72173JS00005B/1077